Lecture Notes in Computer Science 7112

Commenced Publication in 1973
Founding and Former Series Editors:
Gerhard Goos, Juris Hartmanis, and Jan van Leeuwen

Dániel Marx   Peter Rossmanith (Eds.)

Parameterized and Exact Computation

6th International Symposium, IPEC 2011
Saarbrücken, Germany, September 6-8, 2011
Revised Selected Papers

 Springer

Volume Editors

Dániel Marx
MTA SZTAKI
Lágymányosi u. 11
1111 Budapest, Hungary
E-mail: dmarx@cs.bme.hu

Peter Rossmanith
RWTH Aachen
Lehrgebiet Theoretische Informatik
Ahornstraße 55
52056 Aachen, Germany
E-mail: rossmani@cs.rwth-aachen.de

ISSN 0302-9743   e-ISSN 1611-3349
ISBN 978-3-642-28049-8   e-ISBN 978-3-642-28050-4
DOI 10.1007/978-3-642-28050-4
Springer Heidelberg Dordrecht London New York

Library of Congress Control Number: 2011945507

CR Subject Classification (1998): F.2, G.1-2, I.3.5, F.1, G.4, E.1, I.2.8

LNCS Sublibrary: SL 1 – Theoretical Computer Science and General Issues

Typesetting: Camera-ready by author, data conversion by Scientific Publishing Services, Chennai, India

Printed on acid-free paper

Springer is part of Springer Science+Business Media (www.springer.com)

Preface

The International Symposium on Parameterized and Exact Computation (IPEC, formerly IWPEC) is an international symposium series that covers research in all aspects of parameterized and exact algorithms and complexity. Started in 2004 as a biennial workshop, it became an annual event in 2008.

This volume contains the papers presented at IPEC 2011: the 6th International Symposium on Parameterized and Exact Computation held during September 6–8, 2011 in Saarbrücken. The symposium was part of ALGO 2011, which also hosted the 19th European Symposium on Algorithms (ESA 2011), the 11th Workshop on Algorithms for Bioinformatics (WABI 2011), the 11th Workshop on Algorithmic Approaches for Transportation Modelling, Optimization, and Systems (ATMOS 2011), the 9th Workshop on Approximation and Online Algorithms (WAOA 2011), and the 7th International Symposium on Algorithms for Sensor Systems, Wireless Ad Hoc Networks and Autonomous Mobile Entities (ALGOSENSORS). The five previous meetings of the IPEC/IWPEC series were held in Bergen, Norway (2004), Zürich, Switzerland (2006), Victoria, Canada (2008), Copenhagen, Denmark (2009), and Chennai, India (2010).

The IPEC 2011 plenary keynote talk was given by Martin Grohe (Humboldt-Universität zu Berlin) on "Excluding Topological Subgraphs." We had two additional invited tutorial speakers: Hans L. Bodlaender (Utrecht University, The Netherlands) speaking on kernels and Fedor V. Fomin (University of Bergen, Norway) speaking on width measures. We thank the speakers for accepting our invitation.

In response to the call for papers, 40 papers were submitted. Each submission was reviewed by at least three, and on average 3.8, reviewers. The reviewers were either Program Committee members or invited external reviewers. The Program Committee held electronic meetings using the EasyChair system, went through extensive discussions, and selected 21 of the submissions for presentation at the symposium and inclusion in this LNCS volume. The Program Committee decided to award the Excellent Student Paper Award to the paper "A Faster Algorithm for Dominating Set Analyzed by the Potential Method" by Yoichi Iwata (The University of Tokyo). We thank Frances Rosamond for sponsoring the award.

We are very grateful to the Program Committee, and the external reviewers they called on, for the hard work and expertise which they brought to the difficult selection process. We also wish to thank all the authors who submitted their work for our consideration.

October 2011

Dániel Marx
Peter Rossmanith

Organization

Program Committee

Hans L. Bodlaender	Utrecht University, The Netherlands
Rod Downey	Victoria University of Wellington, New Zealand
David Eppstein	University of California, Irvine, USA
Pinar Heggernes	University of Bergen, Norway
Thore Husfeldt	IT University of Copenhagen, Denmark and Lund University, Sweden
Iyad Kanj	DePaul University, Chicago, USA
Dieter Kratsch	Université Paul Verlaine, Metz, France
Daniel Lokshtanov	University of Bergen, Norway
Dániel Marx (Co-chair)	Budapest University of Technology and Economics, Hungary
Rolf Niedermeier	TU Berlin, Germany
Peter Rossmanith (Co-chair)	RWTH Aachen University, Germany
Ulrike Stege	University of Victoria, Canada
Stéphan Thomassé	Équipe AlGCo, LIRMM-Université Montpellier 2, France
Ryan Williams	IBM Almaden Research Center, USA
Gerhard Woeginger	TU Eindhoven, The Netherlands

Additional Reviewers

Abu-Khzam, Faisal	Hartung, Sepp	Otachi, Yota
Bevern, René van	Hermelin, Danny	Paulusma, Daniël
Bredereck, Robert	Hliněný, Petr	Pilipczuk, Michał
Cao, Yixin	Hof, Pim van 't	Reidl, Felix
Chen, Jiehua	Hüffner, Falk	Rooij, Johan M. M. van
Creignou, Nadia	Jansen, Bart	Saurabh, Saket
Cygan, Marek	Kern, Walter	Scott, Allan
Damaschke, Peter	Komusiewicz, Christian	Sikdar, Somnath
Dell, Holger	Kowaluk, Miroslaw	Sorge, Manuel
Fernau, Henning	Kratsch, Stefan	Szeider, Stefan
Fleischer, Rudolf	Langer, Alexander	Taslaman, Nina
Flum, Jörg	Liedloff, Mathieu	Thilikos, Dimitrios M.
Fomin, Fedor V.	Lonergan, Steven	Wahlström, Magnus
Gaspers, Serge	McCartin, Catherine	Watt, Nathaniel
Golovach, Petr	Mnich, Matthias	Weller, Mathias
Grandoni, Fabrizio	Nichterlein, André	

Table of Contents

On Multiway Cut Parameterized
above Lower Bounds[*]

Marek Cygan[1], Marcin Pilipczuk[1],
Michał Pilipczuk[1], and Jakub Onufry Wojtaszczyk[2]

[1] Institute of Informatics, University of Warsaw, Poland
{cygan@,malcin@,mp248287@students.}mimuw.edu.pl
[2] Google Inc., Cracow, Poland
onufry@google.com

Abstract. In this paper we consider two *above lower bound* parameterizations of the NODE MULTIWAY CUT problem — above the maximum separating cut and above a natural LP-relaxation — and prove them to be fixed-parameter tractable. Our results imply $O^*(4^k)$ algorithms for VERTEX COVER ABOVE MAXIMUM MATCHING and ALMOST 2-SAT as well as an $O^*(2^k)$ algorithm for NODE MULTIWAY CUT with a standard parameterization by the solution size, improving previous bounds for these problems.

1 Introduction

The study of cuts and flows is one of the most active fields in combinatorial optimization. However, while the simplest case, where we seek a cut separating two given vertices of a graph, is algorithmically tractable, the problem becomes hard as soon as one starts to deal with multiple terminals. For instance, given three vertices in a graph it is NP-hard to decide what is the smallest size of a cut that separates every pair of them (see [4]). The generalization of this problem — the well-studied NODE MULTIWAY CUT problem — asks for the size of the smallest set separating a given set of terminals. The formal definition is as follows:

NODE MULTIWAY CUT
Input: A graph $G = (V, E)$, a set $T \subseteq V$ of *terminals* and an integer k.
Question: Does there exist a set $X \subseteq V \setminus T$ of size at most k such that any path between two different terminals intersects X?

For various approaches to this problem we refer the reader for instance to [8,2,4,13].

Before describing our results, let us discuss the methodology we will be working with. We will be studying NODE MULTIWAY CUT (and several other problems) from the parameterized complexity point of view. Note that since the

---

[*] The first two authors were partially supported by National Science Centre grant no. N206 567140 and Foundation for Polish Science.

solution to our problem is a set of k vertices and it is easy to verify whether a solution is correct, we can solve the problem by enumerating and verifying all the $O(|V|^k)$ sets of size k. Therefore, for every fixed value of k, our problem can be solved in polynomial time. This approach, however, is not feasible even for, say, $k = 10$. The idea of parameterized complexity is to try to split the (usually exponential) dependency on k from the (hopefully uniformly polynomial) dependency on $|V|$ — so we look for an algorithm where the degree of the polynomial does not depend on k, e.g., an $O(C^k|V|^{O(1)})$ algorithm for a constant C.

Formally, a parameterized problem Q is a subset of $\Sigma^* \times \mathbb{N}$ for some finite alphabet Σ, where the integer is the *parameter*. We say that the problem is *fixed parameter tractable* (*FPT*) if there exists an algorithm solving any instance (x, k) in time $f(k)\text{poly}(|x|)$ for some (usually exponential) computable function f. It is known that a decidable problem is FPT iff it is kernelizable: a kernelization algorithm for a problem Q takes an instance (x, k) and in time polynomial in $|x| + k$ produces an equivalent instance (x', k') (i.e., $(x, k) \in Q$ iff $(x', k') \in Q$) such that $|x'| + k' \leq g(k)$ for some computable function g. The function g is the *size of the kernel*, and if it is polynomial, we say that Q admits a polynomial kernel. The reader is invited to refer to now classical books by Downey and Fellows [5], Flum and Grohe [7] and Niedermeier [15].

The typical parameterization takes the solution size as the parameter. For instance, Chen et al. [2] have shown an algorithm solving NODE MULTIWAY CUT in time $O(4^k n^{O(1)})$, improving upon the previous result of Daniel Marx [13]. However, in many cases it turns out we have a natural lower bound on the solution size — for instance, in the case of the VERTEX COVER problem the cardinality of the maximal matching is such a lower bound. It can happen that this lower bound is large — rendering algorithms parameterized by the solution size impractical. For some problems, better answers have been obtained by introducing the so called *parameterization above guaranteed value*, i.e. taking as the parameter the difference between the expected solution size and the lower bound. The idea was first proposed in [12]. An overview of this currently active research area can be found in the introduction to [10].

We will consider two natural lower bounds for NODE MULTIWAY CUT — the separating cut and the LP-relaxation solution. Let $I = (G, T, k)$ be a NODE MULTIWAY CUT instance and let $s = |T|$. By a *minimum solution* to I we mean a set $X \subseteq V \setminus T$ of minimum cardinality that disconnects the terminals, even if $|X| > k$.

For a terminal $t \in T$ a set $S \subseteq V \setminus T$ is a *separating cut* (also called an *isolating cut*) of t if t is disconnected from $T \setminus \{t\}$ in $G[V \setminus S]$ (the subgraph induced by $V \setminus S$). Let $m(I, t)$ be the size of a minimum isolating cut of t. Notice that for any t the value $m(I, t)$ can be found in polynomial time using standard max-flow techniques. Moreover, the maximum of these values over all t is a lower bound for the size of the minimum solution to I — any solution X has, in particular, to separate t from all the other terminals.

Now we consider a different approach to the problem, stemming from linear programming. Let $\mathcal{P}(I)$ denote the set of all simple paths connecting two different terminals in G. Garg et al. [8] gave a 2-approximation algorithm for NODE MULTIWAY CUT using the following natural LP-relaxation:

$$\text{minimize} \quad \sum_{v \in V \setminus T} d_v \tag{1}$$

$$\text{subject to} \quad \sum_{v \in P \cap (V \setminus T)} d_v \geq 1 \qquad \forall P \in \mathcal{P}(I)$$

$$d_v \geq 0 \qquad \forall v \in V \setminus T$$

In other words, the LP-relaxation asks to assign for each vertex $v \in V \setminus T$ a non-negative weight d_v, such that the distance between pair of terminals, with respect to the weights d_v, is at least one. This is indeed a relaxation of the original problem — if we restrict the values d_v to be integers, we obtain the original NODE MULTIWAY CUT .

The above LP-relaxation has exponential number of constraints, as $\mathcal{P}(I)$ can be exponentially big in the input size. However, the optimal solution for this LP-relaxation can be found in polynomial time either using separation oracle and ellipsoid method or by solving an equivalent linear program of polynomial size (see [8] for details). By $LP(I)$ we denote the cost of the optimal solution of the LP-relaxation (1). As the LP-relaxation is less restrictive than the original NODE MULTIWAY CUT problem, $LP(I)$ is indeed a lower bound on the size of the minimum solution.

We can now define two *above lower bound* parameters: $L(I) = k - LP(I)$ and $C(I) = k - \max_{t \in T} m(I, t)$, and denote by NMWC-A-LP (NODE MULTIWAY CUT ABOVE LP-RELAXATION) and NMWC-A-CUT (NODE MULTIWAY CUT ABOVE MAXIMUM SEPARATING CUT) the NODE MULTIWAY CUT problem parameterized by $L(I)$ and $C(I)$, respectively.

We say that a parameterized problem Q is in XP, if there exists an algorithm solving any instance (x, k) in time $|x|^{f(k)}$ for some computable function f, i.e., polynomial for any constant value of k. The NMWC-A-CUT problem was defined and shown to be in XP by Razgon in [17].

Our results. In Section 2, using the ideas of Xiao [20] and building upon analysis of the LP relaxation by Guillemot [9], we prove a NODE MULTIWAY CUT instance I can be solved in $O^*(4^{L(I)})$ time[1], which easily yields an $O^*(2^{C(I)})$-time algorithm. Both algorithms run in polynomial space. Consequently we prove both NMWC-A-LP and NMWC-A-CUT problems to be FPT, solving an open problem of Razgon [17]. Observe that if $C(I) > k$ the answer is trivially negative, hence as a by-product we obtain an $O^*(2^k)$ time algorithm for the NODE MULTIWAY CUT problem, improving the previously best known $O^*(4^k)$ time algorithm by Chen et al. [2].

By considering a line graph of the input graph, it is easy to see that an edge-deletion variant of MULTIWAY CUT is easier than the node-deletion one, and our results hold also for the edge-deletion variant. We note that the edge-deletion

---

[1] $O^*()$ is the $O()$ notation with suppressed factors polynomial in the size of the input.

variant, parameterized above maximum separating cut, was implicitly proven to be FPT by Xiao [20].

Furthermore we observe that VERTEX COVER ABOVE MAXIMUM MATCHING is a special case of NMWC-A-LP, while it is known that VERTEX COVER ABOVE MAXIMUM MATCHING is equivalent to ALMOST 2-SAT from the point of view of parameterized complexity [11,19]. The question of an FPT algorithm for those two problems was a long-standing open problem until Razgon and O'Sullivan gave an $O^*(15^k)$-time algorithm in 2008, improved recently by Raman et al. [16] to $O^*(9^k)$. Our results improve those bounds to $O^*(4^k)$ for both VERTEX COVER ABOVE MAXIMUM MATCHING and ALMOST 2-SAT. The details are gathered in Section 3.

One of the major open problems in kernelization is the question of a polynomial kernel for NODE MULTIWAY CUT , parameterized by the solution size. Our results show that the number of terminals can be reduced to $2k$ in polynomial-time, improving a quadratic bound due to Razgon [18]. Moreover, our algorithm includes a number of polynomial-time reduction rules, that may be of some interest from the point of view of kernelization.

Finally, we consider the NODE MULTICUT problem, a generalization of NODE MULTIWAY CUT , which was recently proven to be FPT when parameterized by the solution size [14,1]. In Section 4 we show that NODE MULTICUT, when parameterized above a natural LP-relaxation, is significantly more difficult and even not in XP.

Notation. Let us introduce some notation. All considered graphs are undirected and simple. Let $G = (V, E)$ be a graph. For $v \in V$ by $N(v)$ we denote the set of neighbours of v, $N(v) = \{u \in V : uv \in E\}$, and by $N[v]$ the closed neighbourhood of v, $N[v] = N(v) \cup \{v\}$. We extend this notation to subsets of vertices $S \subseteq V$, $N[S] = \bigcup_{v \in S} N[v]$, $N(S) = N[S] \setminus S$. By *removing* a vertex v we mean transforming G to $(V \setminus v, E \setminus \{uv, vu : u \in V\})$. The resulting graph is denoted by $G \setminus v$. By *contracting an edge* uv we mean the following operation: we remove vertices u and v, introduce a new vertex x_{uv} and connect it to all vertices previously connected to u or v. The resulting graph is denoted by G/uv. If $u \in T$ and $v \notin T$, we somewhat abuse the notation and identify the new vertex x_{uv} with u, so that the terminal set remains unchanged. In this paper we do not contract any edge that connects two terminals.

2 Algorithms for MULTIWAY CUT

Let $I = (G, T, k)$, where $G = (V, E)$, be a NODE MULTIWAY CUT instance. First, let us recall the two known facts about the LP-relaxation (1).

Definition 1 ([8,9]). *Let $(d_v)_{v \in V \setminus T}$ be a feasible solution to the LP-relaxation (1) of I. For a terminal t, the zero area of t, denoted by U_t, is the set of vertices within distance zero from t with respect to weights d_v.*

Lemma 2 ([8]). *Given an optimal solution $(d_v^*)_{v \in V \setminus T}$ to the LP-relaxation (1), let us construct an assignment $(d_v)_{v \in V \setminus T}$ as follows. First, for each terminal t compute its zero area U_t with respect to weights $(d_v^*)_{v \in V \setminus T}$. Second, for $v \in V \setminus T$ we take $d_v = 1$ if $v \in N(U_t)$ for at least two terminals t, $d_v = 1/2$ if $v \in N(U_t)$ for exactly one terminal t, and $d_v = 0$ otherwise. Then $(d_v)_{v \in V \setminus T}$ is also an optimal solution to the LP-relaxation (1).*

Lemma 3 ([9], Lemma 3). *Let $(d_v^*)_{v \in V \setminus T}$ be any optimal solution to the LP-relaxation (1) of I. Then there is a minimum solution to I that is disjoint with $\bigcup_{t \in T} U_t$.*

Our algorithm consists of two parts. The first part is a set of several polynomial-time reduction rules. At any moment, we apply the lowest-numbered applicable rule. We shall prove that the original instance I is a YES–instance if and only if the reduced instance is a YES–instance, we will say this means the reduction is *sound*. We prove that no reduction rule increases the parameter $L(I)$ or the graph size. If no reduction rule can be applied, we proceed to the branching rule. The branching rule outputs two subcases, each with the parameter $L(I)$ decreased by at least $1/2$ and a smaller graph. If the answer to any of the two subcases is YES, we return YES from the original instance, otherwise we return NO. As the parameter $L(I)$ decreases by at least $1/2$ with each branching, and we can trivially return NO if $L(I)$ is negative, we obtain the claimed $O^*(4^{L(I)})$ running time.

Reduction 1. If two terminals are connected by an edge or $L(I) < 0$, return NO.

The first part of the above rule is obviously sound, as we only remove vertices, not edges. The second part is sound as the optimal cost of the LP-relaxation (1) is a lower bound for the size of the minimum solution to the instance I.

Reduction 2. If there exists a vertex $w \in V \setminus T$ that is adjacent to two terminals $t_1, t_2 \in T$, remove w from G and decrease k by one.

The above rule is sound, as such a vertex w has to be included in any solution to I. Let us now analyse how the parameter $L(I)$ is influenced by this rule. Let $I' = (G \setminus w, T, k - 1)$ be the output instance. Notice that any feasible solution $(d_v)_{v \in V \setminus (T \cup \{w\})}$ to I' can be extended to a feasible solution of I by putting $d_w = 1$. Thus $LP(I) \leq LP(I') + 1$, and we infer $L(I) \geq L(I')$.

Reduction 3. Let $w \in V \setminus T$ be a neighbour of a terminal $t \in T$. Let $(d_v^\circ)_{v \in V \setminus T}$ be a solution to the LP-relaxation (1) with an additional constraint $d_w = 0$. If the cost of the solution $(d_v^\circ)_{v \in V \setminus T}$ is equal to $LP(I)$, contract the edge tw.

As $(d_v^\circ)_{v \in V \setminus T}$ is a feasible solution to the LP-relaxation (1), its cost is at least $LP(I)$. If the rule is applicable, $(d_v^\circ)_{v \in V \setminus T}$ is an optimal solution to the LP-relaxation (1) and $w \in U_t$. The soundness of Reduction 3 follows from Lemma 3. Moreover, note that if I' is the output instance of Reduction 3, we have

$LP(I) = LP(I')$, as $(d_v^\circ)_{v \in V \setminus (T \cup \{w\})}$ is a feasible solution to the LP-relaxation (1) for the instance I'. We infer that $L(I) = L(I')$.

The following lemma summarizes properties of an instance, assuming none of the above reduction rules is applicable.

Lemma 4. *If Reductions 1, 2 and 3 are not applicable, then:*

1. *An assignment $(d_v)_{v \in V \setminus T}$ that assigns $d_v = 1/2$ if $v \in N(T)$ and $d_v = 0$ otherwise is an optimal solution to the LP-relaxation (1).*
2. *For each terminal $t \in T$, the set $N(t)$ is the unique minimum separating cut of t.*

Proof. Let $(d_v^*)_{v \in V \setminus T}$ be any optimal solution to the LP-relaxation (1). As Reduction 3 is not applicable, $d_w^* > 0$ for any $w \in N(T)$. As Reduction 2 is not applicable, if we invoke Lemma 2 on the assignment $(d_v^*)_{v \in V \setminus T}$, we obtain the assignment $(d_v)_{v \in V \setminus T}$. Thus the first part of the lemma is proven.

For the second part, obviously $N(t)$ is a separating cut of t. Let $C \subseteq V \setminus T$ be any other separating cut of t and assume $|C| \leq |N(t)|$. Let $d_v' = d_v + 1/2$ if $v \in C \setminus N(t)$, $d_v' = d_v - 1/2$ if $v \in N(t) \setminus C$ and $d_v' = d_v$ otherwise. It is easy to see that d_v' is a feasible solution to the LP-relaxation (1). As $|C| \leq |N(t)|$, $\sum_{v \in V \setminus T} d_v' \leq \sum_{v \in V \setminus T} d_v$ and we infer that $(d_v')_{v \in V \setminus T}$ is an optimal solution to the LP-relaxation (1). However, $d_v' = 0$ for $v \in N(t) \setminus C$, since $d_v = 1/2$ in this case. Therefore Reduction 3 would be applicable.

Branching Rule. Let $w \in V \setminus T$ be a neighbour of a terminal $t \in T$. Branch into two subcases, either w is included in a solution to the NODE MULTIWAY CUT instance I or not. In the first branch, we remove w from the graph and decrease k by one. In the second one, we contract the edge tw.

The soundness of the branching rule is straightforward. We now prove that in both subcases the parameter $L(I)$ drops by at least $1/2$. Let $I_1 = (G \setminus w, T, k-1)$ and $I_2 = (G/tw, T, k)$ be the output instances in the first and second cases, respectively.

In the first subcase, it is sufficient to prove that $LP(I_1) \geq LP(I) - 1/2$, i.e., that the cost of the optimal solution to the LP-relaxation (1) drops by at most half. Assume the contrary, that $LP(I_1) < LP(I) - 1/2$. Let $(d_v)_{v \in V \setminus T \setminus \{w\}}$ be a half-integral optimal solution to the LP-relaxation (1) for I_1, as asserted by Lemma 2. Note that if we put $d_w = 1$, then $(d_v)_{v \in V \setminus T}$ is a feasible solution to the LP-relaxation (1) for I, and $LP(I_1) \geq LP(I) - 1$. By half-integrality, $LP(I_1) = LP(I) - 1$ and $(d_v)_{v \in V \setminus T}$ is an optimal half-integral solution to the LP-relaxation (1) for I. As Reduction 3 is not applicable, $d_v > 0$ for all $v \in N(T)$. As $(d_v)_{v \in V \setminus T}$ is half-integral, $d_v \geq 1/2$ for all $v \in N(T)$. However, the assignment given by Lemma 4 has strictly smaller cost than $(d_v)_{v \in V \setminus T}$ (as $d_w = 1$), a contradiction to the fact that $(d_v)_{v \in V \setminus T}$ is an optimal solution to the LP-relaxation (1) for I. Thus $LP(I_1) \geq LP(I) - 1/2$ and $L(I_1) \leq L(I) - 1/2$.

In the second subcase note that, as Reduction 3 is not applicable, $LP(I_2) > LP(I)$. As the LP-relaxation (1) has half-integral solutions, we have $LP(I_2) \geq LP(I) + 1/2$. This implies that $L(I_2) \leq L(I) - 1/2$.

Since Reduction 1 stops when the parameter $L(I)$ becomes negative, we obtain the following theorem.

Theorem 5. *There exists an algorithm that solves a* NODE MULTIWAY CUT *instance I in $O^*(4^{L(I)})$ time.*

To solve NODE MULTIWAY CUT parameterized by $C(I)$, we introduce one more reduction rule. Recall s denotes the number of terminals.

Reduction 4. If $C(I) \geq \frac{s-2}{s-1} \cdot k$ or $C(I) \leq 2L(I)$, return YES.

Now we show that Reduction 4 is sound. Let $t_0 \in T$ be the terminal with the largest separating cut, i.e., $m(I, t_0) = \max_{t \in T} m(I, t)$. Let $X = N(T \setminus \{t_0\})$. Obviously no two terminals are in the same connected component of $G[V \setminus X]$. We claim that $|X| \leq k$.

If $C(I) \geq \frac{s-2}{s-1} \cdot k$, $|N(t)| = m(I, t)$ by Lemma 4, and:

$$|X| = \sum_{t \in T \setminus \{t_0\}} m(I, t) \leq (s-1)m(I, t_0) = (s-1)(k - C(I)) \leq k.$$

In the second case, the condition $C(I) \leq 2L(I)$ is equivalent to $2LP(I) - m(I, t_0) \leq k$. From the structure of the optimum half-integral solution given by Lemma 4, we infer that $2LP(I) \geq |N(T)|$. By Lemma 4, $|N(t_0)| = m(I, t_0)$. Since Reduction 2 is not applicable, $N(t_0) \cap N(t) = \emptyset$ for $t \in T \setminus \{t_0\}$. We infer that $|X| = 2LP(I) - m(I, t_0) \leq k$, and Reduction 4 is sound.

Corollary 6. *There exists an algorithm that solves a* NODE MULTIWAY CUT *instance I in $O^*(2^{\min(C(I), \frac{s-2}{s-1} \cdot k)})$ time. In the case of three terminals, this yields a $O^*(2^{k/2})$-time algorithm.*

Finally, we would like to note that all our reduction rules are polynomial-time and could be used in a hypothetical algorithm to find a polynomial kernel for NODE MULTIWAY CUT . Let us supply them with one additional clean-up rule.

Reduction 5. If there exists a connected component of G with at most one terminal, remove it.

The following lemma shows that our reductions improve the quadratic bound on the number of terminals due to Razgon [18].

Lemma 7. *If Reductions 1, 2, 3 and 5 are not applicable, then $|T| \leq 2k$.*

Proof. As noted before, the optimal half-integral solution given by Lemma 4 implies that $|N(T)| = 2LP(I)$. However, if Reduction 5 is not applicable, $N(t) \neq \emptyset$ for any $t \in T$, and $|T| \leq |N(T)|$ by Reduction 2. We infer that $2LP(I) \geq |T|$. If $|T| > 2k$, then $L(I) < 0$ and Reduction 1 would return NO.

3   From NODE MULTIWAY CUT to ALMOST 2-SAT

We start with problem definitions. For a graph G by $\mu(G)$ we denote the size of a maximum matching in G.

VERTEX COVER ABOVE MAXIMUM MATCHING   **Parameter:** k
Input: A graph $G = (V, E)$ and an integer k.
Question: Does there exist a vertex cover in G of size at most $\mu(G) + k$?

ALMOST 2-SAT   **Parameter:** k
Input: A 2-SAT formula Φ and an integer k.
Question: Does there exist a set X of at most k clauses of Φ, whose deletion makes Φ satisfiable?

Now we prove that VERTEX COVER ABOVE MAXIMUM MATCHING is a special case of NMWC-A-LP.

Theorem 8. *There exists an algorithm that solves* VERTEX COVER ABOVE MAXIMUM MATCHING *in* $O^*(4^k)$ *time.*

Proof. Let $I = (G = (V, E), k)$ be a VERTEX COVER ABOVE MAXIMUM MATCHING instance. We construct a NODE MULTIWAY CUT instance $I' = (G', T, k')$ as follows. For each $v \in V$ we create a terminal t_v and connect it to v, thus $T = \{t_v : v \in V\}$ and each terminal in G' is of degree one. Moreover we take $k' = \mu(G) + k$.

We claim that $X \subseteq V$ is a vertex cover in G if and only if each connected component of $G'[(V \setminus X) \cup T]$ contains at most one terminal. If $X \subseteq V$ is a vertex cover in G, $G[V \setminus X]$ is an independent set, thus every edge in $G'[(V \setminus X) \cup T]$ is of type $t_v v$. In the other direction, note that if $uv \in E$ and $u, v \notin X$, then t_u and t_v are connected in $G'[(V \setminus X) \cup T]$.

We now show that $LP(I') \geq \mu(G)$. Let M be a maximum matching in G and let $(d_v)_{v \in V}$ be an optimal solution to the LP-relaxation (1) for I'. For each $uv \in M$, the path consisting of vertices t_u, u, v and t_v is in $\mathcal{P}(I')$, thus $d_u + d_v \geq 1$. As M is a matching, we infer that $\sum_{v \in V} d_v \geq |M| = \mu(G)$.

Since $LP(I') \geq \mu(G)$ and $k' = k + \mu(G)$, we have $L(I') \leq k$. We apply algorithm from Theorem 5 to the instance I' and the time bound follows.

We now reproduce the reduction from ALMOST 2-SAT to VERTEX COVER ABOVE MAXIMUM MATCHING to prove the following theorem.

Theorem 9. *There exists an algorithm that solves* ALMOST 2-SAT *in* $O^*(4^k)$ *time.*

Proof. Let $I = (\Phi, k)$ be an ALMOST 2-SAT instance. First, we replace each clause $C \in \Phi$ that consists of a single literal l with a clause $(l \vee l)$. From now we assume that each clause of Φ consists of two, possibly equal, literals.

Let x be a variable of Φ. By $n(x)$ we denote the number of occurrences of the variable x in the formula Φ (if $l = x$ or $l = \neg x$, a clause $(l \vee l)$ counts as two occurrences). Let us arbitrarily number those occurrences and for any $1 \leq i \leq n(x)$, by $C(x, i)$ we denote the clause where x occurs the i-th time.

We now construct a VERTEX COVER ABOVE MAXIMUM MATCHING instance $I' = (G, k)$. For each variable x and for each $1 \leq i \leq n(x)$ we create two vertices $v(x, i)$ and $v(\neg x, i)$. For $l \in \{x, \neg x\}$ we denote $V(l) = \{v(l, i) : 1 \leq i \leq n(x)\}$. For each variable x and for each $1 \leq i, j \leq n(x)$ we connect $v(x, i)$ and $v(\neg x, j)$ by an edge, i.e., we make a full bipartite subgraph with sides $V(x)$ and $V(\neg x)$.

Furthermore, if $C(x, i) = C(y, j)$ for some variables x, y and indices $1 \leq i \leq n(x)$, $1 \leq j \leq n(y)$ (possibly $x = y$, but $(x, i) \neq (y, j)$), we introduce an edge $v(l_x, i)v(l_y, j)$, where $C(x, i) = C(y, j) = (l_x \vee l_y)$, l_x is the i-th occurrence of x and l_y is the j-th occurrence of y. Such an edge is called a *clause edge*. Note that we introduce exactly one clause edge for each clause of Φ and no two clause edges share an endpoint in G.

We claim that I is an ALMOST 2-SAT YES-instance if and only if I' is a VERTEX COVER ABOVE MAXIMUM MATCHING YES-instance. First note that G has a perfect matching consisting of all edges of the type $v(x, i)v(\neg x, i)$.

Assume I is a YES-instance. Let $X \subseteq \Phi$ be a set of clauses, such that there exists a truth assignment ϕ of all variables of Φ that satisfies all clauses of $\Phi \setminus X$. We now construct a vertex cover Y of G. For each variable x and for each index $1 \leq i \leq n(x)$, we take into Y the vertex $v(x, i)$ if x is true in the assignment ϕ, and $v(\neg x, i)$ otherwise. Moreover, for each clause $C \in X$ we take into Y any endpoint of the clause edge for C.

Clearly $|Y| \leq \mu(G) + |X|$. Each non-clause edge $v(x, i)v(\neg x, j)$ is covered by Y, as $v(x, i) \in Y$ if x is true in ϕ, and $v(\neg x, j) \in Y$ otherwise. Let $e_C = v(l_x, i)v(l_y, j)$ be a clause edge for clause $C = (l_x \vee l_y)$. If $C \in X$, then one of the endpoints of e_C is chosen into Y. Otherwise, l_x or l_y is true in ϕ and the corresponding vertex is chosen into Y.

In the other direction, let us assume that I' is a YES-instance and let Y be a vertex cover of G. We construct a truth assignment ϕ as follows. Let x be a variable of Φ. Recall that G has a complete bipartite subgraph with sides $V(x)$ and $V(\neg x)$. Thus $V(l) \subseteq Y$ for some $l \in \{x, \neg x\}$, and we take l to be true in ϕ (if $V(x) \cup V(\neg x) \subseteq Y$, we choose whether x is true or false arbitrarily). Let X be the set of clauses of Φ that are not satisfied by ϕ. We claim that $|X| \leq |Y| - \mu(G)$.

Let Y_1 be the union of all sets $V(l)$ for which l is true under ϕ. Obviously $Y_1 \subseteq Y$ and $|Y_1| = \mu(G)$. Let $Y_2 = Y \setminus Y_1$. Take any $C \in X$. As C is not satisfied by ϕ, the clause edge e_C corresponding to C does not have an endpoint in Y_1. Since Y is a vertex cover in G, e_C has an endpoint in Y_2. Finally, recall that no two clause edges share an endpoint. This implies that $|Y_2| \geq |X|$ and $|X| \leq |Y| - \mu(G)$.

We infer that the instances I and I' are equivalent. As the above construction can be done in polynomial time, the running time follows from Theorem 8.

4   Hardness of NODE MULTICUT Parameterized above LP-Relaxation

Recall the definition of NODE MULTICUT, which is a natural generalization of NODE MULTIWAY CUT .

NODE MULTICUT
Input: A graph $G = (V, E)$, a set T of pairs of terminals, and an integer k.
Question: Does there exist a set X of at most k non-terminal vertices, whose removal disconnects all pairs of terminals in T?

The LP-relaxation (1) for NODE MULTIWAY CUT naturally generalizes to NODE MULTICUT as follows. Let T be the set of all terminals in the given NODE MULTICUT instance $I = (G, T, k)$. In the LP-relaxation we ask for an assignment of non-negative weights $(d_v)_{v \in V \setminus T}$, such that for each pair $(s, t) \in T$ the distance between s and t with respect to the weights $(d_v)_{v \in V \setminus T}$ is at least one. Clearly, if X is a solution to I, an assignment that takes $d_v = 1$ if $v \in X$ and $d_v = 0$ otherwise, is a feasible solution to the LP-relaxation. Let $LP(I)$ be the cost of an optimal solution to this LP-relaxation. We denote by NMC-A-LP the NODE MULTICUT problem parameterized by $L(I) = k - LP(I)$, i.e., parameterized above LP lower bound.

In this section we prove that NMC-A-LP does not even belong to XP, by the following lemma.

Lemma 10. NMC-A-LP, *restricted to instances where $L(I) = 0$, is NP-hard.*

Proof. We reduce from MULTICOLOURED INDEPENDENT SET which is NP-complete (see [6]). In this problem we are given a graph $G = (V, E)$ together with a partition of the vertex set into sets $V_1, V_2, \ldots, V_r$, such that $G[V_i]$ is a clique for $1 \leq i \leq r$, and we are to decide whether G contains an independent set of size r. Note that such an independent set needs to take exactly one vertex from each set V_i. W.l.o.g. we may assume that $|V_i| \geq 2$ for each $1 \leq i \leq r$. Let $|V| = n$ and let I be the given MULTICOLOURED INDEPENDENT SET instance.

We construct a NODE MULTICUT instance $I' = (G', T, n)$ as follows. We start with the graph G. Then, for each $v \in V$ we create a vertex v' and connect it to v. For each set V_i, we connect the vertices $\{v' : v \in V_i\}$ into a path P_i in an arbitrary order. We now add terminal pairs. Each terminal will be of degree one in the graph G'.

First, for each $v \in V$ we create a terminal t_v connected to v and we include in T all pairs (t_v, t_u) for $u, v \in V$, $u \neq v$. Second, for each $v \in V$ we create terminals s_v and s'_v, connected to v and v' respectively, and include (s_v, s'_v) in T. Finally, for each set V_i, we create terminals a_i and b_i, connected to two endpoints of the path P_i, and include (a_i, b_i) in T. This finishes the construction of the instance I'.

First note that for each $(s, t) \in T$, we have $N(s) \cap N(t) = \emptyset$, due to the assumption $|V_i| \geq 2$ for each $1 \leq i \leq r$. Thus an assignment that takes $d_v = d_{v'} = 1/2$ for each $v \in V$ is a feasible solution to the LP-relaxation of cost n.

Moreover, it is an optimal solution, as $d_v + d_{v'} \geq 1$ for each $v \in V$ due to the terminal pair (s_v, s'_v). Thus $LP(I') = n$.

Assume I is a YES-instance and let $X \subseteq V$ be an independent set of size r in G. Take $X' = \{v' : v \in X\}$ and $Y = (V \setminus X) \cup X'$. Clearly $|Y| = n$. To see that Y is a solution to the instance I' observe that $V \setminus X$ is a vertex cover of G.

In the other direction, let Y be a solution to the instance I'. Y needs to include v or v' for each $v \in V$, due to the terminal pair (s_v, s'_v). Thus $|Y| = n$ and Y includes exactly one vertex from the set $\{v, v'\}$ for each $v \in V$. Moreover, for each V_i, if $u', v' \in Y$, $u, v \in V_i$, then Y does not disconnect t_u from t_v. On the other hand, if $V_i \subseteq Y$, then Y does not intersect P_i and the pair (a_i, b_i) is not disconnected by Y. We infer that for each $1 \leq i \leq r$ there exists a vertex $v_i \in V_i$, such that $(V_i \setminus \{v_i\}) \cup \{v'_i\} \subseteq Y$. Moreover, if $v_i v_j \in E$ for some $1 \leq i < j \leq r$, then the pair (t_{v_i}, t_{v_j}) is not disconnected by Y. We infer that $\{v_i : 1 \leq i \leq r\}$ is an independent set in G, and the instances I and I' are equivalent.

5   Conclusions

In this paper, building upon work of Xiao [20] and Guillemot [9], we show that NODE MULTIWAY CUT is fixed-parameter tractable when parameterized above two lower bounds: largest isolating cut and the cost of the optimal solution of the LP-relaxation. We also believe that our results may be of some importance in resolving the question of an existence of a polynomial kernel for NODE MULTIWAY CUT.

One of the tools used in the parameterized complexity is the notion of important separators introduced by Marx in 2004 [13]. From that time important separators were used for proving several problems to be in FPT, including MULTIWAY CUT [13], DIRECTED FEEDBACK VERTEX SET [3], ALMOST 2-SAT [19] and MULTICUT [14,1]. In this paper we show that in the NODE MULTIWAY CUT problem half-integral solutions of the natural LP-relaxation of the problem can be even more useful than important separators. Is it possible to use linear programming in other graph separation problems, for example to obtain a $O^*(c^k)$ algorithm for DIRECTED FEEDBACK VERTEX SET?

We have shown that NODE MULTICUT parameterized above LP-relaxation is not in XP. Is the edge-deletion variant similarly difficult?

Acknowledgements. We thank Saket Saurabh for pointing us to [9].

References

1. Bousquet, N., Daligault, J., Thomassé, S.: Multicut is FPT. In: Proc. of STOC 2011 (to appear, 2011)
2. Chen, J., Liu, Y., Lu, S.: An improved parameterized algorithm for the minimum node multiway cut problem. Algorithmica 55(1), 1–13 (2009)
3. Chen, J., Liu, Y., Lu, S., O'Sullivan, B., Razgon, I.: A fixed-parameter algorithm for the directed feedback vertex set problem. In: Proc. of STOC 2008, pp. 177–186 (2008)

4. Dahlhaus, E., Johnson, D.S., Papadimitriou, C.H., Seymour, P.D., Yannakakis, M.: The complexity of multiterminal cuts. SIAM J. Comput. 23(4), 864–894 (1994)
5. Downey, R.G., Fellows, M.R.: Parameterized Complexity. Springer, Heidelberg (1999), citeseer.ist.psu.edu/downey98parameterized.html
6. Fellows, M.R., Hermelin, D., Rosamond, F.A., Vialette, S.: On the parameterized complexity of multiple-interval graph problems. Theor. Comput. Sci. 410, 53–61 (2009)
7. Flum, J., Grohe, M.: Parameterized Complexity Theory. Texts in Theoretical Computer Science, 1st edn. An EATCS Series. Springer, Heidelberg (2006), http://www.worldcat.org/isbn/3540299521
8. Garg, N., Vazirani, V.V., Yannakakis, M.: Multiway cuts in node weighted graphs. J. Algorithms 50(1), 49–61 (2004)
9. Guillemot, S.: FPT Algorithms for Path-Transversals and Cycle-Transversals Problems in Graphs. In: Grohe, M., Niedermeier, R. (eds.) IWPEC 2008. LNCS, vol. 5018, pp. 129–140. Springer, Heidelberg (2008)
10. Gutin, G., van Iersel, L., Mnich, M., Yeo, A.: All Ternary Permutation Constraint Satisfaction Problems Parameterized above Average have Kernels with Quadratic Numbers of Variables. In: de Berg, M., Meyer, U. (eds.) ESA 2010, Part I, LNCS, vol. 6346, pp. 326–337. Springer, Heidelberg (2010)
11. Gutin, G., Kim, E.J., Lampis, M., Mitsou, V.: Vertex cover problem parameterized above and below tight bounds. Theory Comput. Syst. 48(2), 402–410 (2011)
12. Mahajan, M., Raman, V.: Parameterizing above guaranteed values: Maxsat and maxcut. J. Algorithms 31(2), 335–354 (1999)
13. Marx, D.: Parameterized graph separation problems. Theor. Comput. Sci. 351(3), 394–406 (2006)
14. Marx, D., Razgon, I.: Fixed-parameter tractability of multicut parameterized by the size of the cutset. In: Proc. of STOC 2011 (to appear, 2011)
15. Niedermeier, R.: Invitation to Fixed Parameter Algorithms. Oxford Lecture Series in Mathematics and Its Applications. Oxford University Press, USA (2006), http://www.worldcat.org/isbn/0198566077
16. Raman, V., Ramanujan, M.S., Saurabh, S.: Paths, Flowers and Vertex Cover. In: Demetrescu, C., Halldórsson, M.M. (eds.) ESA 2011. LNCS, vol. 6942, pp. 382–393. Springer, Heidelberg (2011)
17. Razgon, I.: Computing multiway cut within the given excess over the largest minimum isolating cut. CoRR abs/1011.6267 (2010)
18. Razgon, I.: Large isolating cuts shrink the multiway cut. CoRR abs/1104.5361 (2011)
19. Razgon, I., O'Sullivan, B.: Almost 2-SAT is fixed-parameter tractable. J. Comput. Syst. Sci. 75(8), 435–450 (2009)
20. Xiao, M.: Simple and improved parameterized algorithms for multiterminal cuts. Theory Comput. Syst. 46(4), 723–736 (2010)

Parameterized Complexity
of Firefighting Revisited

Marek Cygan[1], Fedor V. Fomin[2], and Erik Jan van Leeuwen[2]

[1] Institute of Informatics, University of Warsaw, Poland
cygan@mimuw.edu.pl
[2] Department of Informatics, University of Bergen, Norway
{fedor.fomin,E.J.van.Leeuwen}@ii.uib.no

Abstract. The FIREFIGHTER problem is to place firefighters on the vertices of a graph to prevent a fire with known starting point from lighting up the entire graph. In each time step, a firefighter may be permanently placed on an unburned vertex and the fire spreads to its neighborhood in the graph in so far no firefighters are protecting those vertices. The goal is to let as few vertices burn as possible. This problem is known to be NP-complete, even when restricted to bipartite graphs or to trees of maximum degree three. Initial study showed the FIREFIGHTER problem to be fixed-parameter tractable on trees in various parameterizations. We complete these results by showing that the problem is in FPT on general graphs when parameterized by the number of burned vertices, but has no polynomial kernel on trees, resolving an open problem. Conversely, we show that the problem is W[1]-hard when parameterized by the number of unburned vertices, even on bipartite graphs. For both parameterizations, we additionally give refined algorithms on trees, improving on the running times of the known algorithms.

1 Introduction

The FIREFIGHTER problem concerns a deterministic model of fire spreading through a graph via its edges. The problem has recently received considerable attention [9,14]. In the model, we are given a graph G with a vertex $s \in V(G)$. At time $t = 0$, the fire breaks out at s and vertex s starts burning. At each step $t \geq 1$, first the firefighter protects one vertex not yet on fire—this vertex remains permanently protected—and the fire then spreads from burning vertices to all unprotected neighbors of these vertices. The process stops when the fire cannot spread anymore. The goal is to find a strategy for the firefighter that minimizes the amount of burned vertices, or, equivalently, maximizes the number of saved, i.e. not burned, vertices.

It is known that the FIREFIGHTER problem is NP-hard, even when restricted to bipartite graphs [14] or trees of maximum degree three [10]. However, it is polynomial-time solvable on such trees if the root has degree two [14]. We refer to the survey [11] for an overview of further combinatorial results on the problem. The study of the problem from the perspective of parameterized complexity

D. Marx and P. Rossmanith (Eds.): IPEC 2011, LNCS 7112, pp. 13–26, 2012.
© Springer-Verlag Berlin Heidelberg 2012

was initiated by Cai, Verbin, and Yang [6]. They considered the following parameterized versions of the problem and obtained a number of parameterized algorithms on trees.

The first parameterization considered by Cai et al. in [6] is by the number of saved vertices.

SAVING k VERTICES   **Parameter:** k

Input: An undirected graph G, a vertex s, and an integer k.

Question: Is there a strategy to save at least k vertices when a fire breaks out at s?

Cai et al. proved that SAVING k VERTICES on trees has a kernel with $O(k^2)$ vertices. They also gave a randomized algorithm solving SAVING k VERTICES on trees in time $O(4^k + n)$, which can be derandomized to a $O(n + 2^{O(k)})$-time algorithm.

The second parameterization is by the number of burned vertices.

SAVING ALL BUT k VERTICES   **Parameter:** k

Input: An undirected n-vertex graph G, a vertex s, and an integer k.

Question: Is there a strategy to save at least $n - k$ vertices when a fire breaks out at s?

For SAVING ALL BUT k VERTICES on trees, Cai et al. gave a randomized algorithm of running time $O(4^k n)$, which can be derandomized to a $O(2^{O(k)} n \log n)$-time algorithm. They left as an open problem whether SAVING ALL BUT k VERTICES has a polynomial kernel on trees.

The last parameterization is by the number of protected vertices, i.e. the number of vertices occupied by firefighters.

MAXIMUM k-VERTEX PROTECTION   **Parameter:** k

Input: An undirected graph G, a vertex s, and an integer k.

Question: What is a strategy that saves the maximum number of vertices by protecting k vertices when a fire breaks out at s?

For MAXIMUM k-VERTEX PROTECTION on trees, Cai et al. gave a randomized algorithm of running time $O(k^{O(k)} n)$, which can be derandomized to a $O(k^{O(k)} n \log n)$-time algorithm. They left open whether the problem has a polynomial kernel on trees, and asked whether there is an algorithm solving the problem on trees in time $2^{o(k \log k)} n^{O(1)}$.

We will sometimes consider the decision variant of MAXIMUM k-VERTEX PROTECTION.

---

k-VERTEX PROTECTION   **Parameter:** k

Input: An undirected graph G, a vertex s, an integer k, and an integer K.

Question: Is there a strategy that saves at least K vertices by protecting k vertices when a fire breaks out at s?

The unparameterized version of this problem is obviously NP-hard on trees of maximum degree three from the hardness of the FIREFIGHTER problem.

Our Results. We resolve several open questions of Cai, Verbin, and Yang [6]. We also refine and extend some of the results of [6].

- In Section 2, we give a deterministic algorithm solving SAVING k VERTICES on trees in time $O(2^k k^3 + n)$, improving the running time $O(4^k + n)$ of the randomized algorithm from [6]. We also observe that on general graphs the problem is W[1]-hard, which was independently observed by Cai (private communication), but is in FPT when parameterized by k and the treewidth of a graph. Based on that we derive that SAVING k VERTICES is in FPT on graphs of bounded local treewidth, including planar graphs, graphs of bounded genus, apex-minor-free graphs, and graphs of bounded maximum vertex degree.
- In Section 3, we provide deterministic algorithms solving SAVING ALL BUT k VERTICES in time $O(2^k n)$ on trees, and in time $O(3^k n)$ on general graphs. The algorithm on trees improves the $O(4^k n)$ running time of the randomized algorithm from [6]. We also answer the open question of Cai et al. by showing that SAVING ALL BUT k VERTICES has no polynomial kernel on trees of maximum vertex degree four.
- For MAXIMUM k-VERTEX PROTECTION, we answer both open questions of Cai et al.: We give a deterministic algorithm solving MAXIMUM k-VERTEX PROTECTION on trees in time $O(2^k kn)$ in Section 2, and show that the problem has no polynomial kernel on trees in Section 3. The no-poly-kernel result was independently obtained by Yang [15]. Based on the parameterized algorithm, we also give an exact subexponential-time algorithm, solving the FIREFIGHTER problem on an n-vertex tree in time $O(2^{\sqrt{2n}} n^{3/2})$, thus improving on the $2^{O(\sqrt{n} \log n)}$ running time from [6]. On general graphs, we show that the MAXIMUM k-VERTEX PROTECTION problem is W[1]-hard, but is solvable in $f(k, t) \cdot n^{O(1)}$ time, where t is the treewidth of the graph.

Recently, and independent of our work, Bazgan, Chopin, and Fellows [2] proved several of the results mentioned above. This includes the W[1]-hardness of SAVING k VERTICES, as well as its membership of FPT on graphs of bounded treewidth, and the membership of FPT of SAVING ALL BUT k VERTICES on general graphs, as well as it not having a polynomial kernel on trees. In addition, they consider the parameterization by the vertex cover number of a graph, and the extension of the problem to being able to protect b vertices at any time step. However, they do not consider MAXIMUM k-VERTEX PROTECTION, algorithms on trees, or exact algorithms.

2   Saving and Protecting Vertices

In this section, we consider the complexity of SAVING k VERTICES and MAXIMUM k-VERTEX PROTECTION. These problems are known to be fixed-parameter tractable on trees, but their complexity on general graphs was hitherto unknown. We resolve this open problem by giving a W[1]-hardness result for both problems. At the other end of the spectrum, we extend the boundary where SAVING k VERTICES and MAXIMUM k-VERTEX PROTECTION remain fixed-parameter tractable by giving parameterized algorithms on graphs of bounded treewidth. Finally, we improve the algorithms known to exist for trees.

2.1 W[1]-Hardness on General Graphs

We show that SAVING k VERTICES and the decision variant of MAXIMUM k-VERTEX PROTECTION are W[1]-hard, even on bipartite graphs. We reduce from the k-CLIQUE problem, which is well known to be W[1]-hard [7].

Theorem 1. SAVING k VERTICES *is W[1]-hard, even on bipartite graphs.*

Proof. Let (G, k) be an instance of k-CLIQUE. We can assume that G has at least $k + 1$ vertices that are not isolated, or we can easily output a trivial YES- or NO-instance. We construct the following bipartite graph G' (see Figure 1). For each edge $(u, v) \in E(G)$, we add a vertex s_{uv}, and for each vertex $v \in V(G)$, we add a vertex s_v. Call these two sets of vertices E and V respectively. Now add an edge from s_{uv} to both s_u and s_v for each $(u, v) \in E(G)$. Add a root vertex s, and add vertices $a_{i,j}$ for all $1 \leq i \leq k - 1$ and $1 \leq j \leq k$. Connect $a_{i,j}$ to $a_{i',j'}$ $(i' = i + 1)$ for all i, j, j', connect $a_{1,j}$ to s for all j, and connect $a_{k-1,j}$ to each vertex of V for all j. Now set $k' = k + \binom{k}{2} + 1$.

We claim that SAVING k VERTICES on (G', s, k') is a YES-instance if and only if k-CLIQUE on (G, k) is a YES-instance. Suppose that G has a k-clique K. Then the strategy that protects the vertices s_v for all $v \in K$ saves the vertices s_{uv} for all $u, v \in K$. Since K is a clique, these vertices s_{uv} are indeed present in G'. Additionally, we can protect (and thus save) a vertex s_{xy} for some edge $xy \notin E(G[K])$. This edge exists, as G is assumed to have at least $k+1$ nonisolated vertices. It follows that this strategy saves at least k' vertices.

Suppose that $P = \{p_1, \ldots, p_\ell\}$ is a strategy for (G', s, k') that chooses vertex p_t at time t and saves at least k' vertices. First observe that if $p_t = a_{i,j}$ for some i, j, then this vertex is not helpful, as there is always a vertex $a_{i,j'}$ that will be burned at time t and has the same neighborhood as $a_{i,j}$. Hence we can assume that no vertex $a_{i,j}$ is protected by the strategy. This implies that all vertices of V will be burned, except those that are protected by the strategy. But then protecting vertices of E does not save any further vertices. Since the fire will reach V in k time steps, and thus E in $k + 1$ time steps, the vertices in $S \cap V$ are responsible for saving $\binom{k}{2}$ vertices, which is only possible if the vertices of $S \cap V$ induce a k-clique in G.   □

Observe that essentially the same construction works for the decision variant of MAXIMUM k-VERTEX PROTECTION.

Theorem 2. k-VERTEX PROTECTION *is W[1]-hard, even on bipartite graphs.*

Proof. We again reduce from k-CLIQUE and construct the same bipartite graph as in the proof of Theorem 1. We set $k' = k + 1$ and $K' = k + \binom{k}{2} + 1$. Correctness now follows straightforwardly from the arguments in the proof of Theorem 1.   □

The above reduction also implies an NP-hardness reduction, which is simpler than the earlier reduction for the FIREFIGHTER problem on bipartite graphs [14].

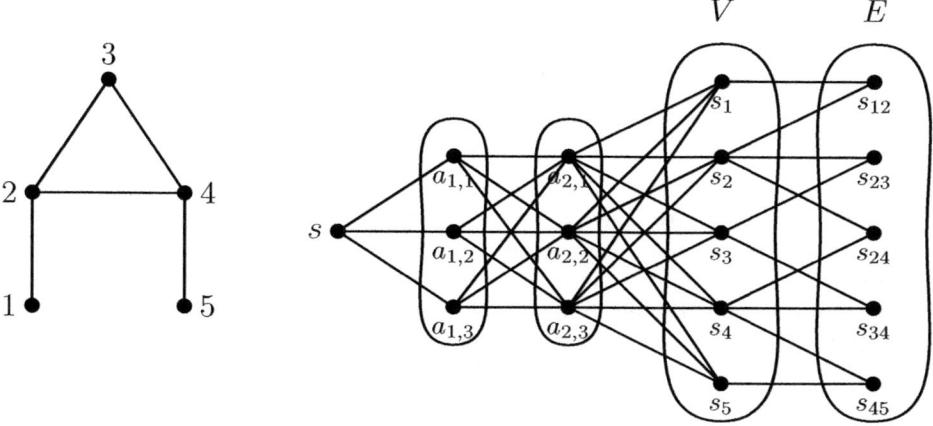

Fig. 1. An instance of k-Clique and the corresponding graph G' constructed in the proof of Theorem 1 for $k = 3$

2.2   Improved Algorithm on Trees

We show that Saving k Vertices and Maximum k-Vertex Protection have a deterministic $O(n + 2^k k^3)$ and $O(2^k kn)$ algorithm, respectively, on trees. This resolves an open question of Cai et al. [6]. As a consequence, we also obtain a refined subexponential algorithm for the Firefighter problem on trees, running in time $O(2^{\sqrt{2n}} n^{3/2})$.

The following observation is by MacGillivray and Wang [14, Sect. 4.1].

Lemma 1. *For any optimum strategy for an instance of the* Firefighter *problem on trees, there is an integer ℓ such that all protected vertices have depth at most ℓ, exactly one vertex p_i at each depth $1 \le i \le \ell$ is protected, and all ancestors of each p_i are burned.*

We need the following notation. Let T be any rooted tree. Use a pre-order traversal of T to number the vertices of T from 1 to n. We say that $u \in V(T)$ is *to the left* of $v \in V(T)$ if the number assigned to u is not greater than the number of v in the order. It is then easy to define what the *leftmost* or *rightmost* vertex is.

Theorem 3. Maximum k-Vertex Protection *has an $O(2^k kn)$-time algorithm on trees.*

Proof. Let (T, s, k) be an instance of Maximum k-Vertex Protection on a tree T. Assume that T is rooted at s. By Lemma 1, we can define a characteristic vector χ_v of length k for each vertex v of the tree, which has a 1 at position i if and only if the optimal strategy protects a vertex at depth i in the part of the tree to the left of v. We use these vectors as the basis for a dynamic programming procedure. However, the vector cannot ensure that no ancestors of a protected vertex will be protected. To ensure this, we add another dimension to our dynamic programming procedure. The pre-order numbering ensures that no descendant is protected.

The dynamic programming algorithm is then as follows. Let L be the set of vertices in T that are at depth at most k. For each $v \in L$, let P_v denote the path in T between v and s. For each vector $\chi \subseteq \{0,1\}^k$ and each integer $0 \le i \le k$, we compute $A_v(\chi, i)$, the maximum number of vertices one can save when protecting at most one vertex at depth j for each j for which $\chi(j) = 1$ and no vertex otherwise, where protected vertices must lie to the left of v but at depth greater than i when lying on P_v, and no protected vertex is an ancestor of another. Observe that s is the leftmost vertex of L. Now set $A_s(\chi, i) = 0$ for any χ and i. Then

$$
\begin{aligned}
A_v(\chi, i) = \max \big\{ \ & A_{l(v)}(\chi, \min\{\text{depth}(v) - 1, i\}), \\
& [\chi(\text{depth}(v)) = 1 \wedge \text{depth}(v) > i] \cdot \\
& (r(v) + A_{l(v)}(\chi^v, \text{depth}(v) - 1)) \big\}
\end{aligned}
$$

Here depth(v) is the depth of a vertex v, $l(v)$ is the rightmost vertex in L which has strictly smaller value in the pre-order than v, and $r(x)$ is the number of vertices saved when protecting only x. Moreover, χ^v is the 0-1 vector obtained from χ by setting the number at the index of χ corresponding to depth(v) to 0. In the formula we use Iverson's bracket notation, where $[\phi]$ is equal to one if ϕ is true and zero otherwise.

To see that the above formula is correct, observe that we can either protect the considered vertex v or not. If we do not protect v, then we must ensure that the value for the second dimension of our dynamic programming procedure does not exceed the length of P_v, yet still captures the same forbidden part of P_v. Correctness then follows from the fact that the parent of v is always on $P_{l(v)}$. If we do protect v, we can protect v only if we are allowed to do so, i.e. if $\chi(\text{depth}(v)) = 1$ and depth$(v) > i$. Furthermore, we need to ensure that no ancestor of v is protected later. Therefore, we set the value for the second dimension of our dynamic programming procedure to depth$(v) - 1$.

To get the solution for the whole tree T, return $A_{v^*}(1^k, 0)$, where v^* is the rightmost vertex of L. To obtain the claimed running time, first find L, and then $l(v)$ for each vertex $v \in L$. This can be done in linear time by a depth-first search. We can also compute $r(x)$ for each $x \in V(T)$ in linear time, as $r(x)$ equals one plus the number of descendants of x. By traversing the vertices of L from left to right, the total running time is $O(2^k kn)$. $\square$

Corollary 1. SAVING k VERTICES *has an* $O(2^k kn)$-*time algorithm on trees.*

Proof. Let (T, s, k) be an instance of SAVING k VERTICES on a tree T. We run the above algorithm for MAXIMUM k-VERTEX PROTECTION for all $k' = 1, \ldots, k$. Observe that it is possible to save k vertices of the tree if and only if the algorithm saves at least k vertices for some value of k'. Furthermore, we note that

$$\sum_{i=1}^{k}(2^i in) \ \leq \ kn\sum_{i=1}^{k} 2^i \ = \ (2^{k+1} - 2)kn,$$

implying that the worst-case running time of the algorithm is $O(2^k kn)$. □

Using the known kernel of size $O(k^2)$ [6], we can improve running time of the above algorithm for SAVING k VERTICES to $O(n + 2^k k^3)$.

To obtain a good subexponential algorithm, we use the following lemma. A similar idea appeared independently in [12].

Lemma 2. *If a vertex at depth d burns in an optimum strategy for an instance of the* FIREFIGHTER *problem on trees, then at least $\frac{1}{2}(d^2 + d)$ vertices are saved.*

Proof. Let (T, s) be an instance of the FIREFIGHTER problem on trees, and let v be a vertex of depth d that burns in an optimum strategy. Then the strategy protects a vertex at depth d, and by Lemma 1 it thus protects a vertex p_i at each depth i for $1 \leq i \leq d$. For any i, the subtree rooted at p_i should contain at least $d - i + 1$ vertices, or it would have been better to protect the vertex at depth i that is on the path from v to s. But then the strategy saves at least $\sum_{i=1}^{d}(d - i + 1) = \frac{1}{2}(d^2 + d)$ vertices. □

Theorem 4. *The* FIREFIGHTER *problem has an $O(2^{\sqrt{2n}}n^{3/2})$-time algorithm on trees.*

Proof. Let (T, s) be an instance of the FIREFIGHTER problem on trees. Suppose that a vertex v at depth $\sqrt{2n}$ burns in an optimum strategy. Then, by Lemma 2, the strategy saves at least $n + \sqrt{n/2} > n$ vertices, which is not possible. It follows that all vertices at depth $\sqrt{2n}$ are saved in any optimum strategy. Since in any optimum strategy every protected vertex has a burned ancestor by Lemma 1, all protected vertices are at depth at most $\sqrt{2n}$. Hence there is an optimum strategy that protects at most $\sqrt{2n}$ vertices, and we can find the optimum strategy by running the algorithm of Theorem 3 with $k = \sqrt{2n}$. □

2.3   Tractability on Graphs of Bounded Treewidth

We generalize the above results by showing that MAXIMUM k-VERTEX PROTECTION and SAVING k VERTICES remain fixed-parameter tractable when parameterized by k and the treewidth of the underlying graph. To this end, we use Monadic Second Order Logic (MSOL). The syntax of MSOL of graphs includes the logical connectives $\vee, \wedge, \neg, \Leftrightarrow, \Rightarrow$, variables for vertices, edges, sets of vertices, and sets of edges, the quantifiers $\forall, \exists$ that can be applied to these variables, and the following four binary relations:

1. $u \in U$, where u is a vertex variable and U is a vertex set variable.
2. $d \in D$, where d is an edge variable and D is an edge set variable.

3. $\text{adj}(u, v)$, where u, v are vertex variables, and the interpretation is that u and v are adjacent.
4. Equality, $=$, of variables representing vertices, edges, sets of vertices, and sets of edges.

For MAXIMUM k-VERTEX PROTECTION, we actually need Linear Extended MSOL [1], which allows the maximization over a linear combination of the size of unbound set variables in the MSOL formula. (The definition of LEMSOL in [1] is slightly more general, but this suffices for our purposes.)

Theorem 5. MAXIMUM k-VERTEX PROTECTION *is solvable in $f(k,t) \cdot n^{O(1)}$ time, where t is the treewidth of the graph.*

Proof. Let (G, s, k) be an instance of MAXIMUM k-VERTEX PROTECTION such that the treewidth of G is t. Use Bodlaender's Algorithm [3] to find a tree decomposition of G of width at most t. Consider the following MSOL formulae.

$$\text{NextBurn}(B_{i-1}, B_i, p_1, \ldots, p_i) :=$$
$$\forall v \left(\left(v \in B_{i-1} \vee \exists u \left(u \in B_{i-1} \wedge \text{adj}(u, v) \wedge \left(\bigwedge_{1 \le j \le i} v \neq p_j \right) \right) \right) \Leftrightarrow v \in B_i \right)$$

This expresses is that if the vertices of B_{i-1} are burning by time step $i-1$, then the vertices of B_i burn by time step i, assuming that vertices $p_1, \ldots, p_i$ have been protected so far.

$$\text{Saved}(S, B, p_1, \ldots, p_\ell) :=$$
$$\forall u \left(u \in S \Rightarrow \left(u \notin B \wedge \forall v \left(\text{adj}(u, v) \Rightarrow v \in S \vee \bigvee_{1 \le i \le \ell} p_i = u \right) \right) \right)$$

This expresses that S is a set of saved vertices when B is a set of burned vertices and vertices $p_1, \ldots, p_\ell$ are protected.

$$\text{Protect}(S, \ell) := \exists p_1, \ldots, p_\ell \ \exists B, B_0, \ldots, B_{\ell-1}$$

$$\forall u \ (u \in B_0 \Leftrightarrow u = s) \tag{1}$$

$$\wedge \bigwedge_{1 \le i \le \ell - 1} \text{NextBurn}(B_{i-1}, B_i, p_1, \ldots, p_i) \tag{2}$$

$$\wedge \bigwedge_{1 \le i \le \ell} p_i \notin B_{i-1} \tag{3}$$

$$\wedge \forall u \left(\left(\bigvee_{0 \le i \le \ell - 1} u \in B_i \right) \Rightarrow u \in B \right) \tag{4}$$

$$\wedge \text{Saved}(S, B, p_1, \ldots, p_\ell) \tag{5}$$

This expresses that S can be saved by protecting ℓ vertices. The sets B_i contain all vertices that are burned by time step i, which is ensured by the formulas in lines 1 and 2. The set B contains vertices that are not saved (line 5) and all vertices of the sets B_i (line 4). The vertices $p_1, \ldots, p_\ell$ are the vertices that are protected. Line 3 ensures that the vertices we want to protect are not burned by the time we pick them. Then we want to find the largest set S such that

$$\text{Protect}_k(S) := \bigvee_{1 \leq \ell \leq k} \text{Protect}(S, \ell)$$

is true. Following a result of Arnborg, Lagergren, and Seese [1], this can be done in $f(k, t) \cdot n^{O(1)}$ time using the above formula. $\square$

In the same way as Corollary 1, we then obtain the following.

Corollary 2. SAVING k VERTICES *is in FPT when parameterized by k and the treewidth of the graph.*

Observe that this algorithm also works on graphs of bounded local treewidth, because if the graph has a vertex at distance more than k from the root, then any strategy that protects a vertex at distance i from the root in time step i will save at least k vertices, and we can answer YES immediately.

Corollary 3. SAVING k VERTICES *is in FPT on graphs of bounded local treewidth.*

The class of graphs having bounded local treewidth coincides with the class of apex-minor-free graphs [8], which includes the class of planar graphs.

Corollary 4. SAVING k VERTICES *is in FPT on planar graphs.*

3 Burning Vertices

In this section, we consider the FIREFIGHTER problem when parameterized by the number of burned vertices, which we call the SAVING ALL BUT k VERTICES problem. We improve on results of Cai et al. [6] by showing an $O(2^k n)$-time deterministic algorithm for trees and an $O(3^k n)$-time deterministic algorithm for general graphs. Furthermore, we prove that the SAVING ALL BUT k VERTICES problem has no polynomial kernel for trees, resolving an open problem from [6].

3.1 Algorithms

In this subsection, we show an $O(2^k n)$-time algorithm for the SAVING ALL BUT k VERTICES problem on trees, and an $O(3^k n)$-time algorithm on general graphs.

Theorem 6. *The* SAVING ALL BUT k VERTICES *problem for trees can be solved in $O(2^k n)$ time and polynomial space.*

Proof. If the root s has at most one child, then we immediately answer YES. We may assume that the root has exactly $a \geq 2$ children, and $k \geq a - 1$ since otherwise we simply answer NO. We use Lemma 1 and branch on every child of the root s. In each branch, we cut the subtree rooted at the protected vertex, identify all the vertices that are on fire after the first round, and decrease the

parameter by $a - 1$. In this way, we obtain a new instance of the SAVING ALL
BUT k VERTICES problem with parameter value equal to $k - (a - 1)$. The time
bound follows from the inequality

$$T(k) \leq aT(k - (a - 1)) + O(n)$$

which is worst when $a = 2$.   □

Before we present the algorithm on general graphs, we need to reformulate the
FIREFIGHTER problem to an equivalent version. Consider a different version of
the FIREFIGHTER problem, where in each round an arbitrary number of vertices
may be protected under the following restrictions:

- each protected vertex must have a neighbor which is on fire,
- after i rounds of the process at most i vertices are protected.

By SAVING ALL BUT k VERTICES II we denote the SAVING ALL BUT k VER-
TICES problem where vertices are protected subject to the above rules.

Lemma 3. *An instance* (G, s, k) *of the* SAVING ALL BUT k VERTICES *problem
is a* YES-*instance if and only if it is a* YES-*instance of the* SAVING ALL BUT k
VERTICES II *problem.*

Proof. Assume that (G, s, k) is a YES-instance of the SAVING ALL BUT k VER-
TICES problem. Let P be the set of protected vertices of an optimum strategy
S. We construct a strategy S', which in the i-th round of SAVING ALL BUT k
VERTICES II protects exactly those vertices of P which have a neighbor which
is on fire. Clearly after i rounds at most i vertices will be protected, since each
vertex of P is protected by the strategy S' not earlier than by the strategy S.

   In the other direction assume that (G, s, k) is a YES-instance of the SAVING
ALL BUT k VERTICES II problem and P is the set of protected vertices of an
optimum strategy S'. We construct a strategy S as follows. Let $(v_1, \ldots, v_{|P|})$ be
a sequence of vertices of P sorted by the round in which a vertex is protected
by S' (breaking ties arbitrarily). In the i-th round of strategy S we protect the
vertex v_i. The vertex v_i is not on fire in the i-th round, because in the strategy
S' it is protected not earlier than in the i-th round.   □

Theorem 7. *There is an* $O(3^k n)$-*time and polynomial-space algorithm for the*
SAVING ALL BUT k VERTICES II *problem on general graphs.*

Proof. We present a simple branching algorithm. Assume that we are in the i-th
time step and let B be the set of vertices which are currently on fire. Moreover,
let P be the set of already protected vertices (in the first round we have $B = \{s\}$
and $P = \emptyset$). Let $a = i - |P|$ and $r = |N(B) \setminus P|$. The algorithm does the following:

1. If $|B| > k$, then we immediately answer NO.
2. Observe that in the i-th round we are allowed to protect at most $\min(a, r)$
   vertices. If $a \geq r$, then we can greedily protect the whole set $N(B) \setminus P$. Hence
   in this case we answer YES.

3. In the last case, when $a < r$, we branch on all subsets of $N(B) \setminus P$ of size at most a. Observe that the number of branches is equal to $\sum_{j=0}^{a} \binom{r}{j} \leq 2^r - 1$, since we have $a < r$.

The running time of the algorithm is as follows. We introduce a measure $\alpha = (k - |B|) + (i - |P|)$ which we use in our time bound. At the beginning of the first round of the burning process we have $\alpha = (k-1) + (1-0) = k$. By $T(\alpha)$ we denote the upper bound on the number of steps that our algorithm requires for a graph with measure value α. Observe that for $\alpha \leq 0$, we have $T(\alpha) = O(n)$. Let us assume that the algorithm did not stop in step 1 or 2, and it branches into at most $2^r - 1$ choices of protected vertices. Observe that no matter how many vertices the algorithm decides to protect, the value of the measure decreases by exactly $r - 1$. Consequently, we have the inequality $T(\alpha) \leq (2^r - 1)T(\alpha - r + 1) + O(n)$. Since the algorithm did not stop in steps 1 or 2, we infer that $r \geq 2$. The time bound follows from the fact that the worst case for the inequality occurs when $r = 2$. □

Corollary 5. *There is an $O(3^k n)$-time and polynomial-space algorithm for the* SAVING ALL BUT k VERTICES *problem on general graphs.*

3.2 No Poly-kernel for Trees

The aim of this subsection is to prove the following theorem.

Theorem 8. *Unless $NP \subseteq coNP/poly$, there is no polynomial kernel for the* SAVING ALL BUT k VERTICES *problem, even if the input graph is a tree of maximum degree four.*

Before we prove Theorem 8 we describe the necessary tools. We use the cross-composition technique introduced by Bodlaender et al. [5], which is based on the previous results of Bodlaender et al. [4] and Fortnow and Santhanam [13]. We recall the crucial definitions.

Definition 1 (Polynomial Equivalence Relation [5]). *An equivalence relation $\mathcal{R}$ on Σ^* is called a* polynomial equivalence relation *if (1) there is an algorithm that given two strings $x, y \in \Sigma^*$ decides whether $\mathcal{R}(x, y)$ in $(|x| + |y|)^{O(1)}$ time; (2) for any finite set $S \subseteq \Sigma^*$ the equivalence relation $\mathcal{R}$ partitions the elements of S into at most $(\max_{x \in S} |x|)^{O(1)}$ classes.*

Definition 2 (Cross-Composition [5]). *Let $L \subseteq \Sigma^*$, and let $Q \subseteq \Sigma^* \times \mathbb{N}$ be a parameterized problem. We say that L* cross-composes *into Q if there is a polynomial equivalence relation $\mathcal{R}$ and an algorithm which, given t strings $x_1, x_2, \ldots x_t$ belonging to the same equivalence class of $\mathcal{R}$, computes an instance $(x^*, k^*) \in \Sigma^* \times \mathbb{N}$ in time polynomial in $\sum_{i=1}^{t} |x_i|$ such that (1) $(x^*, k^*) \in Q$ iff $x_i \in L$ for some $1 \leq i \leq t$; (2) k^* is bounded polynomially in $\max_{i=1}^{t} |x_i| + \log t$.*

Theorem 9 ([5], Theorem 9). *If $L \subseteq \Sigma^*$ is NP-hard under Karp reductions and L cross-composes into the parameterized problem Q that has a polynomial kernel, then $NP \subseteq coNP/poly$.*

We apply Theorem 9, where as the language L we use SAVING ALL BUT k VERTICES in trees of maximum degree three, which is NP-complete [10]. To finish the proof of Theorem 8, we present a cross-composition algorithm.

Lemma 4. *The unparameterized version of the* SAVING ALL BUT k VERTICES *problem on trees with maximum degree three cross-composes to* SAVING ALL BUT k VERTICES *on trees with maximum degree four.*

Proof. Observe that any polynomial equivalence relation is defined on all words over the alphabet Σ and for this reason we should also define how the relation behaves on words that do not represent instances of the problem. For the equivalence relation $\mathcal{R}$ we take a relation that puts all malformed instances into one equivalence class and all well-formed instances are grouped according to the number of vertices we are allowed to burn.

If we are given malformed instances, we simply output a trivial NO-instance. Thus in the rest of the proof we assume we are given a sequence of instances $(T_i, s_i, k)_{i=1}^t$ of the FIREFIGHTER problem, where each T_i is of maximum degree three. Observe that in all instances we have the same value of the parameter k. W.l.o.g. we assume that $t = 2^h$ for some integer $h \geq 1$. Otherwise we can duplicate an appropriate number of instances (T_i, s_i, k).

We create a new tree T' as follows. Let T' be a full binary tree with exactly t leaves rooted at a vertex s'. Now for each $i = 1, \ldots, t$, we replace the i-th leaf of the tree by tree T_i rooted at s_i. Finally, we set $k' = k + h = k + \log_2 t$. Observe that since each tree T_i is of maximum degree three, the tree T' is of maximum degree four. To prove correctness, it is enough to show that any strategy that minimizes the number of burned vertices protects exactly one vertex at each depth $1, \ldots, h$, which follows from Lemma 1.

Hence in any strategy that minimizes the number of burned vertices, there will be exactly one vertex s_i which is on fire after h rounds. $\square$

We can obtain a similar result for the decision variant of MAXIMUM k-VERTEX PROTECTION.

Theorem 10. *Unless $NP \subseteq coNP/poly$, there is no polynomial kernel for the k-VERTEX PROTECTION problem, even if the input graph is a tree of maximum degree four.*

Proof. There are only two differences compared to the proof for SAVING ALL BUT k VERTICES.

- For the equivalence relation $\mathcal{R}$, we take a relation that puts all malformed instances into one equivalence class, and all well-formed instances are grouped according to the number of vertices of the tree, the parameter value k, and the value K.
- The value of k' for the tree T' is $k + h$, and the value of K' is equal to $K + (t - 1)n + (t - h - 1)$, where n is the number of vertices in each of the trees T_i. The additional summands are derived from the fact that any optimal strategy will ensure that after h rounds exactly one vertex s_i will

be on fire and hence we save $t - 1$ subtrees rooted at s_i, each containing n vertices, and $t - h - 1$ vertices of the full binary tree.

This completes the proof. $\square$

4   Open Problems

In this paper, we refined and extended several parameterized algorithmic and complexity results about different parameterizations of the FIREFIGHTER problem. We conclude with the following open problems.

- We have shown that SAVING k VERTICES is in FPT on graphs of bounded local treewidth , and thus on planar graphs, by showing that the problem is in FPT parameterized by k and the treewidth of a graph. While MAXIMUM k-VERTEX PROTECTION is also solvable in $f(k,t) \cdot n^{O(1)}$ time, where t is the treewidth of the graph, we do not know if the problem has an $f(k) \cdot n^{O(1)}$-time algorithm on planar graphs, and leave it as an open problem.
- The FIREFIGHTER problem is solvable in subexponential time on trees. Is it solvable in time $2^{o(n)}$ on n-vertex planar graphs? Even the case of outerplanar graphs is open.
- Finally, we do not know if any of the three parameterized versions of the problem is solvable in parameterized subexponential time $2^{o(k)} n^{O(1)}$ on trees.

Acknowledgement. We thank Leizhen Cai for pointing us to [15] and for sending us the full version of [6]. We also acknowledge the support of Schloss Dagstuhl for Seminar 11071 (GRASTA 2011 - Theory and Applications of Graph Searching Problems). Research of Marek Cygan was partially supported by National Science Centre grant no. N206 567140 and Foundation for Polish Science. Research of Fedor Fomin was supported by the European Research Council (ERC) grant Rigorous Theory of Preprocessing, reference 267959.

References

1. Arnborg, S., Lagergren, J., Seese, D.: Easy problems for tree-decomposable graphs. J. Algorithms 12, 308–340 (1991)
2. Bazgan, C., Chopin, M., Fellows, M.R.: Parameterized Complexity of the Firefighter Problem. In: Asano, T., Nakano, S.-i., Okamoto, Y., Watanabe, O. (eds.) ISAAC 2011. LNCS, vol. 7074, pp. 643–652. Springer, Heidelberg (2011)
3. Bodlaender, H.L.: A linear-time algorithm for finding tree-decompositions of small treewidth. SIAM J. Computing 25(6), 1305–1317 (1996)
4. Bodlaender, H.L., Downey, R.G., Fellows, M.R., Hermelin, D.: On problems without polynomial kernels. J. Comput. Syst. Sci. 75(8), 423–434 (2009)
5. Bodlaender, H.L., Jansen, B.M.P., Kratsch, S.: Cross-composition: A new technique for kernelization lower bounds. CoRR abs/1011.4224 (2010)
6. Cai, L., Verbin, E., Yang, L.: Firefighting on Trees (1-1/e)-Approximation, Fixed Parameter Tractability and a Subexponential Algorithm. In: Hong, S.-H., Nagamochi, H., Fukunaga, T. (eds.) ISAAC 2008. LNCS, vol. 5369, pp. 258–269. Springer, Heidelberg (2008)

7. Downey, R.G., Fellows, M.R.: Parameterized complexity. Springer, New York (1999)
8. Eppstein, D.: Diameter and treewidth in minor-closed graph families. Algorithmica 27(3), 275–291 (2000)
9. Finbow, S., Hartnell, B., Li, Q., Schmeisser, K.: On minimizing the effects of fire or a virus on a network. J. Combin. Math. Combin. Comput. 33, 311–322 (2000)
10. Finbow, S., King, A., MacGillivray, G., Rizzi, R.: The firefighter problem for graphs of maximum degree three. Discrete Math. 307(16), 2094–2105 (2007)
11. Finbow, S., MacGillivray, G.: The firefighter problem: a survey of results, directions and questions. Australas. J. Combin. 43, 57–77 (2009)
12. Floderus, P., Lingas, A., Persson, M.: Towards more efficient infection and fire fighting. In: CATS 2011: 17th Computing: The Australasian Theory Symposium. CRPIT, vol. 119, pp. 69–74. Australian Computer Society (2011)
13. Fortnow, L., Santhanam, R.: Infeasibility of instance compression and succinct PCPs for NP. In: STOC 2008: Proceedings of the 40th Annual ACM Symposium on Theory of Computing, pp. 133–142. ACM (2008)
14. MacGillivray, G., Wang, P.: On the firefighter problem. J. Combin. Math. Combin. Comput. 47, 83–96 (2003)
15. Yang, L.: Efficient Algorithms on Trees. M. Phil thesis, Department of Computer Science and Engineering. The Chinese University of Hong Kong (2009)

Parameterized Complexity in Multiple-Interval Graphs: Domination

Minghui Jiang[1,*] and Yong Zhang[2]

[1] Department of Computer Science, Utah State University,
Logan, UT 84322, USA
mjiang@cc.usu.edu

[2] Department of Computer Science, Kutztown University of PA,
Kutztown, PA 19530, USA
zhang@kutztown.edu

Abstract. We show that several variants of the problem k-DOMINATING SET, including k-CONNECTED DOMINATING SET, k-INDEPENDENT DOMINATING SET, k-DOMINATING CLIQUE, d-DISTANCE k-DOMINATING SET, k-PERFECT CODE and d-DISTANCE k-PERFECT CODE, when parameterized by the solution size k, remain W[1]-hard in either multiple-interval graphs or their complements or both.

1 Introduction

We introduce some basic definitions. The *intersection graph* $\Omega(\mathcal{F})$ of a family of sets $\mathcal{F} = \{S_1, \ldots, S_n\}$ is the graph with $\mathcal{F}$ as the vertex set and with two different vertices S_i and S_j adjacent if and only if $S_i \cap S_j \neq \emptyset$; the family $\mathcal{F}$ is called a *representation* of the graph $\Omega(\mathcal{F})$. Let $t \geq 2$ be an integer. A *t-interval graph* is the intersection graph of a family of t-intervals, where each *t-interval* is the union of t disjoint intervals in the real line. A *t-track interval graph* is the intersection graph of a family of t-track intervals, where each *t-track interval* is the union of t disjoint intervals on t disjoint parallel lines called tracks, one interval on each track. Note that the t disjoint tracks for a t-track interval graph can be viewed as t disjoint "host intervals" in the real line for a t-interval graph. Thus t-track interval graphs are a subclass of t-interval graphs. If a t-interval graph has a representation in which all intervals have unit lengths, then the graph is a *unit t-interval graph*. If a t-interval graph has a representation in which the t disjoint intervals of each t-interval have the same length (although the intervals from different t-intervals may have different lengths), then the graph is a *balanced t-interval graph*. Similarly we define unit t-track interval graphs and balanced t-track interval graphs.

As generalizations of the ubiquitous interval graphs, multiple-interval graphs such as t-interval graphs and t-track interval graphs have numerous applications, traditionally to scheduling and resource allocation [13,1], and more recently to bioinformatics [4,8]. For this reason, a systematic study of various classical optimization

---

* Supported in part by NSF grant DBI-0743670.

D. Marx and P. Rossmanith (Eds.): IPEC 2011, LNCS 7112, pp. 27–40, 2012.
© Springer-Verlag Berlin Heidelberg 2012

problems in multiple-interval graphs has been undertaken by several groups of researchers. In terms of approximability, Bar-Yehuda et al. [1] presented a $2t$-approximation algorithm for MAXIMUM INDEPENDENT SET in t-interval graphs, and Butman et al. [2] presented approximation algorithms for MINIMUM VERTEX COVER, MINIMUM DOMINATING SET, and MAXIMUM CLIQUE in t-interval graphs with approximation ratios $2 - 1/t$, t^2, and $(t^2 - t + 1)/2$, respectively.

Fellows et al. [7] initiated the study of multiple-interval graph problems from the perspective of parameterized complexity. In general graphs, the four problems k-VERTEX COVER, k-INDEPENDENT SET, k-CLIQUE, and k-DOMINATING SET, parameterized by the solution size k, are exemplary problems in parameterized complexity theory [6]: it is well-known that k-VERTEX COVER is in FPT, k-INDEPENDENT SET and k-CLIQUE are W[1]-hard, and k-DOMINATING SET is W[2]-hard. Since t-interval graphs are a special class of graphs, all FPT algorithms for k-VERTEX COVER in general graphs immediately carry over to t-interval graphs. On the other hand, the parameterized complexities of k-INDEPENDENT SET, k-CLIQUE, and k-DOMINATING SET in t-interval graphs are not at all obvious. Indeed, in general graphs, k-INDEPENDENT SET and k-CLIQUE are essentially the same problem (the problem k-INDEPENDENT SET in any graph G is the same as the problem k-CLIQUE in the complement graph $\overline{G}$), but in t-interval graphs, they manifest different parameterized complexities. Fellows et al. [7] showed that k-INDEPENDENT SET in t-interval graphs is W[1]-hard for any $t \geq 2$, then, in sharp contrast, gave an FPT algorithm for k-CLIQUE in t-interval graphs parameterized by both k and t. Fellows et al. [7] also showed that k-DOMINATING SET in t-interval graphs is W[1]-hard for any $t \geq 2$. Recently, Jiang [9] strengthened the two hardness results for t-interval graphs, and showed that k-INDEPENDENT SET and k-DOMINATING SET remain W[1]-hard even in unit t-track interval graphs for any $t \geq 2$. In particular, we have the following theorem on the parameterized complexity of k-DOMINATING SET in unit 2-track interval graphs:

Theorem 1 (Jiang 2010 [9]). k-DOMINATING SET *in unit 2-track interval graphs is W[1]-hard with parameter* k.

The lack of symmetry in the parameterized complexities of k-INDEPENDENT SET and k-CLIQUE in multiple-interval graphs and their complements leads to a natural question about k-DOMINATING SET, which is known to be W[1]-hard in multiple-interval graphs: Is it still W[1]-hard in the complements of multiple-interval graphs? Our following theorem (here "co-3-track interval graphs" denotes "complements of 3-track interval graphs") gives a positive answer:

Theorem 2. k-DOMINATING SET *in co-3-track interval graphs is W[1]-hard with parameter* k.

A *connected dominating set* in a graph G is a dominating set S in G such that the induced subgraph $G(S)$ is connected. An *independent dominating set* in a graph G is both a dominating set and an independent set in G. A *dominating clique* in a graph G is both a dominating set and a clique in G. With connectivity taken in account, the problem k-DOMINATING SET has three important variants: k-CONNECTED DOMINATING SET, k-INDEPENDENT DOMINATING SET, and k-DOMINATING CLIQUE. Recall the sharp contrast in parameterized complexities of the two problems k-INDEPENDENT SET and k-CLIQUE in multiple-interval graphs and their complements. This leads to more natural questions about k-DOMINATING SET: Are the two problems k-INDEPENDENT DOMINATING SET and k-DOMINATING CLIQUE still W[1]-hard in multiple-interval graphs and their complements? Also, without veering to either extreme, how about k-CONNECTED DOMINATING SET?

We show that our FPT reduction for the W[1]-hardness of k-DOMINATING SET in co-3-track interval graphs in Theorem 2 also establishes the following theorem:

Theorem 3. k-CONNECTED DOMINATING SET *and* k-DOMINATING CLIQUE *in co-3-track interval graphs are both W[1]-hard with parameter* k.

Similarly, it is not difficult to verify that the FPT reduction for the W[1]-hardness of k-DOMINATING SET in unit 2-track interval graphs [9] also establishes the following theorem:

Theorem 4. k-INDEPENDENT DOMINATING SET *in unit 2-track interval graphs is W[1]-hard with parameter* k.

For the two problems k-CONNECTED DOMINATING SET and k-DOMINATING CLIQUE in multiple-interval graphs, we obtain a weaker result:

Theorem 5. k-CONNECTED DOMINATING SET *and* k-DOMINATING CLIQUE *in unit 3-track interval graphs are both W[1]-hard with parameter* k.

Another important variant (indeed a generalization) of k-DOMINATING SET is called d-DISTANCE k-DOMINATING SET, where each vertex is able to dominate all vertices within a threshold distance d. Note that k-DOMINATING SET is simply d-DISTANCE k-DOMINATING SET with $d = 1$. For this distance variant of k-DOMINATING SET, we obtain the following theorem:

Theorem 6. d-DISTANCE k-DOMINATING SET *for any* $d \geq 2$ *in balanced 3-interval graphs is W[1]-hard with parameter* k.

The last variant of k-DOMINATING SET that we study in this paper is called k-PERFECT CODE. A *perfect code* in a graph $G = (V, E)$, also known as a *perfect dominating set* or an *efficient dominating set*, is a subset of vertices $V' \subseteq V$ that includes exactly one vertex from the closed neighborhood of each vertex $u \in V$. Recall that the *open neighborhood* of u is $N(u) = \{v \mid \{u, v\} \in E\}$, and that the *closed neighborhood* of u is $N[u] = N(u) \cup \{u\}$. The problem k-PERFECT CODE is that of deciding whether a given graph G has a perfect code of size exactly k. It is known to be W[1]-complete with parameter k in general graphs [5,3]. Since every graph of maximum degree 3 is the intersection graph of a family of unit 2-track intervals [10, Theorem 4], it follows that k-PERFECT CODE is NP-complete in unit 2-track interval graphs. In the following theorem, we show that k-PERFECT CODE is indeed W[1]-hard in unit 2-track interval graphs:

Theorem 7. k-PERFECT CODE *in unit 2-track interval graphs is W[1]-hard with parameter* k.

The distance variant of k-PERFECT CODE, denoted as d-DISTANCE k-PERFECT CODE, is also studied in the literature [12]. We show that d-DISTANCE k-PERFECT CODE is also W[1]-hard in unit 2-track interval graphs:

Theorem 8. d-DISTANCE k-PERFECT CODE *for any* $d \geq 2$ *in unit 2-track interval graphs is W[1]-hard with parameter* k.

We refer to [11] for some related results. All proofs of W[1]-hardness in this paper are based on FPT reductions from the W[1]-complete problem k-MULTICOLORED CLIQUE [7]: Given a graph G of n vertices and m edges, and a vertex-coloring κ : $V(G) \rightarrow \{1, 2, \ldots, k\}$, decide whether G has a clique of k vertices containing exactly one vertex of each color. Without loss of generality, we assume that no edge in G connects two vertices of the same color.

2   Dominating Set

In this section we prove Theorem 2. We show that k-DOMINATING SET in co-3-track interval graphs is W[1]-hard by an FPT reduction from the W[1]-complete problem k-MULTICOLORED CLIQUE [7].

Let (G, κ) be an instance of k-MULTICOLORED CLIQUE. We will construct a family $\mathcal{F}$ of 3-track intervals such that G has a clique of k vertices containing exactly one vertex of each color if and only if the complement of the intersection graph $G_{\mathcal{F}}$ of $\mathcal{F}$ has a dominating set of k' vertices, where $k' = k + \binom{k}{2}$.

Vertex selection: Let $v_1, \ldots, v_n$ be the set of vertices in G, sorted by color such that the indices of all vertices of each color are contiguous. For each color i, $1 \leq i \leq k$, let $V_i = \{v_p \mid s_i \leq p \leq t_i\}$ be the set of vertices v_p of color i. For each vertex v_p, $1 \leq p \leq n$, let $\langle v_p \rangle$ be a *vertex 3-track interval* consisting of the following three intervals on the three tracks:

$$\langle v_p \rangle = \begin{cases} \text{track } 1 : (p - 1, p) \\ \text{track } 2 : (p - 1 + m + 1, p + m + 1) \\ \text{track } 3 : (p - 1 + m + 1, p + m + 1). \end{cases}$$

For each color i, $1 \leq i \leq k$, let $\langle V_i \rangle$ be the following 3-track interval:

$$\langle V_i \rangle = \begin{cases} \text{track } 1 : (t_i, m + n + 1) \\ \text{track } 2 : (0, s_i - 1 + m + 1) \\ \text{track } 3 : (m, m + 1). \end{cases}$$

Edge selection: Let $e_1, \ldots, e_m$ be the set of edges in G, also sorted by color such that the indices of all edges of each color pair are contiguous. For each pair of distinct colors i and j, $1 \leq i < j \leq k$, let $E_{ij} = \{e_r \mid s_{ij} \leq r \leq t_{ij}\}$ be the set of edges $v_p v_q$ such that v_p has color i and v_q has color j. For each edge e_r, $1 \leq r \leq m$, let $\langle e_r \rangle$ be an *edge 3-track interval* consisting of the following three intervals on the three tracks:

$$\langle e_r \rangle = \begin{cases} \text{track } 1 : (r - 1 + n + 1, r + n + 1) \\ \text{track } 2 : (r - 1, r) \\ \text{track } 3 : (r - 1, r). \end{cases}$$

For each pair of distinct colors i and j, $1 \leq i < j \leq k$, let $\langle E_{ij} \rangle$ be the following 3-track interval:

$$\langle E_{ij} \rangle = \begin{cases} \text{track 1}: (0, s_{ij} - 1 + n + 1) \\ \text{track 2}: (t_{ij}, n + m + 1) \\ \text{track 3}: (m, m + 1). \end{cases}$$

Validation: For each edge $e_r = v_p v_q$ such that v_p has color i and v_q has color j, let $\langle v_p e_r \rangle$ and $\langle v_q e_r \rangle$ be the following 3-track intervals:

$$\langle v_p e_r \rangle = \begin{cases} \text{track 1}: (p, s_{ij} - 1 + n + 1) \\ \text{track 2}: (t_{ij}, p - 1 + m + 1) \\ \text{track 3}: (r - 1, r), \end{cases} \quad \langle v_q e_r \rangle = \begin{cases} \text{track 1}: (q, s_{ij} - 1 + n + 1) \\ \text{track 2}: (t_{ij}, q - 1 + m + 1) \\ \text{track 3}: (r - 1, r). \end{cases}$$

Let $\mathcal{F}$ be the following family of $n + m + k + \binom{k}{2} + 2m$ 3-track intervals:

$$\mathcal{F} = \left\{ \langle v_p \rangle \mid 1 \leq p \leq n \right\} \cup \left\{ \langle e_r \rangle \mid 1 \leq r \leq m \right\}$$
$$\cup \left\{ \langle V_i \rangle \mid 1 \leq i \leq k \right\} \cup \left\{ \langle E_{ij} \rangle \mid 1 \leq i < j \leq k \right\}$$
$$\cup \left\{ \langle v_p e_r \rangle, \langle v_q e_r \rangle \mid e_r = v_p v_q \in E_{ij}, 1 \leq i < j \leq k \right\}.$$

This completes the construction. We refer to Figure 1 for an example. The following five properties of the construction can be easily verified:

1. For each color i, $1 \leq i \leq k$, all 3-track intervals $\langle v_p \rangle$ for $v_p \in V_i$ are pairwise-disjoint.
2. For each color i, $1 \leq i \leq k$, $\langle V_i \rangle$ intersects all other 3-track intervals except the vertex 3-track intervals $\langle v_p \rangle$ for $v_p \in V_i$.
3. For each pair of distinct colors i and j, $1 \leq i < j \leq k$, all 3-track intervals $\langle e_r \rangle$ for $e_r \in E_{ij}$ are pairwise-disjoint.
4. For each pair of distinct colors i and j, $1 \leq i < j \leq k$, $\langle E_{ij} \rangle$ intersects all other 3-track intervals except the edge 3-track intervals $\langle e_r \rangle$ for $e_r \in E_{ij}$.
5. For each pair of distinct colors i and j, $1 \leq i < j \leq k$, for each edge $e_r \in E_{ij}$ and each vertex v_p incident to e_r, $\langle v_p e_r \rangle$ intersects all other 3-track intervals except the vertex 3-track interval $\langle v_p \rangle$ and the edge 3-track intervals for the edges in E_{ij} other than $\langle e_r \rangle$.

Lemma 1. *G has a k-multicolored clique if and only if $\overline{G_{\mathcal{F}}}$ has a k'-dominating set, where $k' = k + \binom{k}{2}$.*

Proof. For the direct implication, if $K \subseteq V(G)$ is a k-multicolored clique in G, then the following subset $\mathcal{D} \subseteq \mathcal{F}$ of 3-track intervals is a k'-dominating set in $\overline{G_{\mathcal{F}}}$:

$$\mathcal{D} = \left\{ \langle v_p \rangle \mid v_p \in K \right\} \cup \left\{ \langle e_r \rangle \mid v_p, v_q \in K, e_r = v_p v_q \right\}.$$

To verify this, check that each $\langle v_p \rangle \notin \mathcal{D}$ is dominated by $\langle v_{p'} \rangle \in \mathcal{D}$ for some vertex $v_{p'}$ of the same color as v_p (Property 1), each $\langle e_r \rangle \notin \mathcal{D}$ is dominated by $\langle e_{r'} \rangle \in \mathcal{D}$ for some edge $e_{r'}$ of the same color pair as e_r (Property 3), each $\langle V_i \rangle$ is dominated by $\langle v_p \rangle \in \mathcal{D}$ for some $v_p \in V_i$ (Property 2), each $\langle E_{ij} \rangle$ is dominated by $\langle e_r \rangle \in \mathcal{D}$ for some $e_r \in E_{ij}$

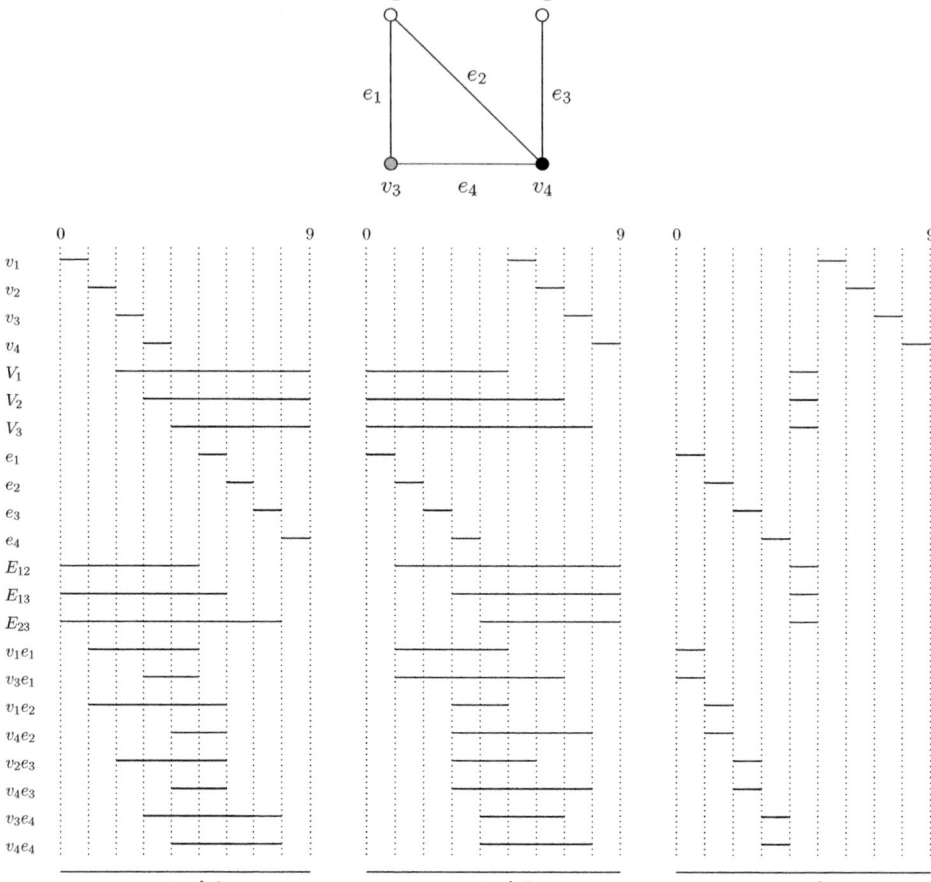

Fig. 1. Top: A graph G of $n = 4$ vertices v_1, v_2, v_3, v_4 and $m = 4$ edges $e_1 = v_1v_3, e_2 = v_1v_4, e_3 = v_2v_4, e_4 = v_3v_4$, with $k = 3$ colors $\kappa(v_1) = \kappa(v_2) = 1, \kappa(v_3) = 2$, and $\kappa(v_4) = 3$. $V_1 = \{v_1, v_2\}, V_2 = \{v_3\}, V_3 = \{v_4\}$; $E_{12} = \{e_1\}, E_{13} = \{e_2, e_3\}, E_{23} = \{e_4\}$. $K = \{v_1, v_3, v_4\}$ is a 3-multicolored clique. Bottom: A family $\mathcal{F}$ of $n + m + k + \binom{k}{2} + 2m = 22$ 3-track intervals. $\mathcal{D} = \{\langle v_1 \rangle, \langle v_3 \rangle, \langle v_4 \rangle, \langle e_1 \rangle, \langle e_2 \rangle, \langle e_4 \rangle\}$ is a 6-dominating set in the complement of the intersection graph of $\mathcal{F}$.

(Property 4), and each $\langle v_p e_r \rangle$ is dominated either by $\langle v_p \rangle \in \mathcal{D}$, when $v_p \in K$, or by $\langle e_{r'} \rangle \in \mathcal{D}$ for some edge $e_{r'}$ of the same color pair as e_r, when $v_p \notin K$ (Property 5).

For the reverse implication, suppose that $\mathcal{D} \subseteq \mathcal{F}$ is a k'-dominating set in $\overline{G_{\mathcal{F}}}$. We will find a k-multicolored clique $K \subseteq V(G)$ in G. For each color i, $1 \leq i \leq k$, $\mathcal{D}$ must include either $\langle V_i \rangle$ or at least one of its neighbors in $\overline{G_{\mathcal{F}}}$. Thus by Properties 1 and 2, we can assume without loss of generality that $\mathcal{D}$ does not include $\langle V_i \rangle$ but includes at least one vertex 3-track interval $\langle v_p \rangle$ for some $v_p \in V_i$. Similarly, for each pair of distinct colors i and j, $1 \leq i < j \leq k$, we can assume by Properties 3 and 4 that $\mathcal{D}$ does not include $\langle E_{ij} \rangle$ but includes at least one edge 3-track interval $\langle e_r \rangle$ for some $e_r \in E_{ij}$.

Since $k' = k + \binom{k}{2}$, it follows that $\mathcal{D}$ includes exactly one vertex 3-track interval of each color, and exactly one edge 3-track interval of each color pair. For each pair of distinct colors i and j, $1 \leq i < j \leq k$, let $e_r = v_p v_q$ be the edge whose 3-track interval $\langle e_r \rangle$ is included in $\mathcal{D}$. By Property 5 of the construction, the two 3-track intervals $\langle v_p e_r \rangle$ and $\langle v_q e_r \rangle$ cannot be dominated by $\langle e_r \rangle$ and hence must be dominated by $\langle v_p \rangle$ and $\langle v_q \rangle$, respectively. Therefore the vertex selection and the edge selection are consistent, and the set of k vertex 3-track intervals in $\mathcal{D}$ corresponds to a k-multicolored clique K in G.

3 Connected Dominating Set, Independent Dominating Set, and Dominating Clique

In this section we prove Theorems 3, 4, and 5.

For Theorem 3, to show the W[1]-hardness of k-CONNECTED DOMINATING SET and k-DOMINATING CLIQUE in co-3-track interval graphs, let us review our FPT reduction for Theorem 2, in particular, the proof of Lemma 1, in the previous section. Observe that for the direct implication of Lemma 1, our proof composes a dominating set $\mathcal{D}$ of pairwise-disjoint 3-track intervals, and that for the reverse implication of Lemma 1, our proof uses only the fact that $\mathcal{D}$ is a dominating set without any assumption about its connectedness. This implies that our FPT reduction for Theorem 2 also establishes Theorem 3. By a similar argument, it is not difficult to verify that the FPT reduction for the W[1]-hardness of k-DOMINATING SET in unit 2-track interval graphs [9] also establishes the W[1]-hardness of k-INDEPENDENT DOMINATING SET in unit 2-track interval graphs in Theorem 4.

For Theorem 5, to show the W[1]-hardness of k-CONNECTED DOMINATING SET and k-DOMINATING CLIQUE in unit 3-track interval graphs, we use the same construction as in the previous FPT reduction for the W[1]-hardness of k-DOMINATING SET in unit 2-track interval graphs [9] for the first two tracks. Then, on track 3, we use the same (coinciding) unit interval for all multiple-intervals in

$$\mathcal{F}' = \left\{ \widehat{u_i} \mid u \in V_i, 1 \leq i \leq k \right\} \cup \left\{ \widehat{u_i v_j}_{\text{left}}, \widehat{u_i v_j}_{\text{right}} \mid uv \in E_{ij}, 1 \leq i < j \leq k \right\},$$

and use a distinct unit interval disjoint from all other unit intervals for each of the remaining multiple-intervals. Now the dominating set composed in the direct implication of the proof in [9] becomes a clique. Since the reverse implication of the proof in [9] does not depend on the additional intersections between the multiple-intervals in $\mathcal{F}'$, the modified reduction establishes Theorem 5.

4 Distance Dominating Set

In this section we prove Theorem 6. We show that d-DISTANCE k-DOMINATING SET is W[1]-hard in 3-interval graphs for any $d \geq 2$ by an FPT reduction again from k-MULTICOLORED CLIQUE.

First we consider the case $d = 2$. Let (G, κ) be an instance of k-MULTICOLORED CLIQUE. We will construct a family $\mathcal{F}$ of 3-intervals as illustrated in Figure 2 such that G has a k-multicolored clique if and only if the intersection graph $G_{\mathcal{F}}$ of $\mathcal{F}$ has a 2-distance k'-dominating set, where $k' = k + \binom{k}{2}$. For convenience, we specify some 3-intervals in $\mathcal{F}$ as 2-intervals or intervals, and assume an implicit extension of each 2-interval or interval to a 3-interval by adding extra intervals that are disjoint from the other intervals in $\mathcal{F}$. We use (u, v, w) to denote a 3-interval that is the union of three disjoint intervals u, v, w, in no particular order. Similarly, we use (u, v) for a 2-interval.

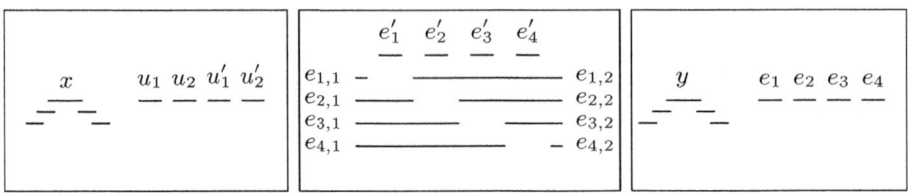

Fig. 2. The vertex gadget for V_i (left) is connected to the edge gadget for E_{ij} (right) by a validation gadget (middle)

Vertex selection: For each color i, $1 \le i \le k$, let V_i be the set of vertices of color i. Write $|V_i| = \phi$. There are $2\phi+1$ disjoint intervals labeled with $x, u_1, \ldots, u_\phi, u'_1, \ldots, u'_\phi$ in the vertex selection gadget for V_i. For each vertex $u = u_s \in V_i$, we add two 2-intervals $\langle u \rangle_1 = (x, u_s)$ and $\langle u \rangle_2 = (u_s, u'_s)$ to $\mathcal{F}$. We also add four dummy intervals to $\mathcal{F}$: two dummy intervals intersect with x; the other two dummy intervals intersect with the first two dummy intervals, respectively.

Edge Selection: For each pair of distinct colors i and j, $1 \le i < j \le k$, let E_{ij} be the set of edges uv such that u has color i and v has color j. Write $|E_{ij}| = \psi$. There are $\psi + 1$ disjoint intervals labeled with $y, e_1, \ldots, e_\psi$ in the edge selection gadget for E_{ij}. For each edge $e = e_s \in E_{ij}$, we add a 2-interval $\langle e \rangle = (y, e_s)$ to $\mathcal{F}$. We also add four dummy intervals to $\mathcal{F}$ in the same way as in each vertex selection gadget.

Validation: For each pair of distinct colors i and j, $1 \le i < j \le k$, we construct two validation gadgets that connect the two vertex gadgets for V_i and V_j, respectively, to the edge gadget for E_{ij}. In the following we describe the validation gadget between the vertex gadget for V_i and the edge gadget for E_{ij}; the construction of the other validation gadget is similar. Write $|E_{ij}| = \psi$. There are 3ψ intervals in this validation gadget. First, there are ψ disjoint intervals labeled with $e'_1, \ldots, e'_\psi$. Then, for each e'_s, there are two disjoint intervals $e_{s,1}$ and $e_{s,2}$ intersecting with all intervals e'_t with $t \ne s$. For each edge $e = e_s \in E_{ij}$, we add a 2-interval $\langle e, i \rangle = (e_s, e'_s)$ to $\mathcal{F}$. For each vertex $u = u_t \in V_i$ incident to some edge $e = e_s \in E_{ij}$, we add a 3-interval $\langle u, e \rangle = (u'_t, e_{s,1}, e_{s,2})$ to $\mathcal{F}$.

In summary, the construction gives us the following family $\mathcal{F}$ of 3-intervals:

$$\mathcal{F} = \big\{ \langle u \rangle_1, \langle u \rangle_2 \mid u \in V_i, 1 \le i \le k \big\} \cup \big\{ \langle e \rangle \mid e \in E_{ij}, 1 \le i < j \le k \big\}$$
$$\cup \big\{ \langle e, i \rangle, \langle e, j \rangle, \langle u, e \rangle, \langle v, e \rangle \mid e = uv \in E_{ij}, 1 \le i < j \le k \big\} \cup \text{DUMMIES},$$

where DUMMIES is the set of $4k + 4\binom{k}{2}$ dummy intervals.

Observe that for each pair of disjoint intervals $e_{s,1}$ and $e_{s,2}$ in the validation gadgets, we can extend $e_{s,1}$ to the left and extend $e_{s,2}$ to the right until they have the same length. This does not change the intersection pattern of the intervals. Therefore $\mathcal{F}$ can be transformed into a family of balanced 3-intervals, where $\langle e \rangle, \langle e, i \rangle, \langle e, j \rangle$ and DUMMIES use intervals of length 1, and $\langle u \rangle_1, \langle u \rangle_2, \langle u, e \rangle$ use intervals of length m, where m is the number of edges in G.

Lemma 2. *G has a k-multicolored clique if and only if $G_{\mathcal{F}}$ has a 2-distance k'-dominating set, where $k' = k + \binom{k}{2}$.*

Proof. We first prove the direct implication. Suppose G has a k-multicolored clique $K \subseteq V(G)$, then it is easy to verify the following subfamily $\mathcal{D}$ of 3-intervals is a 2-distance k'-dominating set in $G_{\mathcal{F}}$:

$$\mathcal{D} = \big\{ \langle u \rangle_1 \mid u \in K \big\} \cup \big\{ \langle e \rangle \mid e = uv,\ u, v \in K \big\}.$$

We next prove the reverse implication. Suppose that $\mathcal{D}$ is a 2-distance k'-dominating set in $G_{\mathcal{F}}$. In order to dominate the dummies we can assume without loss of generality that $\mathcal{D}$ includes at least one $\langle u \rangle_1$ from each vertex gadget and at least one $\langle e \rangle$ from each edge gadget. Since $\mathcal{D}$ has size $k' = k + \binom{k}{2}$, we must have exactly one $\langle u \rangle_1$ from each vertex gadget and exactly one $\langle e \rangle$ from each edge gadget in $\mathcal{D}$. Consider $\langle e \rangle$ from the edge gadget for E_{ij}, where $e = uv$. Note that $\langle e \rangle$ dominates all multiple-intervals in the two validation gadgets for E_{ij} except $\langle u, e \rangle$ and $\langle v, e \rangle$, which must be dominated by $\langle u \rangle$ and $\langle v \rangle$, respectively, in the corresponding vertex gadgets. Therefore the subset of vertices $K = \{v \in V(G) \mid \langle v \rangle_1 \in \mathcal{D}\}$ is a k-multicolored clique in G. □

The above construction can be easily generalized to handle the case $d > 2$. To do this, extend each vertex gadget to include d pairs of dummy intervals instead of two pairs, and to include d disjoint intervals for each vertex u (instead of only the two labeled with u and u') such that there is a path of length $d - 1$ from $\langle u \rangle_1$ to $\langle u \rangle_d$ in $G_{\mathcal{F}}$. Extend each edge gadget in a similar way. Then the same argument applies.

5 Perfect Code

In this section we prove Theorem 7. We show that k-PERFECT CODE in unit 2-track interval graphs is W[1]-hard by a reduction from k-MULTICOLORED CLIQUE.

Let (G, κ) be an instance of k-MULTICOLORED CLIQUE. We will construct a family $\mathcal{F}$ of unit 2-track intervals such that G has a k-multicolored clique if and only if the intersection graph $G_{\mathcal{F}}$ of $\mathcal{F}$ has a k'-perfect code, where $k' = k + 2\binom{k}{2}$.

Vertex selection: For each color i, $1 \leq i \leq k$, let V_i be the set of vertices of color i. We construct a vertex-selection gadget for V_i as illustrated in Figure 3. Write $|V_i| = \phi$. On each track, we start with 2ϕ unit intervals arranged in ϕ rows and two (slanted) columns. The ϕ intervals in each column are pairwise-intersecting. The two intervals in each row slightly overlap such that each interval in the left column intersects with all intervals in the same or higher rows in the right column. For the rth vertex u in V_i, $1 \leq r \leq \phi$, we add a *vertex 2-track interval* $\langle u \rangle = (u_1, u_2)$ to $\mathcal{F}$, where u_1 and u_2

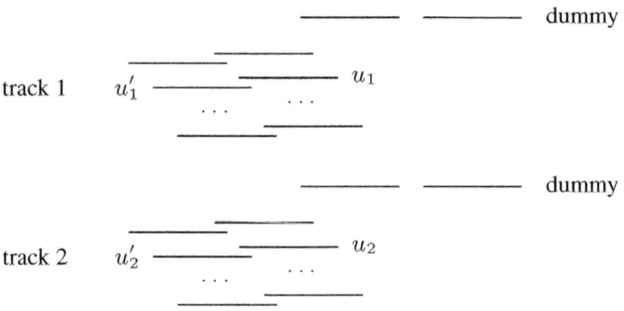

Fig. 3. An illustration of a vertex-selection gadget

are the intervals in the rth row and the right column on tracks 1 and 2, respectively. Denote by u'_1 and u'_2 the intervals in the rth row and the left column on tracks 1 and 2, respectively; they will be used for validation. Besides the ϕ vertex 2-track intervals $\langle u \rangle$, we also add two dummy 2-track intervals to $\mathcal{F}$. The first (resp. second) dummy 2-interval consists of a unit interval on track 1 (resp. track 2) that intersects all intervals in the right column and no interval in the left column, and a unit interval on track 2 (resp. track 1) that is disjoint from all other intervals.

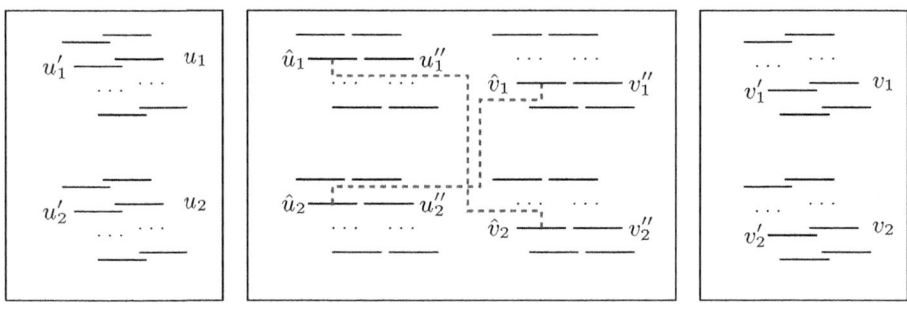

Fig. 4. An illustration of an edge-selection gadget (middle) and the corresponding vertex-selection gadgets (left and right). Two edge 2-track intervals $(\hat{u}_1, \hat{v}_2)$ and $(\hat{u}_2, \hat{v}_1)$ are represented by dashed lines. Dummy 2-track intervals are omitted from the figure.

Edge selection: For each pair of distinct colors i and j, $1 \leq i < j \leq k$, let E_{ij} be the set of edges uv such that u has color i and v has color j. We construct an edge selection gadget for E_{ij} as illustrated in Figure 4. We start with four disjoint groups of intervals, two groups on each track, with two columns of intervals in each group. Write $|V_i| = \phi_i$ and $|V_j| = \phi_j$. The two groups on the left correspond to color i and have ϕ_i rows; the two groups on the right correspond to color j and have ϕ_j rows. Different from the formation in the vertex selection gadgets, here in each group each interval in the left column intersects with all intervals in higher rows in the right column but not the interval in the same row. In the two groups on the left, for the rth vertex $u \in V_i$, $1 \leq r \leq \phi_i$, denote by $\hat{u}_1$ and $\hat{u}_2$ the intervals in the rth row and the left column on tracks 1 and 2, respectively, and denote by u_1'' and u_2'' the intervals in the rth row and the right column on tracks 1 and 2, respectively. Similarly, for each vertex $v \in V_j$, denote by $\hat{v}_1, \hat{v}_2, v_1', v_2'$ the corresponding intervals in the two groups on the right. For each edge $uv \in E_{ij}$, we add two *edge 2-track intervals* $\langle uv \rangle_1 = (\hat{u}_1, \hat{v}_2)$ and $\langle uv \rangle_2 = (\hat{u}_2, \hat{v}_1)$ to $\mathcal{F}$. Besides these edge 2-track intervals, we also add four dummy 2-track intervals to $\mathcal{F}$, one for each group of intervals. The dummy 2-track interval for each group consists of a unit interval that intersects all intervals in the left column and no interval in the right column in the group, and a unit interval on the other track that is disjoint from all other intervals.

Validation: For each pair of distinct colors i and j, $1 \leq i < j \leq k$, we add $2|V_i| + 2|V_j|$ *validation 2-track intervals* to $\mathcal{F}$ as illustrated in Figure 4. Specifically, for each vertex $u \in V_i$, we add $\langle u*_{ij} \rangle_1 = (u_1', u_2'')$ and $\langle u*_{ij} \rangle_2 = (u_2', u_1'')$, and for each vertex $v \in V_j$, we add $\langle *v_{ij} \rangle_1 = (v_1', v_2'')$ and $\langle *v_{ij} \rangle_2 = (v_2', v_1'')$.

In summary, the construction gives us the following family $\mathcal{F}$ of unit 2-track intervals:

$$\mathcal{F} = \big\{ \langle u \rangle \mid u \in V_i, 1 \leq i \leq k \big\} \cup \big\{ \langle uv \rangle_1, \langle uv \rangle_2 \mid uv \in E_{ij}, 1 \leq i < j \leq k \big\}$$
$$\cup \big\{ \langle u*_{ij} \rangle_1, \langle u*_{ij} \rangle_2, \langle *v_{ij} \rangle_1, \langle *v_{ij} \rangle_2 \mid u \in V_i, v \in V_j, 1 \leq i < j \leq k \big\}$$
$$\cup \text{DUMMIES},$$

where DUMMIES is the set of $2k + 4\binom{k}{2}$ dummy 2-track intervals.

Lemma 3. *G has a k-multicolored clique if and only if $G_{\mathcal{F}}$ has a k'-perfect code, where $k' = k + 2\binom{k}{2}$.*

Proof. We first prove the direct implication. Suppose G has a k-multicolored clique $K \subseteq V(G)$, then it is easy to verify that the following subfamily $\mathcal{D}$ of unit 2-track intervals is a k'-perfect code in $G_{\mathcal{F}}$:

$$\mathcal{D} = \big\{ \langle u \rangle \mid u \in K \big\} \cup \big\{ \langle uv \rangle_1, \langle uv \rangle_2 \mid u, v \in K \big\}.$$

We next prove the reverse implication. Suppose $\mathcal{D}$ is a k'-perfect code in $G_{\mathcal{F}}$. Observe that the dummy 2-track intervals in our construction are pairwise-disjoint. Moreover, the two dummies in each vertex gadget share the same open neighborhood which is not empty, and the same is true about the two dummies associated with the two groups of intervals, the left group on track 1 and the right group on track 2 (resp. the right group on track 1 and the left group on track 2) of each edge gadget. It follows that these dummies cannot be included in $\mathcal{D}$. In order to perfectly dominate the dummies, $\mathcal{D}$ must include exactly one vertex 2-track interval $\langle u \rangle$ from each vertex selection gadget and two edge 2-track intervals $\langle uv \rangle_1$ and $\langle xy \rangle_2$ from each edge selection gadget. Consider an edge 2-track interval $\langle uv \rangle_1 = (\hat{u}_1, \hat{v}_2)$ from the edge selection gadget for E_{ij}, and observe the validation 2-track intervals dominated by $\langle uv \rangle_1$. To perfectly dominate the validation 2-track intervals $\langle w*_{ij} \rangle_2$ for all $w \in V_i$, $\mathcal{D}$ must include $\langle u \rangle$ from the vertex selection gadget for V_i. Similarly, to perfectly dominate the validation 2-track intervals $\langle *w_{ij} \rangle_1$

for all $w \in V_j$, $\mathcal{D}$ must include $\langle v \rangle$ from the vertex selection gadget for V_j. Then, to perfectly dominate the validation 2-track intervals $\langle w*_{ij} \rangle_1$ for all $w \in V_i$, and $\langle *w_{ij} \rangle_2$ for all $w \in V_j$, the two intervals $\hat{u}_2$ and $\hat{v}_1$ must be used. This implies that the other edge 2-track interval from the same edge selection gadget must be $\langle uv \rangle_2 = (\hat{u}_2, \hat{v}_1)$. Therefore the subset of vertices $K = \{u \in V(G) \mid \langle u \rangle \in \mathcal{D}\}$ is a k-multicolored clique in G. $\square$

6   Distance Perfect Code

In this section we prove Theorem 8. We show that d-DISTANCE k-PERFECT CODE is W[1]-hard in unit 2-interval graphs for any $d \geq 2$ by an FPT reduction from k-MULTICOLORED CLIQUE.

We consider the case $d = 2$ first. Let (G, κ) be an instance of k-MULTICOLORED CLIQUE. We will construct a family $\mathcal{F}$ of unit 2-intervals as illustrated in Figure 5 such that G has a k-multicolored clique if and only if the intersection graph $G_{\mathcal{F}}$ of $\mathcal{F}$ has a 2-distance k'-perfect code, where $k' = k + \binom{k}{2}$.

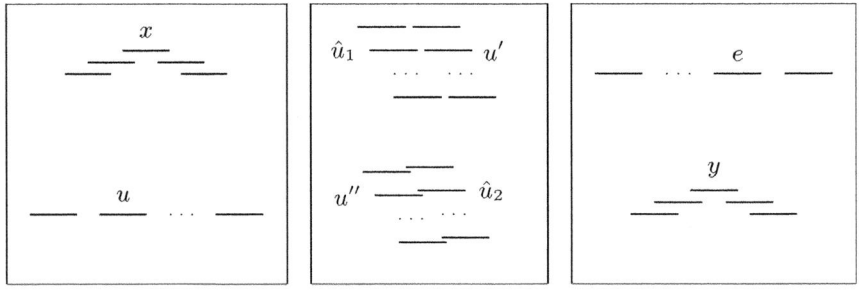

Fig. 5. The vertex gadget for V_i (left) is connected to the edge gadget for E_{ij} (right) by a validation gadget (middle)

Vertex selection: For each color i, $1 \leq i \leq k$, let V_i be the set of vertices of color i. We construct a vertex-selection gadget for V_i as illustrated in Figure 5. Write $|V_i| = \phi$. On track 1 there is an interval labeled by x. On track 2 there are ϕ disjoint intervals, one for each vertex in V_i. For the rth vertex u in V_i, $1 \leq r \leq \phi$, we add a 2-track interval $\langle u \rangle = (x, u)$ to $\mathcal{F}$. We also add four dummy 2-track intervals to $\mathcal{F}$: two dummy 2-track intervals intersect with x; the other two dummy 2-track intervals intersect with the first two dummy 2-track intervals, respectively. In figure 5, only one interval (on track 1) of each dummy 2-track intervals is drawn.

Edge selection: For each pair of distinct colors i and j, $1 \leq i < j \leq k$, let E_{ij} be the set of edges uv such that u has color i and v has color j. Write $|E_{ij}| = \psi$. There are ψ disjoint intervals on track 1, one for each edge in E_{ij}. There is an interval labeled by y on track 2. For each edge $e \in E_{ij}$, add a 2-track interval $\langle e \rangle = (y, e)$ to $\mathcal{F}$. We also add four dummy 2-track intervals to $\mathcal{F}$ in the similar way as in each vertex selection gadget.

Validation selection: For each pair of distinct colors i and j, $1 \leq i < j \leq k$, we construct two validation gadgets that connect the two vertex gadgets for V_i and V_j, respectively, to the edge gadget for E_{ij}. First we describe the validation gadget between the vertex gadget for V_i and the edge gadget for E_{ij}. Write $|V_i| = \phi$ and $|E_{ij}| = \psi$. On track 1, there are 2ϕ interval arranged in ϕ rows and two (slanted) columns. The ϕ intervals in each column are pairwise-intersecting. Moreover, each interval in the left column intersects with all intervals in higher rows in the right column but not the interval in the same row. For the rth vertex $u \in V_i$, $1 \leq r \leq \phi$, denote by $\hat{u}_1$ and u' the left and right intervals, respectively, in the rth row. On track 2, the arrangement of the 2ϕ intervals are similar except that each interval in the left column intersects with all intervals in the higher rows *and* the interval in the same row. Denote by u'' and $\hat{u}_2$ the left and right intervals, respectively, in the rth row. We add $2\phi + \psi$ validation 2-track intervals to $\mathcal{F}$. For each vertex $u \in V_i$, add $\langle u*_{ij} \rangle_1 = (u, u')$ and $\langle u*_{ij} \rangle_2 = (\hat{u}_1, \hat{u}_2)$ to $\mathcal{F}$. For each edge $e = uv \in E_{ij}$, add $\langle u, e \rangle = (e, u'')$ to $\mathcal{F}$.

The validation gadget between the vertex gadget for V_j and the edge gadget for E_{ij} (not shown in Figure 5) is constructed similarly. For each vertex $v \in V_j$, we add $\langle *v_{ij} \rangle_1 = (v, v')$ and $\langle *v_{ij} \rangle_2 = (\hat{v}_1, \hat{v}_2)$ to $\mathcal{F}$. For each edge $e = uv \in E_{ij}$, we add $\langle v, e \rangle = (e, v'')$ to $\mathcal{F}$.

In summary, the construction gives us the following family $\mathcal{F}$ of unit 2-track intervals:

$$\mathcal{F} = \big\{ \langle u \rangle \mid u \in V_i, 1 \leq i \leq k \big\} \cup \big\{ \langle e \rangle \mid e \in E_{ij}, 1 \leq i < j \leq k \big\}$$
$$\cup \big\{ \langle u*_{ij} \rangle_1, \langle u*_{ij} \rangle_2, \langle *v_{ij} \rangle_1, \langle *v_{ij} \rangle_2 \mid u \in V_i, v \in V_j, 1 \leq i < j \leq k \big\}$$
$$\cup \big\{ \langle u, e \rangle, \langle v, e \rangle \mid e = uv \in E_{ij}, 1 \leq i < j \leq k \big\} \cup \text{DUMMIES},$$

where DUMMIES is the set of $4k + 4\binom{k}{2}$ dummy 2-track intervals.

Lemma 4. *G has a k-multicolored clique if and only if $G_{\mathcal{F}}$ has a 2-distance k'-perfect code, where $k' = k + \binom{k}{2}$.*

Proof. We first prove the direct implication. Suppose G has a k-multicolored clique $K \subseteq V(G)$, then one can verify that the following subfamily $\mathcal{D}$ of 2-track intervals is a 2-distance k'-perfect code in $G_{\mathcal{F}}$:

$$\mathcal{D} = \big\{ \langle u \rangle \mid u \in K \big\} \cup \big\{ \langle e \rangle \mid e = uv, u, v \in K \big\}.$$

We next prove the reverse implication. Suppose that $\mathcal{D}$ is a 2-distance k'-perfect code in $G_{\mathcal{F}}$. By a similar argument as in the proof of Lemma 3, the dummies cannot be included in $\mathcal{D}$. In order to perfectly dominate the dummies, $\mathcal{D}$ must include exactly one $\langle u \rangle$ from each vertex gadget and exactly one $\langle e \rangle$ from each edge gadget. For the rth vertex u and tth vertex w in V_i, we write $u \leq_i w$ if $r \leq t$ and $u >_i w$ if $r > t$. Consider $\langle e \rangle$ from the edge gadget for E_{ij}, where $e = uv$. Observe that in the validation gadget between the vertex gadget for V_i and the edge gadget for E_{ij}, the 2-track intervals $\{ \langle w*_{ij} \rangle_2 \mid w \in V_i, w \leq_i u \}$ are within distance 2 from $\langle e \rangle$. Then, to perfectly dominate the 2-track intervals $\{ \langle w*_{ij} \rangle_2 \mid w \in V_i, w >_i u \}$, the 2-track interval $\langle u \rangle$ from the vertex gadget for V_i must be included in $\mathcal{D}$. Similarly, to perfectly dominate the 2-track intervals $\langle *w_{ij} \rangle_2$ in the other validation gadget, the 2-track interval $\langle v \rangle$ from the vertex gadget for V_j must also be included in $\mathcal{D}$. Therefore the subset of vertices $K = \{u \in V(G) \mid \langle u \rangle \in \mathcal{D}\}$ is a k-multicolored clique in G.   □

The above construction can be generalized to handle the case $d > 2$. We postpone the details to the full version of this paper.

Concluding Remarks. A general direction for extending our work is to strengthen the existing W[1]-hardness results for more restricted graph classes. For example, we showed in Theorem 2 that k-DOMINATING SET in co-3-track interval graphs is W[1]-hard with parameter k. Is it still W[1]-hard in co-2-track interval graphs or co-unit 3-track interval graphs? Many questions can be asked in the same spirit. In particular, are k-INDEPENDENT DOMINATING SET and k-PERFECT CODE W[1]-hard in co-t-interval graphs for some constant t?

References

1. Bar-Yehuda, R., Halldórsson, M.M., Naor, J., Shachnai, H., Shapira, I.: Scheduling split intervals. SIAM Journal on Computing 36, 1–15 (2006)
2. Butman, A., Hermelin, D., Lewenstein, M., Rawitz, D.: Optimization problems in multiple-interval graphs. ACM Transactions on Algorithms 6, 40 (2010)
3. Cesati, M.: Perfect Code is W[1]-complete. Information Processing Letters 81, 163–168 (2002)
4. Chen, Z., Fu, B., Jiang, M., Zhu, B.: On recovering syntenic blocks from comparative maps. Journal of Combinatorial Optimization 18, 307–318 (2009)
5. Downey, R.G., Fellows, M.R.: Fixed-parameter tractability and completeness II: on completeness for W[1]. Theoretical Computer Science 141, 109–131 (1995)
6. Downey, R.G., Fellows, M.R.: Parameterized Complexity. Springer, Heidelberg (1999)
7. Fellows, M.R., Hermelin, D., Rosamond, F., Vialette, S.: On the parameterized complexity of multiple-interval graph problems. Theoretical Computer Science 410, 53–61 (2009)
8. Jiang, M.: Approximation algorithms for predicting RNA secondary structures with arbitrary pseudoknots. IEEE/ACM Transactions on Computational Biology and Bioinformatics 7, 323–332 (2010)
9. Jiang, M.: On the parameterized complexity of some optimization problems related to multiple-interval graphs. Theoretical Computer Science 411, 4253–4262 (2010)
10. Jiang, M.: Recognizing d-Interval Graphs and d-Track Interval Graphs. In: Lee, D.-T., Chen, D.Z., Ying, S. (eds.) FAW 2010. LNCS, vol. 6213, pp. 160–171. Springer, Heidelberg (2010)
11. Jiang, M., Zhang, Y.: Parameterized Complexity in Multiple-Interval Graphs: Partition, Separation, Irredundancy. In: Fu, B., Du, D.-Z. (eds.) COCOON 2011. LNCS, vol. 6842, pp. 62–73. Springer, Heidelberg (2011)
12. Kratochvíl, J.: Perfect codes in general graphs. Akademia Praha (1991)
13. Roberts, F.S.: Graph Theory and Its Applications to Problems of Society. SIAM (1987)

A Faster Algorithm for Dominating Set Analyzed by the Potential Method

Yoichi Iwata

Department of Computer Science,
Graduate School of Information Science and Technology,
The University of Tokyo,
Tokyo, Japan
y.iwata@is.s.u-tokyo.ac.jp

Abstract. Measure and Conquer is a recently developed technique to analyze worst-case complexity of backtracking algorithms. The traditional measure and conquer analysis concentrates on one branching at once by using only small number of variables. In this paper, we extend the measure and conquer analysis and introduce a new analyzing technique named "potential method" to deal with consecutive branchings together. In potential method, the optimization problem becomes sparse; therefore, we can use large number of variables. We applied this technique to the minimum dominating set problem and obtained the current fastest algorithm that runs in $O(1.4864^n)$ time and polynomial space. We also combined this algorithm with a precalculation by dynamic programming and obtained $O(1.4689^n)$ time and space algorithm. These results show the power of the potential method and possibilities of future applications to other problems.

1 Introduction

Backtracking is one of the ways to solve NP-hard problems exactly and is widely used for various problems. Despite the practical efficiency of backtracking algorithms, the analysis of their worst-case complexity is difficult, and there is a large gap between the proved upper and lower bounds of the complexity for many backtracking algorithms.

Worst-case complexity of backtracking algorithms is typically analyzed by solving linear recurrences on problem size. Recently Fomin, Grandoni, and Kratsch have made a breakthrough by developing a new analyzing technique named "Measure and Conquer" [2]. In measure and conquer analysis, we define the problem size as the sum of local weights by using variables, and then solve an optimization problem to minimize the proved upper bound. This can be considered as the extension of univariate linear recurrences to multivariate linear recurrences [1].

In this paper, we extend the measure and conquer analysis and introduce a new analyzing technique named "potential method." In measure and conquer analysis, linear recurrences are solved for each branchings separately. We noticed that some of worst-case branchings cannot occur consecutively in many

D. Marx and P. Rossmanith (Eds.): IPEC 2011, LNCS 7112, pp. 41–54, 2012.
© Springer-Verlag Berlin Heidelberg 2012

applications. Therefore, if we deal with the consecutive branchings together, we can prove better complexity. In our method, we introduce variables that capture the state of the algorithm, and for each branchings, we consider not only the reduction of the problem size but also the transition of the state. We call these type of variables "potentials." This can be considered as an extension to systems of linear recurrences. Because each subproblem is assigned only one of the potentials, the optimization problem becomes sparse; therefore, we can use large number of variables. There are several papers using variables in a similar way. For example, Robson [8] and Wahlstrom [13] used potentials to deal with consecutive branchings for independent set and minimum transversals, respectively, without using measure and conquer analysis. For max-2-CSP, Gaspers [5] used potentials corresponding to the maximum degree and regularity with measure and conquer analysis. However, ours can use many more variables and capture the state of the algorithm in more detail. Actually, we used about a thousand variables while in the previous researches, only about ten variables are used.

We apply this method to the minimum dominating set problem. This is a well-studied NP-hard graph optimization problem on which many papers have been published.

In 2004, the first algorithms breaking the trivial 2^n-barrier were proposed by three sets of authors [4,6,7]. Among them, Grandoni's algorithm [6] is based on the reduction to the minimum set cover problem and runs in $O(1.9053^n)$ time. He applied the memorization to this branching algorithm and proved the running time of $O(1.8019^n)$ using exponential space. In 2005, Fomin, Grandoni, and Kratsch applied measure and conquer technique to Grandoni's algorithm and proved the running time of $O(1.5259^n)$ using polynomial space and $O(1.5132^n)$ using exponential space [2]. This result was improved by van Rooij and Bodlaender to $O(1.4969^n)$ time and polynomial space [9,10,11]. They obtained this result by introducing the reduction rules to deal with the smallest worst case efficiently, and they concluded that it is difficult to improve the complexity under $O(1.4952^n)$ by this approach. We note that this algorithm cannot be improved by the memorization. There is another branching algorithm combined with inclusion/exclusion which can count the number of dominating set for each size in $O(1.5048^n)$ time [12].

Our algorithm is also based on Grandoni's algorithm. We modified the algorithm to branch on the same element consecutively. By using potential method, we can deal with consecutive branchings together, and obtained the current fastest algorithm that runs in $O(1.4864^n)$ time and polynomial space. This result breaks the limit of $O(1.4952^n)$ [11], and if we analyze the same algorithm by the traditional measure and conquer without potentials, we can only obtain the running time of $O(1.5040^n)$. Therefore we can conclude that the potential method has the power to refine the analysis and break the limit of traditional measure and conquer analysis.

We also combined the algorithm with dynamic programming on elements to precalculate the solutions of subproblems with small number of elements. In the previous researches, the algorithm memorizes the solutions of subproblems with respect to the remaining sets and elements. In this approach, however, we can focus on the elements only. This combined version of the algorithm can be analyzed to run in $O(1.4689^n)$ but needs exponential space.

2   Definitions

Let $G = (V, E)$ be an undirected graph, where V is the vertex set and E is the edge set. The *neighborhood* of a vertex v is denoted by $N(v) = \{u \in V \mid \{u, v\} \in E\}$, and the *closed neighborhood* by $N[v] = N(v) \cup \{v\}$. A subset $D \subseteq V$ is called a *dominating set* if $\bigcup_{v \in D} N[v] = V$. In the minimum dominating set problem, we are given an undirected graph G and asked to find a dominating set of minimum cardinality.

A set cover instance consists of a collection of sets $\mathcal{S}$. We call $\mathcal{U}_\mathcal{S} = \bigcup_{S \in \mathcal{S}} S$ as a *universe* of $\mathcal{S}$. A subset $\mathcal{C} \subseteq \mathcal{S}$ is called a *set cover* if $\bigcup_{S \in \mathcal{C}} S = \mathcal{U}_\mathcal{S}$. By imposing $\mathcal{S} = \{N[v] \mid v \in V\}$, a minimum dominating set instance with n vertices can be reduced to a minimum set cover instance with n sets and n elements.

We denote sets containing an element $e \in \mathcal{U}_\mathcal{S}$ by $S(e) = \{S \in \mathcal{S} \mid e \in S\}$, the *frequency* of an element $e \in \mathcal{U}_\mathcal{S}$ by $|e| = |S(e)|$, and the *deletion* of $S \in \mathcal{S}$ from $\mathcal{S}$ by $\mathrm{del}(\mathcal{S}, S) = \{R \setminus S \mid R \in \mathcal{S}, R \not\subseteq S\}$.

3   Potential Method

3.1   Main Idea

In the measure and conquer analysis by Fomin et al. [2], the worst-case recurrences corresponds to very special cases such as branching on a set S of cardinality 5 and all of the elements of S have frequency 6 and all of the sets intersecting with S have cardinality 5. Because this is a feasible condition, we cannot neglect it; however, we can avoid the consecutive worst-case branchings. Thus it seems that better complexity can be proved by dealing with consecutive branchings together.

One of the ways to deal with consecutive branchings is to merge the recurrences. For example, consider that there are N worst-case branchings, which generates 2 subproblems with the reduction of the problem size by x_i and y_i. Let $T(k)$ be the maximum number of leaf nodes in a search tree generated from a problem with size k, then for every i, the following recurrences should be satisfied: $T(k) \leq T(k - x_i) + T(k - y_i)$. In addition, assume that in the first subproblems, the algorithm can always do one of the M better branchings with recurrences of $T(k) \leq T(k - x_i') + T(k - y_i')$. We can merge these $N + M$ recurrences to NM recurrences: $T(k) \leq T(k - x_i - x_j') + T(k - x_i - y_j') + T(k - y_i)$. This technique is called "addition of branching vectors" [3]. However, this technique is difficult to combine with measure and conquer analysis. The optimal weights may change by merging the recurrences; therefore, it is not effective to merge the recurrences after assigning values to the weight variables. If we merge the recurrences with weight variables remaining, the number of recurrences grows exponentially and becomes too large to deal with.

This exponential growth can be avoided by introducing one more variable. Consider the previous case. If we want to upper-bound $T(k)$ by λ^k for some real λ, the merged recurrences require the following constraints: $1 \geq \lambda^{-x_i - x'_j} + \lambda^{-x_i - y'_j} + \lambda^{-y_i}$. Now, we introduce a new variable ϕ that satisfies $\lambda^{-\phi} \geq \lambda^{-x'_j} + \lambda^{-y'_j}$ for every j. By using ϕ, we can rewrite the constraints as follows:

$$
\begin{aligned}
1 &\geq \lambda^{-x_i - x'_j} + \lambda^{-x_i - y'_j} + \lambda^{-y_i} \\
&= \lambda^{-x_i} \left(\lambda^{-x'_j} + \lambda^{-y'_j} \right) + \lambda^{-y_i} \\
&\geq \lambda^{-x_i - \phi} + \lambda^{-y_i} \ .
\end{aligned}
$$

As a result, we obtain the following $N + M$ constraints:

$$
\begin{aligned}
1 &\geq \lambda^{-x_i - \phi} + \lambda^{-y_i} \ , \\
1 &\geq \lambda^{-x'_j + \phi} + \lambda^{-y'_j + \phi} \ .
\end{aligned}
$$

We do not know the actual value of ϕ, however, by considering it as a kind of weight variables, we can optimize it with other variables used in measure and conquer analysis.

In measure and conquer analysis, the problem size is measured as the sum of local weights. On the other hand, this variable ϕ can be considered as a global weight corresponding to the state that the algorithm can do a better branching. In our method, we split the states with respect to the branchings the algorithm can do in the next step, and for each state i, we introduce the variable ϕ_i. For each branching, we consider not only the reduction of the problem size but also the transition of the state, and if the state changes from s to t, the reduction of the problem size changes from Δ to $\Delta + \phi_t - \phi_s$. This technique is analogous to the potential method used in the amortized analysis; therefore, we also call this type of variables as potentials and this technique as potential method.

3.2   Algorithm

Before presenting our algorithm, we introduce some notations and lemmas. For a given collection of sets $\mathcal{S}$ and a set $S \in \mathcal{S}$, a set $R \in \mathcal{S} \setminus \{S\}$ is called an *alternative set* of S if R shares an element of frequency two with S, and we denote the alternative sets of S by $\mathcal{A}(S)$. Note that if a set cover of $\mathcal{S}$ doesn't contain S, it must contain all of $\mathcal{A}(S)$. For a given set $S \in \mathcal{S}$, a set $R \in \mathcal{S} \setminus \{S\}$ is called a *mirror* of S if there exists an alternative set $T \in \mathcal{A}(S) \setminus \{R\}$ that is completely covered by S and R. We denote the mirrors of S by $\mathcal{M}(S)$, and the *closed mirrors* of S by $\mathcal{M}[S] = \mathcal{M}(S) \cup \{S\}$. For a given set $S \in \mathcal{S}$, a set $R \in \mathcal{S} \setminus \{S\}$ is called a *quasi-subset* of S if there exists an alternative set $T \in \mathcal{A}(S) \setminus \{R\}$, and R is completely covered by S and T. For mirrors and quasi-subsets, we prove the following lemmas. We denote the cardinality of a minimum set cover of $\mathcal{S}$ by $\alpha(\mathcal{S})$.

Lemma 1 (mirroring). *For any collection of sets $\mathcal{S}$ and for any set $S \in \mathcal{S}$,*

$$\alpha(\mathcal{S}) = \begin{cases} \min(\alpha(\mathcal{S} \setminus \mathcal{M}[S]), 1 + \alpha(\text{del}(\mathcal{S}, S))) & (\mathcal{U}_{\mathcal{S} \setminus \mathcal{M}[S]} = \mathcal{U}_{\mathcal{S}}) \; ; \\ 1 + \alpha(\text{del}(\mathcal{S}, S)) & (\mathcal{U}_{\mathcal{S} \setminus \mathcal{M}[S]} \subset \mathcal{U}_{\mathcal{S}}) \; . \end{cases} \tag{1}$$

Proof. It is sufficient to show that, if S is not contained in any minimum set cover, the same holds for any mirror of S. If S does not have any mirrors, the equation holds. Otherwise, let R be one of the mirrors of S. From the definition of mirror, there is a set $T \in \mathcal{A}(S) \setminus \{R\}$ with $T \subseteq S \cup R$. Assume that S is not contained in any minimum set cover and there is a minimum set cover that contains R. Because this set cover must contain T, by removing T and adding S, we obtain a minimum set cover that contains S. Thus the assumption is false. $\square$

Lemma 2 (quasi-subset). *For any collection of sets $\mathcal{S}$ and for any quasi-subset R, there is a minimum set cover of $\mathcal{S}$ that doesn't contain R.*

Proof. Let $\mathcal{S}' \subseteq \mathcal{S}$ be one of the minimum set cover that contains R. From the definition of quasi-subset, there are sets S and T with $T \in \mathcal{A}(S) \setminus \{R\}$ and $R \subseteq S \cup T$. Because $\mathcal{S}'$ must contain at least one of S and T, $\mathcal{S}' \cup \{S, T\} \setminus \{R\}$ is also a minimum set cover. $\square$

Now, we describe our algorithm. We modified Grandoni's algorithm [6] to extract the power of potential method effectively. This algorithm consists of two functions msc and seq-msc.

Algorithm 1. $\text{msc}(\mathcal{S})$

1: **if** $\exists S, R \in \mathcal{S}$ with $S \subseteq R$ **then**
2:     **return** $\text{msc}(\mathcal{S} \setminus \{S\})$
3: **else if** $\exists e \in \mathcal{U}_{\mathcal{S}}$ with $|e| = 1$ **then**
4:     **return** $1 + \text{msc}(\text{del}(\mathcal{S}, S))$, where $S(e) - \{S\}$
5: **else if** $\forall S \in \mathcal{S}.|S| = 2$ **then**
6:     **return** $\text{poly-msc}(\mathcal{S})$
7: **else if** $\exists e \in \mathcal{U}_{\mathcal{S}}$ contained in at most one set of non-maximum cardinality **then**
8:     choose an element $t \in \mathcal{U}_{\mathcal{S}}$ of the smallest frequency which satisfies the above condition
9:     **return** $\text{seq-msc}(\mathcal{S}, t)$
10: **end if**
11: choose the largest set $S \in \mathcal{S}$
12: **return** $\min(\text{msc}(\mathcal{S} \setminus \{S\}), 1 + \text{msc}(\text{del}(\mathcal{S}, S)))$

The function msc is almost the same as Grandoni's algorithm [6] except for the case where there is an element which is contained in at most one set of non-maximum cardinality (line 7-9). In this case, we can sequentially branch on these sets and reach to the branching on a set containing an element of frequency two. The branching with elements of frequency two is efficient because after the algorithm discards a set S, all of the alternative sets of S must be contained

---

Algorithm 2. seq-msc$(\mathcal{S}, t)$

1: choose the largest set $S \in \mathcal{S}$ with $t \in S$ and $\sum_{e \in S} |e|$ is minimum
2: **if** $|t| \geq 3$ **then**
3:     **return** $\min($seq-msc$(\mathcal{S} \setminus \{S\}, t), 1 + msc(del(\mathcal{S}, S)))$
4: **else if** there exists a quasi-subset R of S **then**
5:     **if** R contains an element of frequency two **then**
6:         **return** msc$(\mathcal{S} \setminus \{R\})$
7:     **else**
8:         **return** seq-msc$(\mathcal{S} \setminus \{R\}, t)$
9:     **end if**
10: **else if** $\mathcal{U}_{\mathcal{S} \setminus \mathcal{M}[S]} \subset \mathcal{U}_{\mathcal{S}}$ **then**
11:     **return** $1 + msc(del(\mathcal{S}, S))$
12: **else**
13:     **return** $\min($msc$(\mathcal{S} \setminus \mathcal{M}[S]), 1 + msc(del(\mathcal{S}, S)))$
14: **end if**

---

in the set cover. Thus we can distribute its efficiency among all the branchings through potentials.

The function seq-msc receives the target element t and sequentially branches on a set containing t. First, the algorithm chooses the largest set S that contains t and has the smallest sum of the frequency (line 1). In the case of $|t| \geq 3$, the algorithm branches on S and generates two subproblems corresponding to the cases where S is discarded or selected, and for the former case, it continues the sequential branchings on t (line 3). Otherwise the frequency of t is two. If there is a quasi-subset R of S, there is a minimum set cover that doesn't contain R from Lemma 2, then the algorithm removes R. If R contains an element of frequency two, the algorithm finishes the sequential branchings and returns to the function msc (line 6). Otherwise, it continues the sequential branchings on t (line 8). If there is no quasi-subset of S, the algorithm branches on S and finishes the sequential branchings. When algorithm discards S, the mirrors of S can be also removed from Lemma 1 (line 11,13).

3.3   Analysis

Theorem 1. *Algorithm* msc *solves the minimum dominating set problem with n vertices in $O(1.4864^n)$ time and polynomial space.*

Proof. In the analysis, we use the following measure:

$$k(\mathcal{S}) = \sum_{S \in \mathcal{S}} \alpha_{|S|} + \sum_{e \in \mathcal{U}_{\mathcal{S}}} \beta_{|e|} , \tag{2}$$

where $\alpha_i, \beta_i \in [0, 1]$ are weight variables. We use the following notations: $\Delta \alpha_i = \alpha_i - \alpha_{i-1}, \Delta \beta_i = \beta_i - \beta_{i-1}$. For the weight variables, we assume the followings:

$$0 \le \Delta\alpha_i \le \Delta\alpha_{i-1} \ ,$$
$$0 \le \Delta\beta_i \ ,$$
$$\alpha_i = \alpha_8, \ \beta_i = \beta_8 \text{ for } i \ge 8 \ ,$$
$$\alpha_8 + \beta_8 = 1 \ .$$

In this measure, an instance of the minimum dominating set problem with n vertices is reduced to an instance of the minimum set cover problem with $k(\mathcal{S}) \le n$.

We introduce the potential variables $\phi_{|S|,|t|,s}$ corresponding to the state where the algorithm runs in seq-msc with $\sum_{e \in S} |e| = s$. For the other states, we set their potentials to 0. For the potentials, we assume the followings:

$$0 \le \phi_{i,j,k} \le \phi_{i,j,k+1} \ ,$$
$$\phi_{i,j,k} = 0 \text{ for } i \ge 9 \text{ or } j \ge 8 \text{ or } k < (j-1)i \ ,$$
$$\phi_{i,j,k} = \phi_{i,j,8i} \text{ for } k \ge 8i \ .$$

These assumptions are used to bound the number of the variables and simplify the analysis.

We denote the reduction of the problem size including the difference of potentials by Δk_{OUT} and Δk_{IN} for discarded and selected cases, respectively. Let r_i be the number of elements of S of frequency i, and s be the sum of the frequency of elements in S.

1. Reduction at Msc. (line 2,4,9) For each case, the size corresponding to the sets and elements does not increase, and the potential does not change for line 2,4, or changes from 0 to nonnegative for line 9. Thus the total problem size never increases.

2. Branching at Msc. (line 12) In this case, S does not contain an element of frequency two and all the elements of S are contained in at least 2 sets of non-maximum cardinality. Therefore the reduction of the problem size due to the reduction of cardinality of the sets intersecting with S is at least $(|e| - 3)\Delta\alpha_{|S|} + 2\Delta\alpha_{|S|-1}$ for each element e of S. Therefore for any $3 \le |S| \le 9$, and $\sum_{i=3}^{9} r_i = |S|$:

$$\Delta k_{OUT} \ge \alpha_{|S|} + \sum_{i=3}^{9} r_i \Delta\beta_i \ ,$$

$$\Delta k_{IN} \ge \alpha_{|S|} + \sum_{i=3}^{9} r_i(\beta_i + (i-3)\Delta\alpha_{|S|} + 2\Delta\alpha_{|S|-1}) \ .$$

Note that it is enough to consider only the cases of $|S| \le 9$ and $|e| \le 9$ by the assumption of α_i and β_i.

3. Reduction at Seq-Msc. (line 6, 8) For line 6, the sequential branchings are finished and the potential changes from $\phi_{i,2,k}$ to 0. By the removal of R, its alternative sets are added to a set cover, then the problem size is reduced by at least $2\alpha_2 + 2\beta_2$. Thus the overall reduction is at least $2\alpha_2 + 2\beta_2 - \phi_{i,2,k}$. For line 8, the sequential branchings continue and the potential changes from $\phi_{i,2,k}$ to $\phi_{i,2,k'}$. By the removal of R, the minimum sum of the frequency is decreased by at most $|R| - 1 \leq i - 1$; therefore, k' is at least $k - i + 1$. Thus the overall reduction is at least $\alpha_2 + \phi_{i,2,k-i+1} - \phi_{i,2,k}$. Therefore if the following constraints are satisfied for any i and k, the problem size does not increase:

$$2\alpha_2 + 2\beta_2 - \phi_{i,2,k} \geq 0 \ ,$$
$$\alpha_2 + \phi_{i,2,k-i+1} - \phi_{i,2,k} \geq 0 \ .$$

4. Branching at Seq-Msc with $|t| \geq 3$. (line 3) For an element e with $|e| < |t|$, it must be contained in at least two sets of non-maximum cardinality; therefore, all the element of S have frequency at least three. When the algorithm discards S, the sequential branchings continue and the potential changes from $\phi_{|S|,|t|,s}$ to $\phi_{|S|,|t|-1,s'}$. Let $c = \max_{R \in S, |R|=|S|} |S \cap R|$. Because the algorithm chose the set S with the smallest sum of the frequency, s' is at least $s - c$. When the algorithm selects S, the sequential branchings are finished and the potential changes from $\phi_{|S|,|t|,s}$ to 0. For an element e with $|e| < |t|$, e is contained in at least two sets of non-maximum cardinality; therefore, the problem size is reduced by at least $(|e| - 3)\Delta\alpha_{|S|} + 2\Delta\alpha_{|S|-1}$. Additionally, because there is a set sharing c elements with S, the problem size is reduced by at least $\alpha_{|S|} - \alpha_{|S|-c}$. In total, for any $3 \leq |S| \leq 9$, $3 \leq |t| \leq 9$, $1 \leq c \leq |S| - 1$, $\sum_{i=3}^{9} r_i = |S|$, and $r_{|t|} \geq 1$:

$$\Delta k_{OUT} \geq \alpha_{|S|} + \sum_{i=3}^{9} r_i \Delta\beta_i + \phi_{|S|,|t|-1,s-c} - \phi_{|S|,|t|,s} \ ,$$

$$\Delta k_{IN} \geq \alpha_{|S|} + \sum_{i=|t|}^{9} r_i(\beta_i + (i-1)\Delta\alpha_{|S|})$$
$$+ \sum_{i=3}^{|t|-1} r_i \left(\beta_i + (i-3)\Delta\alpha_{|S|} + 2\Delta\alpha_{|S|-1}\right)$$
$$+ \alpha_{|S|} - \alpha_{|S|-c} - c\Delta\alpha_{|S|} - \phi_{|S|,|t|,s} \ .$$

5. Branching at Seq-Msc with $|t| = 2$. (line 13) This case is very important because our algorithm is based on the distribution of the efficiency of this case; therefore, we carefully analyze it by considering about the alternative sets and mirrors of S. We denote the reduction due to the alternative sets and the mirrors by $\Delta k'_{OUT}$ and $\Delta k'_{IN}$, and describe Δk_{OUT} and Δk_{IN} as follows for any $3 \leq |S| \leq 9$, $\sum_{i=2}^{9} r_i = |S|$, and $r_2 \geq 1$:

$$\Delta k_{OUT} \geq \alpha_{|S|} + \sum_{i=2}^{9} r_i \Delta \beta_i - \phi_{|S|,2,s} + \Delta k'_{OUT} \ ,$$

$$\Delta k_{IN} \geq \alpha_{|S|} + \sum_{i=2}^{9} r_i \beta_i + \sum_{i=3}^{9} r_i(i-1)\Delta \alpha_{|S|} - \phi_{|S|,2,s} + \Delta k'_{IN} \ .$$

We analyze $\Delta k'_{OUT}$ and $\Delta k'_{IN}$ by dividing into the following subcases.

(a) $\mathcal{A}(S) = \{R\}$ and $|R \setminus S| = 1$,
(b) $\mathcal{A}(S) = \{R\}$ and $|R \setminus S| \geq 2$,
(c) $|\mathcal{A}(S)| \geq 2$ and there exists a set $R \in \mathcal{A}(S)$ with $|R \setminus S| = 1$,
(d) $|\mathcal{A}(S)| = 2$ and for any set $R \in \mathcal{A}(S)$, $|R \setminus S| \geq 2$,
(e) $|\mathcal{A}(S)| \geq 3$ and for any set $R \in \mathcal{A}(S)$, $|R \setminus S| \geq 2$.

Note that if there is an alternative set R with $R \setminus S = \{f\}$, all of the sets which shares f with R are the mirrors of S. Also note that any alternative set of S cannot be a mirror of S, because in this case, there must be a quasi-subset of S. We use the following notation for convenience: [condition] $= 1$ if the condition is true and [condition] $= 0$ otherwise.

(a) $\mathcal{A}(S) = \{R\}$ and $|R \setminus S| = 1$. Let $R \setminus S = \{f\}$. The cardinality of R is at least $r_2 + 1$; therefore, r_2 must be at most $|S| - 1$. In the case of $r_2 = |S| - 1$, the frequency of f must be at least $s - 2r_2$ because the algorithm chose the set S with the smallest sum of the frequency. When the algorithm discards S, R is added to a set cover and the problem size is reduced by at least $\alpha_{r_2+1} + \beta_{|f|}$. Moreover, the mirrors of S are removed. Because all of the set which shares f with R are the mirrors of S, $|\mathcal{M}(S)| = |f| - 1$. Therefore the corresponding reduction is at least $(|f| - 1)\alpha_2$. When the algorithm selects S, R becomes a set of single element. Then, R is removed and the frequency of f is reduced by one. Therefore the problem size is reduced by $\alpha_{r_2+1} + \Delta\beta_{|f|}$. Moreover, in the case of $|f| = 2$, the algorithm must select the remaining set T which shares f with R. T cannot be a single element set, because in this case, T becomes a quasi-subset of S. Therefore this leads to an additional reduction of at least $\alpha_2 + \beta_2$. In total, for any $3 \leq |S| \leq 9$, $1 \leq r_2 \leq |S| - 1$, $|f| \geq 2$, and $r_2 = |S| - 1 \Rightarrow |f| \geq s - 2r_2$:

$$\Delta k'_{OUT} \geq \alpha_{r_2+1} + \beta_{|f|} + (|f| - 1)\alpha_2 \ ,$$
$$\Delta k'_{IN} \geq \alpha_{r_2+1} + \Delta\beta_{|f|} + [|f| = 2](\alpha_2 + \beta_2) \ .$$

(b) $\mathcal{A}(S) = \{R\}$ and $|R \setminus S| \geq 2$. Let $|R \setminus S| = d$. Because the cardinality of R is at least $r_2 + d$, $2 \leq d \leq |S| - r_2$. When the algorithm discards S, R is added to a set cover and the problem size is reduced by at least $\alpha_{r_2+d} + d\beta_2$. In addition, the cardinality of the sets intersecting with R is reduced. Because these sets are not a subset of R, the corresponding reduction is at least $d\Delta\alpha_{|S|}$. Moreover, in the case of $r_2 + d = |S|$, it must be at least $(s - 2r_2 - d)\Delta\alpha_{|S|}$ because the algorithm chose the set S with the smallest sum of the frequency. When the algorithm selects S, R becomes a set of cardinality d, and the problem size is reduced by $\alpha_{r_2+d} - \alpha_d$. In total, for any $3 \leq |S| \leq 9$, $r_2 \geq 1$, and $2 \leq d \leq |S| - r_2$:

$$\Delta k'_{OUT} \geq \alpha_{r_2+d} + d\beta_2$$
$$+ \begin{cases} d\Delta\alpha_{|S|} & (r_2 + d < |S|); \\ (s - 2r_2 - d)\Delta\alpha_{|S|} & (r_2 + d = |S|) \end{cases},$$
$$\Delta k'_{IN} \geq \alpha_{r_2+d} - \alpha_d .$$

(c) $|\mathcal{A}(S)| \geq 2$ and there exists a set $R \in \mathcal{A}(S)$ with $|R \setminus S| = 1$. Let $R \setminus S = \{f\}$. Because there are at least two alternative sets of S, r_2 must be at least 2. When the algorithm discards S, alternative sets of S are added to a set cover. Because each set of $\mathcal{A}(S)$ must contain at least one element outside S, the sum of the cardinality of $\mathcal{A}(S)$ is at least $r_2 + 2$. Therefore the total size of alternative sets is at least $\alpha_2 + \alpha_{r_2}$. Note that f cannot be contained in an alternative set other than R, because in this case, R becomes a quasi-subset of S. Therefore the reduction of the problem size due to the removal of the elements is at least $\beta_{|f|} + \beta_2$. In addition, the mirrors of S are removed and the problem size is reduced by at least $(|f| - 1)\alpha_2$. When the algorithm selects S, almost the same argument as the subcase (a) holds. In this case, there are alternative sets other than R; therefore, the reduction of the problem size due to the decrease of the cardinality is at least $\alpha_2 + (r_2 - 1)\Delta\alpha_{|S|}$. In total, for any $3 \leq |S| \leq 9$, $r_2 \geq 2$, and $|f| \geq 2$:

$$\Delta k'_{OUT} \geq \alpha_2 + \alpha_{r_2} + \beta_{|f|} + \beta_2 + (|f| - 1)\alpha_2 ,$$
$$\Delta k'_{IN} \geq \alpha_2 + (r_2 - 1)\Delta\alpha_{|S|} + \Delta\beta_{|f|} + [|f| = 2](\alpha_2 + \beta_2) .$$

(d) $|\mathcal{A}(S)| = 2$ and for any set $R \in \mathcal{A}(S)$, $|R \setminus S| \geq 2$. Because there are two alternative sets of S, r_2 must be at least 2. Moreover, in the case of $|S| = 3$, r_2 must be 2. Let $\mathcal{A}(S) = \{R_1, R_2\}$. Because R_1 and R_2 must contain at least two elements outside S, $|R_1| + |R_2| \geq r_2 + 4$. Therefore the total size of R_1 and R_2 is at least $\alpha_3 + \alpha_{r_2+1}$. In the case of $r_2 = |S|$, there are no set of cardinality $r_2 + 1$, then the total size is at least $\alpha_4 + \alpha_{r_2}$. When the algorithm discards S, R_1 and R_2 are added to a set cover. R_1 and R_2 must have at least three distinct element outside S, because otherwise R_1 becomes a quasi-subset of S. Therefore the reduction of the problem size due to the removal of the elements is at least $3\beta_2$. In addition, the cardinality of the sets which contains these elements is reduced, and the problem size is reduced by at least $\min(2\Delta\alpha_{|S|}, \alpha_2)$. When the algorithm selects S, the cardinality of R_1 and R_2 is reduced by at least r_2 in total. Therefore the problem size is reduced by at least:

$$\Delta\alpha_{|S|} + \Delta\alpha_{|S|} + \Delta\alpha_{|S|-1} + \ldots + \Delta\alpha_{|S|-\lfloor\frac{r_2-1}{2}\rfloor} = 2\alpha_{|S|} - \alpha_{|S|-\lfloor\frac{r_2}{2}\rfloor} - \alpha_{|S|-\lceil\frac{r_2}{2}\rceil} .$$

In total, for any $3 \leq |S| \leq 9$, $r_2 \geq 2$, and $|S| = 3 \Rightarrow r_2 = 2$:

$$\Delta k'_{OUT} \geq 3\beta_2 + \min(2\Delta\alpha_{|S|}, \alpha_2) + \begin{cases} \alpha_3 + \alpha_{r_2+1} & (r_2 < |S|) ; \\ \alpha_4 + \alpha_{r_2} & (r_2 = |S|) , \end{cases}$$

$$\Delta k'_{IN} \geq 2\alpha_{|S|} - \alpha_{|S|-\lfloor\frac{r_2}{2}\rfloor} - \alpha_{|S|-\lceil\frac{r_2}{2}\rceil} .$$

(e) $|\mathcal{A}(S)| \geq 3$ and for any set $R \in \mathcal{A}(S)$, $|R \setminus S| \geq 2$. Because there are at least three alternative sets of S, r_2 must be at least 3. When the algorithm discards S, alternative sets of S are added to a set cover. Because each set of $\mathcal{A}(S)$ must contain at least two elements outside S, the sum of the cardinality of $\mathcal{A}(S)$ is at least $r_2 + 6$. Therefore the total size of alternative sets is at least $2\alpha_3 + \alpha_{r_2}$. By the same argument as the subcase (d), alternative sets must contain at least three distinct element outside S. Due to the removal of these elements, problem size is reduced by at least $3\beta_2$. When the algorithm selects S, the cardinality of the alternative sets is reduced by at least r_2 in total; therefore, the problem size is reduced by at least $r_2\Delta\alpha_{|S|}$. In total for any $3 \leq |S| \leq 9$, $r_2 \geq 3$:

$$\Delta k'_{OUT} \geq 2\alpha_3 + \alpha_{r_2} + 3\beta_2 ,$$
$$\Delta k'_{IN} \geq r_2\Delta\alpha_{|S|} .$$

We solve the optimization problem to minimize λ under the constraints of $1 \geq \lambda^{-\Delta k_{OUT}} + \lambda^{-\Delta k_{IN}}$ for all reductions $(\Delta k_{IN}, \Delta k_{OUT})$ by implementing Eppstein's smooth quasiconvex programming [1]. Although this optimization problem has many variables (actually it has 908 variables), it is very sparse, and thus the values are converged in a few minutes. The obtained values of α_i and β_i are listed in Table 1. Because there are too many potentials to be listed here, we provide them in our webpage [1]. Also, we provide a program to check the constraints [2]. These values give $\lambda < 1.4864$; therefore, the algorithm runs in $O(1.4864^n)$. □

If we analyze the same algorithm by the traditional measure and conquer without potentials, we obtained the weights listed in Table 2. These weights only give $\lambda = 1.5039\ldots < 1.5040$.

---

[1] http://www-imai.is.s.u-tokyo.ac.jp/~y.iwata/ipec2011/potentials.txt
[2] http://www-imai.is.s.u-tokyo.ac.jp/~y.iwata/ipec2011/check.cpp

Table 1. The obtained values of the weights

i	2	3	4	5	6	7	8
α_i	0.251502	0.459931	0.550662	0.597567	0.620987	0.632065	0.635804
β_i	0.074109	0.263740	0.339850	0.361289	0.364195	0.364196	0.364196

Table 2. The obtained values of the weights without potentials

i	2	3	4	5	6	7	8
α_i	0.227607	0.455213	0.551209	0.596609	0.618020	0.626422	0.627064
β_i	0.066561	0.261058	0.332901	0.360645	0.370450	0.372936	0.372936

4   Exponential Space

4.1   Algorithm

We denote the intersection of a collection of sets $\mathcal{S}$ and a set U by $\mathcal{S}_U = \{S \cap U \mid S \in \mathcal{S}, S \cap U \neq \emptyset\}$, and the cardinality of a minimum set cover of $\mathcal{S}$ by $\alpha(\mathcal{S})$. By dynamic programming on elements, we can compute $\alpha(\mathcal{S}_U)$ for all of the subsets $U \subseteq \mathcal{U}_\mathcal{S}$ of cardinality at most $h(\leq \frac{n}{2})$ in $O^*\left(\binom{n}{h}\right)$ time and space.

We combine the algorithms as follows. We precalculate the solutions of $\mathcal{S}_U$ with small $|U|$ by the dynamic programming. And then, we start to run the algorithm msc. During the execution, if $|\mathcal{U}_{\mathcal{S}'}|$ becomes at most h, the algorithm returns the precalculated solution of $\mathcal{S}_{\mathcal{U}_{\mathcal{S}'}}$. This may not be the solution for $\mathcal{S}'$ because the algorithm may have discarded some of the sets of the minimum set cover of $\mathcal{S}_{\mathcal{U}_{\mathcal{S}'}}$; however, the union of the selected sets and the minimum set cover of $\mathcal{S}_{\mathcal{U}_{\mathcal{S}'}}$ is also a set cover of $\mathcal{S}$; therefore, the algorithm returns the value at least $\alpha(\mathcal{S})$. On the other hand, $\alpha(\mathcal{S}_{\mathcal{U}_{\mathcal{S}'}})$ is at most $\alpha(\mathcal{S}')$; therefore, the returned value is at most $\alpha(\mathcal{S})$. Thus the algorithm computes the cardinality of the minimum set cover correctly. Note that some reduction rules such as "subsumption" used in [11] make this approach impossible because in this case, the union of the selected sets and the minimum set cover of $\mathcal{S}_{\mathcal{U}_{\mathcal{S}'}}$ may not be a set cover of $\mathcal{S}$.

4.2   Analysis

Theorem 2. *Algorithm* msc, *combined with the dynamic programming precalculation, solves the minimum dominating set problem with n vertices in $O(1.4689^n)$ time and space.*

Proof. In the analysis, we use the same weights as the previous section.

We consider the subproblem $\mathcal{S}'$ generated during the execution of the algorithm. If there is a set of cardinality one or an element of frequency one, the algorithm applies one of the reduction rules without branching. Therefore we can assume that all of the sets have cardinality at least two and all of the elements have frequency at least two. Let $h = |\mathcal{U}_{\mathcal{S}'}|$, and we consider about the minimum possible problem size with h elements.

If there is a set of cardinality at least 10, the corresponding recurrences give $\lambda < 1.4434$. Otherwise, all of the sets have cardinality at most 9. If there is a set of cardinality 9, the corresponding recurrences give $\lambda < 1.4744$. Because all of the elements in $\mathcal{S}'$ have frequency at least two, the corresponding size is at least $h\beta_2$. Moreover, because all of the sets have cardinality at most 9, the size corresponding to the sets is at least $\min_{2 \le i \le 9} \frac{2h}{i}\alpha_i$. Thus the overall size is at least:

$$k(\mathcal{S}') \ge h\beta_2 + \min_{2 \le i \le 9} \frac{2h}{i}\alpha_i \ge 0.2153h \ . \tag{3}$$

When all of the sets have cardinality at most 8, the corresponding recurrences give $\lambda < 1.4864$, and in this case, the problem size is at least:

$$k(\mathcal{S}') \ge h\beta_2 + \min_{2 \le i \le 8} \frac{2h}{i}\alpha_i \ge 0.2330h \ . \tag{4}$$

Consider that we precalculate the solutions of $\mathcal{S}_U$ with $|U| \le h$. For this precalculation, it takes $O^*\left(\binom{n}{h}\right)$ time, and for the execution of msc, it takes $O(\max\left\{1.4434^n, 1.4744^{n-0.2153h}, 1.4864^{n-0.2330h}\right\})$ time. By setting $h = 0.1290n$, we obtain $O(1.4689^n)$ of total running time. $\qquad\square$

5   Conclusion

In this paper, we developed a new analyzing technique "potential method." By applying it to the minimum dominating set problem, we obtained the current fastest algorithm that runs in $O(1.4864^n)$. Therefore we conclude that the potential method has the power to refine the analysis of traditional measure and conquer analysis.

In the analysis, we used about a thousand potential variables. If we can use more variables, better complexity can be proved. Therefore a faster implementation of quasi-convex programming solver will be needed. Also, it might be possible to improve the complexity by using a different strategy. By introducing the potential method, more complicated algorithm can be analyzed, and the choice of the strategy grows wider.

Potential method is not a specific tool for dominating set or set cover, and applications to other problems are left as a future work.

References

1. Eppstein, D.: Quasiconvex analysis of multivariate recurrence equations for backtracking algorithms. ACM Transactions on Algorithms 2(4), 492–509 (2006)
2. Fomin, F.V., Grandoni, F., Kratsch, D.: Measure and Conquer: Domination – A Case Study. In: Caires, L., Italiano, G.F., Monteiro, L., Palamidessi, C., Yung, M. (eds.) ICALP 2005. LNCS, vol. 3580, pp. 191–203. Springer, Heidelberg (2005)
3. Fomin, F.V., Kratsch, D.: Exact Exponential Algorithms. Texts in Theoretical Computer Science. An EATCS Series. Springer, Heidelberg (2010)

4. Fomin, F.V., Kratsch, D., Woeginger, G.J.: Exact (Exponential) Algorithms for the Dominating Set Problem. In: Hromkovič, J., Nagl, M., Westfechtel, B. (eds.) WG 2004. LNCS, vol. 3353, pp. 245–256. Springer, Heidelberg (2004)
5. Gaspers, S., Sorkin, G.: A universally fastest algorithm for max 2-sat, max 2-csp, and everything in between. In: Proceedings of the Twentieth Annual ACM-SIAM Symposium on Discrete Algorithms, pp. 606–615. Society for Industrial and Applied Mathematics (2009)
6. Grandoni, F.: Exact algorithms for hard graph problems. Ph.D. thesis, Universitá di Roma "Tor Vergata" (2004)
7. Randerath, B., Schiermeyer, I.: Exact algorithms for MINIMUM DOMINATING SET. Tech. rep., zaik-469, Zentrum für Angewandte Informatik Köln, Germany (2004)
8. Robson, J.: Algorithms for maximum independent sets. Journal of Algorithms 7(3), 425–440 (1986)
9. van Rooij, J.M.M.: Design by measure and conquer: an $O(1.5086^n)$ algorithm for minimum dominating set and similar problems. Master's thesis, Utrecht University (2006)
10. van Rooij, J.M.M., Bodlaender, H.L.: Design by measure and conquer: a faster exact algorithm for dominating set. In: Proceedings of the 25th Annual Symposium on Theoretical Aspects of Computer Science (STACS 2008), pp. 657–668 (2008)
11. van Rooij, J.M.M., Bodlaender, H.L.: Design by measure and conquer: exact algorithms for dominating set. Tech. rep., UU-CS-2009-025, Utrecht University, The Netherlands (2009)
12. van Rooij, J.M.M., Nederlof, J., van Dijk, T.C.: Inclusion/Exclusion Meets Measure and Conquer. In: Fiat, A., Sanders, P. (eds.) ESA 2009. LNCS, vol. 5757, pp. 554–565. Springer, Heidelberg (2009)
13. Wahlstrom, M.: Exact algorithms for finding minimum transversals in rank-3 hypergraphs. Journal of Algorithms 51(2), 107–121 (2004)

Contracting Graphs to Paths and Trees*

Pinar Heggernes[1], Pim van 't Hof[1], Benjamin Lévêque[2],
Daniel Lokshtanov[3], and Christophe Paul[2]

[1] Department of Informatics, University of Bergen, N-5020 Bergen, Norway
{pinar.heggernes,pim.vanthof}@ii.uib.no
[2] CNRS, LIRMM, Université Montpellier 2, Montpellier, France
{leveque,paul}@lirmm.fr
[3] Dept. Computer Science and Engineering, University of California San Diego, USA
dlokshtanov@cs.ucsd.edu

Abstract. Vertex deletion and edge deletion problems play a central role in Parameterized Complexity. Examples include classical problems like FEEDBACK VERTEX SET, ODD CYCLE TRANSVERSAL, and CHORDAL DELETION. The study of analogous edge contraction problems has so far been left largely unexplored from a parameterized perspective. We consider two basic problems of this type: TREE CONTRACTION and PATH CONTRACTION. These two problems take as input an undirected graph G on n vertices and an integer k, and the task is to determine whether we can obtain an acyclic graph or a path, respectively, by a sequence of at most k edge contractions in G. We present an algorithm with running time $4.98^k n^{O(1)}$ for TREE CONTRACTION, based on a variant of the color coding technique of Alon, Yuster and Zwick, and an algorithm with running time $2^{k+o(k)} + n^{O(1)}$ for PATH CONTRACTION. Furthermore, we show that PATH CONTRACTION has a kernel with at most $5k+3$ vertices, while TREE CONTRACTION does not have a polynomial kernel unless NP $\subseteq$ coNP/poly. We find the latter result surprising, because of the connection between TREE CONTRACTION and FEEDBACK VERTEX SET, which is known to have a kernel with $4k^2$ vertices.

1 Introduction

For a graph class Π, the Π-CONTRACTION problem takes as input a graph G and an integer k, and the question is whether there is a graph $H \in \Pi$ such that G can be contracted to H using at most k edge contractions. In early papers by Watanabe et al. [29,30] and Asano and Hirata [2], Π-CONTRACTION was proved to be NP-complete for several classes Π. The Π-CONTRACTION problem fits into a wider and well studied family of graph modification problems, where vertex deletions and edge deletions are two other ways of modifying a graph. Π-VERTEX DELETION and Π-EDGE DELETION are the problems of deciding whether some graph belonging to graph class Π can be obtained from G by at most k vertex deletions or by at most k edge deletions, respectively. All of

* Supported by the Research Council of Norway (project SCOPE, 197548/V30) and the French ANR (project AGAPE, ANR-09-BLAN-0159).

D. Marx and P. Rossmanith (Eds.): IPEC 2011, LNCS 7112, pp. 55–66, 2012.

these problems are shown to be NP-complete for most of the interesting graph classes Π [25,31,32,33]. However, whereas Π-VERTEX DELETION and Π-EDGE DELETION have studied in detail for several graph classes Π with respect to fixed parameter tractability (e.g., [3,5,10,17,19,21,22,23,26,28]), this has not been the case for Π-CONTRACTION. Note that every edge contraction reduces the number of vertices of the input graph by one, which means that the parameter k of Π-CONTRACTION is never more than $n - 1$.

Here we study Π-CONTRACTION when Π is the class of acyclic graphs and when Π is the class of paths. Since edge contractions preserve the number of connected components, we may assume that the input graph is connected, justifying the names TREE CONTRACTION and PATH CONTRACTION. Both problems are NP-complete [2,9]. We find these problems of particular interest, since their vertex deletion versions, widely known as FEEDBACK VERTEX SET and LONGEST INDUCED PATH, are famous and well-studied. These two problems are known to be fixed parameter tractable and have polynomial kernels, when parameterized by the number of deleted vertices.

The question whether a fixed parameter tractable problem has a polynomial kernel or not has attracted considerable attention during the last years, especially after the establishment of methods for proving non-existence of polynomial kernels, up to some complexity theoretical assumptions [6,7,8]. During the last decade, considerable effort has also been devoted to improving the parameter dependence in the running time of classical parameterized problems. Even in the case of a running time which is single exponential in k, lowering the base of the exponential function is considered to be an important challenge. For instance, the running time of FEEDBACK VERTEX SET has been successively improved from $37.7^k n^{O(1)}$ [18] to $10.57^k n^{O(1)}$ [15], $5^k n^{O(1)}$ [12], $3.83^k n^{O(1)}$ [11], and randomized $3^k n^{O(1)}$ [14].

In this paper, we present results along these established lines for TREE CONTRACTION and PATH CONTRACTION. It is easy to see that if a graph G is contractible to a path or a tree (i.e., a graph with treewidth 1) with at most k edge contractions, then the treewidth of G is at most $k + 1$. Consequently, when parameterized by k, fixed parameter tractability of TREE CONTRACTION and PATH CONTRACTION follows from the well known result of Courcelle [13], as both problems are expressible in monadic second order logic. However, this approach yields very unpractical algorithms whose running times involve huge functions of k. Here, we give algorithms with running time $2^{k+o(k)} + n^{O(1)}$ for PATH CONTRACTION, and $4.98^k n^{O(1)}$ for TREE CONTRACTION. To obtain the latter result, we use a variant of the color coding technique of Alon, Yuster and Zwick [1]. Combined with a recent result of Cygan et al. [14], our results also imply a randomized algorithm for TREE CONTRACTION with running time $4^{k+o(k)} n^{O(1)}$. Furthermore, we show that PATH CONTRACTION has a linear vertex kernel. On the negative side, we show that TREE CONTRACTION does not have a polynomial kernel, unless NP $\subseteq$ coNP/poly. This is a contrast compared to the corresponding vertex deletion problem FEEDBACK VERTEX SET, which is known to have a quadratic kernel [27].

2   Definitions and Notation

All graphs in this paper are finite, undirected, and simple, i.e., do not contain multiple edges or loops. Given a graph G, we denote its vertex set by $V(G)$ and its edge set by $E(G)$. We also use the ordered pair $(V(G), E(G))$ to represent G. We let $n = |V(G)|$. Let $G = (V, E)$ be a graph. The *neighborhood* of a vertex v in G is the set $N_G(v) = \{w \in V \mid vw \in E\}$ of *neighbors* of v in G. Let $S \subseteq V$. We write $N_G(S)$ to denote $\bigcup_{v \in S} N_G(v) \setminus S$. We say that S *dominates* a set $T \subseteq V$ if every vertex in T either belongs to S or has at least one neighbor in S. We write $G[S]$ to denote the subgraph of G *induced* by S. We use shorthand notation $G - v$ to denote $G[V \setminus \{v\}]$ for a vertex $v \in V$, and $G - S$ to denote $G[V \setminus S]$ for a set of vertices $S \subseteq V$. A graph is *connected* if it has a path between every pair of its vertices, and is *disconnected* otherwise. The *connected components* of a graph are its maximal connected subgraphs. We say that a vertex subset $S \subseteq V$ is *connected* if $G[S]$ is connected. A *bridge* in a connected graph is an edge whose deletion results in a disconnected graph. A *cut vertex* in a connected graph is a vertex whose deletion results in a disconnected graph. A graph is *2-connected* if it has no cut vertex. A *2-connected component* of a graph G is a maximal 2-connected subgraph of G.

We use P_ℓ to denote the graph isomorphic to a path on ℓ vertices, i.e., the graph with ordered vertex set $\{p_1, p_2, p_3, \ldots, p_\ell\}$ and edge set $\{p_1 p_2, p_2 p_3, \ldots, p_{\ell-1} p_\ell\}$. We will also write $p_1 p_2 \cdots p_\ell$ to denote P_ℓ. A *tree* is a connected acyclic graph. A vertex with exactly one neighbor in a tree is called a *leaf*. A *star* is a tree isomorphic to the graph with vertex set $\{a, v_1, v_2, \ldots, v_s\}$ and edge set $\{av_1, av_2, \ldots, av_s\}$. Vertex a is called the *center* of the star.

The *contraction* of edge xy in G removes vertices x and y from G, and replaces them by a new vertex, which is made adjacent to precisely those vertices that were adjacent to at least one of the vertices x and y. A graph G is *contractible* to a graph H, or H-*contractible*, if H can be obtained from G by a sequence of edge contractions. Equivalently, G is H-contractible if there is a surjection $\varphi : V(G) \to V(H)$, with $W(h) = \{v \in V(G) \mid \varphi(v) = h\}$ for every $h \in V(H)$, that satisfies the following three conditions: (1) for every $h \in V(H)$, $W(h)$ is a connected set in G; (2) for every pair $h_i, h_j \in V(H)$, there is an edge in G between a vertex of $W(h_i)$ and a vertex of $W(h_j)$ if and only if $h_i h_j \in E(H)$; (3) $\mathcal{W} = \{W(h) \mid h \in V(H)\}$ is a partition of $V(G)$. We say that $\mathcal{W}$ is an H-*witness structure* of G, and the sets $W(h)$, for $h \in V(H)$, are called *witness sets* of $\mathcal{W}$.

If a witness set contains more than one vertex of G, then we call it a *big* witness set; a witness set consisting of a single vertex of G is called *small*. We say that G is k-*contractible* to H, with $k \leq n - 1$, if H can be obtained from G by at most k edge contractions. The next observation follows from the above.

Observation 1. *If a graph G is k-contractible to a graph H, then $|V(G)| \leq |V(H)| + k$, and any H-witness structure $\mathcal{W}$ of G satisfies the following three properties: no witness set of $\mathcal{W}$ contains more than $k + 1$ vertices, $\mathcal{W}$ has at most k big witness sets, and all the big witness sets of $\mathcal{W}$ together contain at most $2k$ vertices.*

A 2-*coloring* of a graph G is a function $\phi : V(G) \to \{1, 2\}$. Here, a 2-coloring of G is merely an assignment of colors 1 and 2 to the vertices of G, and should not be confused with a *proper* 2-coloring of G, which is a 2-coloring with the additional property that no two adjacent vertices receive the same color. If all the vertices belonging to a set $S \subseteq V(G)$ have been assigned the same color by ϕ, we say that S is *monochromatic* with respect to ϕ, and we use $\phi(S)$ to denote the color of the vertices of S. Any 2-coloring ϕ of G defines a partition of $V(G)$ into two sets V_ϕ^1 and V_ϕ^2, which are the sets of vertices of G colored 1 and 2 by ϕ, respectively. A set $X \subseteq V(G)$ is a *monochromatic component* of G with respect to ϕ if $G[X]$ is a connected component of $G[V_\phi^1]$ or a connected component of $G[V_\phi^2]$. We say that two different 2-colorings ϕ_1 and ϕ_2 of G *coincide* on a vertex set $A \subseteq V(G)$ if $\phi_1(v) = \phi_2(v)$ for every vertex $v \in A$.

3   TREE CONTRACTION

Asano and Hirata [2] showed that TREE CONTRACTION is NP-complete. In this section, we first show that TREE CONTRACTION does not have a polynomial kernel, unless NP $\subseteq$ coNP/poly. We then present a $4.98^k n^{O(1)}$ time algorithm for TREE CONTRACTION.

A *polynomial parameter transformation* from a parameterized problem Q_1 to a parameterized problem Q_2 is a polynomial time reduction from Q_1 to Q_2 such that the parameter of the output instance is bounded by a polynomial in the parameter of the input instance. Bodlaender et al. [8] proved that if Q_1 is NP-complete, if Q_2 is in NP, if there is a polynomial parameter transformation from Q_1 to Q_2, and if Q_2 has a polynomial kernel, then Q_1 has a polynomial kernel.

Theorem 1. TREE CONTRACTION *does not have a kernel with size polynomial in k, unless NP $\subseteq$ coNP/poly.*

Proof. We give a polynomial parameter transformation from RED-BLUE DOMINATION to TREE CONTRACTION. RED-BLUE DOMINATION takes as input a bipartite graph $G = (A, B, E)$ and an integer t, and the question is whether there exists a subset of at most t vertices in B that dominates A. We may assume that every vertex of A has a neighbor in B, and that $t \leq |A|$. This problem, when parameterized by $|A|$, has been shown not to have a polynomial kernel, unless NP $\subseteq$ coNP/poly [16]. Since TREE CONTRACTION is in NP, the existence of the polynomial parameter transformation described below implies that TREE CONTRACTION does not have a kernel with size polynomial in k, unless NP $\subseteq$ coNP/poly.

Given an instance of RED-BLUE DOMINATION, that is a bipartite graph $G = (A, B, E)$ and an integer t, we construct an instance (G', k) of TREE CONTRACTION with $G' = (A' \cup B', E')$ as follows. To construct G', we first add a new vertex a to A and make it adjacent to every vertex of B. We define $A' = A \cup \{a\}$. We then add, for every vertex u of A, $k + 1$ new vertices to B that are all made adjacent to exactly u and a. The set B' consists of the set B and the $|A|(k + 1)$

newly added vertices. Finally, we set $k = |A| + t$. This completes the construction. Observe that $k \leq 2|A|$, which means that the construction is parameter preserving. In particular, we added $|A|(k+1) + 1 \leq 2|A|^2 + |A| + 1$ vertices to G to obtain G', and we added $|B|$ edges incident to a and then two edges incident to each vertex of $B' \setminus B$. Hence the size of the graph has increased by $O(|B| + |A|^2)$. We show that there is a subset of at most t vertices in B that dominates A in G if and only if G' is k-contractible to a tree.

Assume there exists a set $S \subseteq B$ of size at most t such that S dominates A in G. Vertex a is adjacent to all vertices of S, so the set $X = \{a\} \cup S \cup A$ is connected in G'. Note that all the vertices of G' that do not belong to X form an independent set in G. Consider the unique witness structure of G' that has X as its only big witness set. Contracting all the edges of a spanning tree of $G[X]$ yields a star. Since X has at most $1 + t + |A| = 1 + k$ vertices, any spanning tree of $G[X]$ has at most k edges. Hence G' is k-contractible to a tree.

For the reverse direction, assume that G' is k-contractible to a tree T, and let $\mathcal{W}$ be a T-witness structure of G'. Vertex a is involved in $k+1$ different cycles with each vertex of A through the vertices of $B' \setminus B$. Hence, if a and a vertex u of A appear in different witness sets, we need more than k contractions to kill the $k+1$ cycles containing both a and u. Consequently, there must be a witness set $W \in \mathcal{W}$ that contains all the vertices of $A \cup \{a\}$. Since all the vertices of $G' - W$ belong to B', they form an independent set in G'. This means that W is the only big witness set of $\mathcal{W}$, and T is in fact a star. Since G' is k-contractible to T, we know that $|W| \leq k+1$ by Observation 1. Suppose W contains a vertex $x \in B' \setminus B$. By construction, x is adjacent only to a and exactly one vertex $a' \in A$. Let b' be a neighbor of a' in B. Then we have $N_{G'}(x) \subseteq N_{G'}(b')$, so $W' = (W \setminus \{x\}) \cup \{b'\}$ is connected and $|W'| \leq |W|$. The unique witness structure of G' that has W' as its only big witness set shows that G' can be k-contracted to a tree T' on at least as many vertices as T. Thus we may assume that W contains no vertices of $B' \setminus B$. Let $S = W \setminus A'$. The set W is connected and A' is an independent set, so S dominates A'. Moreover $|S| = |W| - |A| - 1 \leq k - |A| = t$. We conclude that S is a subset of at most t vertices in B that dominates A in G. $\square$

As a contrast to this negative result, we present below an algorithm for TREE CONTRACTION with running time $4.98^k n^{O(1)}$. The straightforward proof of the following lemma has been omitted due to page restrictions.

Lemma 1. *A connected graph is k-contractible to a tree if and only if each of its 2-connected components can be contracted to a tree, using at most k edge contractions in total.*

The main idea for our algorithm for TREE CONTRACTION is to use 2-colorings of the input graph G. Let T be a tree, and let $\mathcal{W}$ be a T-witness structure of G. We say that a 2-coloring ϕ of G is *compatible* with $\mathcal{W}$ (or $\mathcal{W}$-*compatible*) if the following two conditions are both satisfied: **(1)** every witness set of $\mathcal{W}$ is monochromatic with respect to ϕ, and **(2)** if $W(u)$ and $W(v)$ are big witness sets and $uv \in E(T)$, then $\phi(W(u)) \neq \phi(W(v))$. In Lemma 2, we will show that if we are given a 2-coloring ϕ of G that is $\mathcal{W}$-compatible, then we can use the monochromatic components of G with respect to ϕ to compute a T'-witness structure of G, such that T' is a tree with at least as many vertices as T. Informally, we do this by finding a "star-like" partition of each monochromatic component M of G, where one set of the partition is a connected vertex cover of $G[M]$, and all the other sets have size 1. A *connected vertex cover* of a graph G is a subset $V' \subseteq V(G)$ such that $G[V']$ is connected and every edge of G has at least one endpoint in V'.

Proposition 1 ([4]). *Given a graph G, it can be decided in $2.4882^t n^{O(1)}$ time whether G has a connected vertex cover of size at most t. If such a connected vertex cover exists, then it can be computed within the same time.*

Lemma 2. *Let ϕ be a 2-coloring of a 2-connected graph G. If ϕ is compatible with a T-witness structure of G whose largest witness set has size d, where T is a tree, then a T'-witness structure of G can be computed in time $2.4882^d n^{O(1)}$, such that T' is a tree with as at least as many vertices as T.*

Proof. Suppose ϕ is compatible with a T-witness structure $\mathcal{W}$ of G, such that T is a tree, and the largest witness set of $\mathcal{W}$ has size d. The 2-connectedness of G implies that, if a witness set $W(v) \in \mathcal{W}$ is small, then v is a leaf of T.

Let $\mathcal{X}$ be the set of monochromatic components of G with respect to ϕ. Every witness set of $\mathcal{W}$ is monochromatic by property (1) of a $\mathcal{W}$-compatible 2-coloring, and connected by definition. Hence, for every $W \in \mathcal{W}$, there exists an $X \in \mathcal{X}$ such that $W \subseteq X$. Moreover, since every $X \in \mathcal{X}$ is connected, there exists a vertex subset $Y \subseteq V(T)$ such that $T[Y]$ is a connected subtree of T and $X = \bigcup_{y \in Y} W(y)$. Hence, $\mathcal{X}$ is a T''-witness structure of G for a tree T'' that has at most as many vertices as T. We now show how to partition the big witness sets of $\mathcal{X}$ in such a way, that we obtain a T'-witness structure of G for some tree T' with at least as many vertices as T.

Suppose there exists a set $X \in \mathcal{X}$ that contains more than one witness set of $\mathcal{W}$, say $W(v_1), \ldots, W(v_p)$ for some $p \geq 2$. We know that at most one of these sets can be big, due to properties (1) and (2) of a $\mathcal{W}$-compatible 2-coloring and the observation that every small witness set corresponds to a leaf of T. If all the sets $W(v_1), \ldots, W(v_p)$ are small, then all the vertices $v_1, \ldots, v_p$ are leaves in T. This means that $p = 2$ and T consists of only two vertices; a trivial case. Suppose one of the sets, say $W(p_1)$, is big. Since each of the sets $W(v_2), \ldots, W(v_p)$ is small, the vertices $v_2, \ldots, v_p$ are leaves in T. This means that the vertices $v_1, \ldots, v_p$ induce a star in T, with center v_1 and leaves $v_2, \ldots, v_p$. Note that $W(v_1)$ is a connected vertex cover in the graph $G[X]$; this observation will be used in the algorithm below. Also note that the sets $W(v_1), \ldots, W(v_p)$ define an S-witness structure $\mathcal{S}$ of the graph $G[X]$, where S is a star with $p-1$ leaves.

We use the above observations to decide, for each $X \in \mathcal{X}$, if we can partition X into several witness sets. Recall that, given ϕ, we only know $\mathcal{X}$, and not $\mathcal{W}$. We perform the following procedure on each set $X \in \mathcal{X}$ that contains more than one vertex. Let $\hat{X} = X \cap N_G(V \setminus X)$ be the set of vertices in X that have at least one neighbor outside X. A *shatter* of X is a partition of X into sets, such that one of them is a connected vertex cover C of $G[X]$ containing every vertex of $\hat{X}$, and each of the others has size 1. The *size* of a shatter is the size of C. A shatter of X of minimum size can be found as follows. Recall that we assumed the largest witness set of $\mathcal{W}$ to be of size d. Construct a graph G' from the graph $G[X]$ by adding, for each vertex $x \in \hat{X}$, a new vertex x' and an edge xx'. Find a connected vertex cover C of minimum size in G' by applying the algorithm of Proposition 1 for all values of t from 1 to d. Since ϕ is $\mathcal{W}$-compatible and each witness set of $\mathcal{W}$ has size at most d, such a set C will always be found. Observe that a minimum size connected vertex cover of G' does not contain any vertex of degree 1, which implies that $\hat{X} \subseteq C$. Hence C, together with the sets of size 1 formed by each of the vertices of $X \setminus C$, is a minimum size shatter of X. If, in $\mathcal{X}$, we replace X by the sets of this minimum size shatter of X, we obtain a $\tilde{T}$-witness structure of G, for some tree $\tilde{T}$ with at least as many (or strictly more, if $|C| < |X|$) vertices as T''. We point out that the size of C is at most as big as the size of the only possible big witness set of $\mathcal{W}$ that X contains. Hence, after repeating the above procedure on each of the sets of $\mathcal{X}$ that contain more than one vertex, we obtain a desired T'-witness structure of G, where T' is a tree with at least at many vertices as T.

By Proposition 1, we can find a minimum size shatter in $2.4882^d n^{O(1)}$ time for each set of $\mathcal{X}$. Since all the other steps can be performed in polynomial time, the overall running time is $2.4882^d n^{O(1)}$. □

The idea of our algorithm for TREE CONTRACTION is to generate a number of 2-colorings of the input graph G, and to check, using the algorithm described in the proof of Lemma 2, whether any of the generated 2-colorings yields a T-witness structure of G for a tree T on at least $n - k$ vertices. Using the notion of universal sets, defined below, we are able to bound the number of 2-colorings that we need to generate and check.

The *restriction* of a function $f : X \rightarrow Y$ to a set $S \subseteq X$ is the function $f_{|S} : S \rightarrow Y$ such that $f_{|S}(s) = f(s)$ for every $s \in S$. An (n,t)-*universal set* $\mathcal{F}$ is a set of functions from $\{1, 2, \ldots, n\}$ to $\{1, 2\}$ such that, for every $S \subseteq \{1, 2, \ldots, n\}$ with $|S| = t$, the set $\mathcal{F}_{|S} = \{f_{|S} \mid f \in \mathcal{F}\}$ is equal to the set 2^S of all the functions from S to $\{1, 2\}$.

Theorem 2 ([24]). *There is a deterministic algorithm that constructs an (n,t)-universal set $\mathcal{F}$ of size $2^{t+O(\log^2 t)} \log n$ in time $2^{t+O(\log^2 t)} n \log n$.*

Theorem 3. TREE CONTRACTION *can be solved in time $4.98^k n^{O(1)}$.*

Proof. Let G be an n-vertex input graph of TREE CONTRACTION. We assume that G is 2-connected, by Lemma 1. Our algorithm has an outer loop, which iterates over the values of an integer d from 1 to $k + 1$. For each value of d, the algorithm constructs an $(n, 2k - d + 2)$-universal set $\mathcal{F}_d$, and runs an inner loop that iterates over all 2-colorings $\phi \in \mathcal{F}_d$. At each iteration of the inner loop, the algorithm computes a minimum size shatter for each of the monochromatic components of G with respect to ϕ, using the $2.4882^d n^{O(1)}$ time procedure described in the proof of Lemma 2 with the value d determined by the outer loop. If this procedure yields a T'-witness structure of G for a tree T' with at least

$n - k$ vertices at some iteration of the inner loop, then the algorithm outputs YES. If none of the iterations of the inner loop yields a YES-answer, the outer loop picks the next value of d. If none of the iterations of the outer loop yields a YES-answer, then the algorithm returns NO.

To prove correctness of the algorithm, suppose G is k-contractible to a tree T. Let $\mathcal{W}$ be a T-witness structure of G whose largest witness set has size d^*. Note that $d^* \leq k + 1$ by Observation 1. Let ψ be a 2-coloring of G such that each of the big witness sets of $\mathcal{W}$ is monochromatic with respect to ψ, such that $\psi(W(u)) \neq \psi(W(v))$ whenever uv is an edge in T, and such that the vertices in the small witness sets are all colored 1. Observe that ψ is a $\mathcal{W}$-compatible 2-coloring of G, as is any other 2-coloring of G that coincides with ψ on all the vertices of the big witness sets of $\mathcal{W}$. The largest witness set requires $d^* - 1$ edge contractions, after which our remaining budget of edge contractions is $k - (d^* - 1) = k - d^* + 1$. As a result of Observation 1, the total number of vertices contained in big witness sets is thus at most $d^* + 2(k - d^* + 1) = 2k - d^* + 2$. Consequently, if we generate an $(n, 2k - d^* + 2)$-universal set $\mathcal{F}_{d^*}$, then, by Theorem 2, $\mathcal{F}_{d^*}$ contains at least one 2-coloring ϕ of G that coincides with ψ on all the vertices of the big witness sets of $\mathcal{W}$. Note that such a 2-coloring ϕ that is $\mathcal{W}$-compatible. Recall that our algorithm iterates over all values of d from 1 to $k + 1$, and that $d^* \leq k + 1$. Hence, at the correct iteration of the outer loop, i.e., the iteration where $d = d^*$, our algorithm will process ϕ. As a result of Lemma 2, the algorithm will then find a T'-witness structure of G for some tree T' with at least $n - k$ vertices. This means that the algorithm correctly outputs YES if G is k-contractible to a tree. Since the algorithm only outputs YES when it has detected a T'-witness structure for some tree T' with at least $n - k$ vertices, it correctly outputs NO if G is not k-contractible to a tree.

For each d, the size of $\mathcal{F}_d$ is $2^{2k-d+2+\log^2(2k-d+2)} \log n$, and $\mathcal{F}_d$ can be constructed in $2^{2k-d+2+\log^2(2k-d+2)} n \log n$ time, by Theorem 2. Summing $|\mathcal{F}_d| \cdot 2.4882^d n^{O(1)}$ over all values of d from 1 to $k + 1$ shows that this deterministic algorithm runs in time $4.98^k n^{O(1)}$.   □

We would like to remark that due to recent developments in the field, our result in fact implies a randomized $4^{k+o(k)} n^{O(1)}$ time algorithm for TREE CONTRACTION. Cygan et al. [14] give a Monte Carlo algorithm with running time $2^t n^{O(1)}$ for deciding whether a graph on n vertices has a connected vertex cover of size at most t and finding such a set if it exists. Summing $|\mathcal{F}_d| \cdot 2^d n^{O(1)}$ over all values of d from 1 to $k + 1$, as it was done in the last line of the proof of Theorem 3, we obtain total running time $4^{k+o(k)} n^{O(1)}$ for a randomized algorithm.

4   PATH CONTRACTION

Brouwer and Veldman [9] showed that, for every fixed $\ell \geq 4$, it is NP-complete to decide whether a graph can be contracted to the path P_ℓ. This, together with the observation that a graph G is k-contractible to a path if and only if G is contractible to P_{n-k}, implies that PATH CONTRACTION is NP-complete. In this section, we first show that PATH CONTRACTION has a linear vertex kernel. We then present an algorithm with running time $2^{k+o(k)} + n^{O(1)}$ for this problem. Throughout this section, whenever we mention a P_ℓ-witness structure $\mathcal{W} = \{W_1, \ldots W_\ell\}$, it will be implicit that $P_\ell = p_1 \cdots p_\ell$, and $W_i = W(p_i)$ for every $i \in \{1, \ldots, \ell\}$.

Rule 1. *Let (G, k) be an instance of* PATH CONTRACTION. *If G contains a bridge uv such that the deletion of edge uv from G results in two connected components that contain at least $k + 2$ vertices each, then return (G', k), where G' is the graph resulting from the contraction of edge uv.*

The proof of the following lemma has been omitted due to page restrictions.

Lemma 3. *Let (G', k) be an instance of* PATH CONTRACTION *resulting from the application of Rule 1 on (G, k). Then G' is k-contractible to a path if and only if G is k-contractible to a path.*

Theorem 4. PATH CONTRACTION *has a kernel with at most $5k + 3$ vertices.*

Proof. Let (G, k) be an instance of PATH CONTRACTION. We repeatedly test, in linear time, whether Rule 1 can be applied on the instance under consideration, and apply the reduction rule if possible. Each application of Rule 1 strictly decreases the number of vertices. Hence, starting from G, we reach in polynomial time a *reduced* graph, on which Rule 1 cannot be applied anymore. By Lemma 3, we know that the resulting reduced graph is k-contractible to a path if and only if G is k-contractible to a path.

We now assume that G is reduced. We show that if G is k-contractible to a path, then G has at most $5k + 3$ vertices. Let $\mathcal{W} = \{W_1, \ldots, W_\ell\}$ be a P_ℓ-witness structure of G with $\ell \geq n-k$. We first prove that $\ell \leq 4k+3$. Assume that $\ell \geq 2k+4$, and let i be such that $k+2 \leq i \leq \ell-k-2$. Suppose, for contradiction, that both W_i and W_{i+1} are small witness sets, i.e., $W_i = \{u\}$ and $W_{i+1} = \{v\}$ for two vertices u and v of G. Then uv forms a bridge in G whose deletion results in two connected components. Each of these components contains at least all vertices from $W_1, \ldots, W_{k+2}$ or all vertices from $W_{\ell-k-1}, \ldots, W_\ell$. Hence they contain at least $k + 2$ vertices each. Consequently, Rule 1 can be applied, contradicting the assumption that G is reduced. So there are no consecutive small sets among $W_{k+2}, \ldots, W_{\ell-k-1}$. By Observation 1, $\mathcal{W}$ contains at most k big witness sets, so we have $(\ell - k - 1) - (k+2) + 1 \leq 2k+1$ implying $\ell \leq 4k+3$. Combining this with the earlier assumption that $\ell \geq n - k$ yields $n \leq 5k + 3$. $\square$

The existence of a kernel with at most $5k + 3$ vertices easily implies a $32^{k+o(k)} + n^{O(1)}$ time algorithm for PATH CONTRACTION, which tests for each 2-coloring ϕ of the reduced input graph G' whether the monochromatic components of G' with respect to ϕ form a P_ℓ-witness structure of G' for some $\ell \geq n - k$. The natural follow-up question, which we answer affirmatively below, is whether this running time can be significantly improved.

Theorem 5. PATH CONTRACTION *can be solved in time $2^{k+o(k)} + n^{O(1)}$.*

Proof. Given an instance (G, k) of PATH CONTRACTION, our algorithm first constructs an equivalent instance (G', k) such that G' has at most $5k+3$ vertices. This can be done in $n^{O(1)}$ time by Theorem 4. For the rest of the proof, we assume that the input graph G has $n \leq 5k + 3$ vertices. Suppose G is k-contractible to a path P_ℓ, and let $\mathcal{W} = \{W_1, \ldots, W_\ell\}$ be a P_ℓ-witness structure of G. We distinguish two cases, depending on whether or not ℓ is larger than $\sqrt{k}$.

Suppose $\ell \leq \sqrt{k}$. Then $n \leq k + \sqrt{k}$. We define $X^* = W_1 \cup W_3 \cup W_5 \cup \ldots$ and $Y^* = W_2 \cup W_4 \cup \ldots$. Then X^* and Y^* form a 2-partition of $V(G)$, and the connected components of the graphs $G[X^*]$ and $G[Y^*]$ form a P_ℓ-witness structure of G. If we contract every edge of G that has both endpoints in the same connected component of $G[X^*]$ or $G[Y^*]$, we end up with the path P_ℓ. Hence, for every given partition of $V(G)$ into two sets X and Y, we can check in $k^{O(1)}$ time whether the connected components of $G[X]$ and $G[Y]$ constitute a $P_{\ell'}$-witness structure of G for some $\ell' \geq n - k$. Based on this analysis, if G has at most $k + \sqrt{k}$ vertices, the algorithm checks for each 2-partition X, Y of $V(G)$ whether this 2-partition yields a desired witness structure. Note that if G is k-contractible to a path on more than $\sqrt{k}$ vertices, then it is also k-contractible to a path on exactly $\sqrt{k}$ vertices, since G has at most $k + \sqrt{k}$ vertices. Since there are at most $2^{k+\sqrt{k}}$ partitions to consider, the running time of the algorithm in this case is $2^{k+o(k)}k^{O(1)} = 2^{k+o(k)}$.

Now suppose $\ell > \sqrt{k}$. For each integer i with $1 \leq i \leq \lfloor \sqrt{k} \rfloor$, we define $W_i^* = W_i \cup W_{i+\lfloor \sqrt{k} \rfloor} \cup W_{i+2\lfloor \sqrt{k} \rfloor} \cup \ldots$. Since $n \leq 5k + 3$, there is at least one index j such that $|W_j^*| \leq (5k + 3)/\sqrt{k}$. Let $G_1^*, \ldots, G_p^*$ denote the connected components of $G - W_j^*$, where $p \leq (5k + 3)/\sqrt{k}$. Note that each connected component G_i^* has a P^*-witness structure $\mathcal{W}_i^*$ for some path P^* on at most $\sqrt{k} - 1$ vertices, such that the union of these witness structures $\mathcal{W}_i^*$, together with the vertex sets of $G_1^*, \ldots, G_p^*$, forms a P_ℓ-witness structure of G. Moreover, each connected component G_i^* has at most $k + \sqrt{k} - 1$ vertices by Observation 1. Based on this analysis, if G has more than $k + \sqrt{k}$ vertices, the algorithm searches for the correct set W_i^* by generating all subsets $W \subseteq V(G)$ of size at most $(5k + 3)/\sqrt{k}$, and performing the following checks for each subset W. If the graph $G - W$ has more than $(5k + 3)/\sqrt{k}$ connected components, or if one of the connected components has more than $k + \sqrt{k} - 1$ vertices, then W is discarded. For each W that is not discarded, we run the algorithm of the previous case on each connected component G_i of $G - W$ to check whether G_i has a $P_{\ell'}$-witness structure with $\ell' \leq \sqrt{k} - 1$. Since in that algorithm we check every 2-partition of $V(G_i)$, we can check whether G_i has a $P_{\ell'}$-witness structure with the additional constraint that precisely the vertices in the first and the last witness sets have neighbors in the appropriate connected components of $G[W]$, and pick such a $P_{\ell'}$-witness structure $\mathcal{W}_i$ for which ℓ' is a large as possible. Finally, we check if all these witness structures $\mathcal{W}_i$, together with the vertex sets of the connected components of $G[W]$, form a P-witness structure of G for some path P on at least $n - k$ vertices. If so, the algorithm outputs YES. Otherwise, the algorithm tries another subset W, or outputs NO if all subsets W have been considered. For each generated set W, we run the algorithm of the previous case on each of the $O(\sqrt{k})$

connected components of $G - W$, so we can check in $2^{k+o(k)}O(\sqrt{k}) = 2^{k+o(k)}$ time whether we get a desired P-witness structure of G. Since we generate no more than $(5k + 3)^{(5k+3)/\sqrt{k}} = 2^{o(k)}$ subsets W, we get a total running time of $2^{k+o(k)}$ also for this case.   □

5 Concluding Remarks

The number of edges to contract in order to obtain a certain graph property is a natural measure of how close the input graph is to having that property, similar to the more established similarity measures of the number of edges or vertices to delete. The latter measures are well studied when the desired property is being acyclic or being a path, defining some of the most widely known and well studied problems within Parameterized Complexity. Inspired by this, we gave kernelization results and fast fixed parameter algorithms for PATH CONTRACTION and TREE CONTRACTION. We think these results motivate the parameterized study of similar problems, an example of which is INTERVAL CONTRACTION. It is not known whether the vertex deletion variant of this problem, INTERVAL VERTEX DELETION, is fixed parameter tractable. Is INTERVAL CONTRACTION fixed parameter tractable?

Acknowledgments. The authors would like to thank Martin Vatshelle, Saket Saurabh, and Erik Jan van Leeuwen for valuable suggestions and comments.

References

1. Alon, N., Yuster, R., Zwick, U.: Color-coding. J. ACM 42, 844–856 (1995)
2. Asano, T., Hirata, T.: Edge-contraction problems. J. Comput. Syst. Sci. 26, 197–208 (1983)
3. van Bevern, R., Komusiewicz, C., Moser, H., Niedermeier, R.: Measuring Indifference: Unit Interval Vertex Deletion. In: Thilikos, D.M. (ed.) WG 2010. LNCS, vol. 6410, pp. 232–243. Springer, Heidelberg (2010)
4. Binkele-Raible, D., Fernau, H.: Enumerate & Measure: Improving Parameter Budget Management. In: Raman, V., Saurabh, S. (eds.) IPEC 2010. LNCS, vol. 6478, pp. 38–49. Springer, Heidelberg (2010)
5. Bodlaender, H.L.: On disjoint cycles. Int. J. Found. Comput. Sci. 5, 59–68 (1994)
6. Bodlaender, H.L., Fomin, F.V., Lokshtanov, D., Penninkx, E., Saurabh, S., Thilikos, D.M.: (Meta) kernelization. In: Proc. FOCS, pp. 629–638. IEEE (2009)
7. Bodlaender, H.L., Downey, R.G., Fellows, M.R., Hermelin, D.: On problems without polynomial kernels. J. Comput. Syst. Sci. 75(8), 423–434 (2009)
8. Bodlaender, H.L., Thomassé, S., Yeo, A.: Kernel bounds for disjoint cycles and disjoint paths. Theor. Comput. Sci. 412(35), 4570–4578 (2011)
9. Brouwer, A.E., Veldman, H.J.: Contractibility and NP-completeness. Journal of Graph Theory 11(1), 71–79 (1987)
10. Cai, L.: Fixed-parameter tractability of graph modification problems for hereditary properties. Inf. Process. Lett. 58(4), 171–176 (1996)
11. Cao, Y., Chen, J., Liu, Y.: On Feedback Vertex Set New Measure and New Structures. In: Kaplan, H. (ed.) SWAT 2010. LNCS, vol. 6139, pp. 93–104. Springer, Heidelberg (2010)

12. Chen, J., Fomin, F.V., Liu, Y., Lu, S., Villanger, Y.: Improved algorithms for feedback vertex set problems. J. Comput. Syst. Sci. 74(7), 1188–1198 (2008)
13. Courcelle, B.: The monadic second-order logic of graphs. I. Recognizable sets of finite graphs. Information and Computation 85, 12–75 (1990)
14. Cygan, M., Nederlof, J., Pilipczuk, M., Pilipczuk, M., van Rooij, J., Wojtaszczyk, J.O.: Solving connectivity problems parameterized by treewidth in single exponential time. In: Proc. FOCS 2011 (to appear, 2011)
15. Dehne, F.K.H.A., Fellows, M.R., Langston, M.A., Rosamond, F.A., Stevens, K.: An $O(2^{O(k)}n^3)$ FPT algorithm for the undirected feedback vertex set problem. Theory Comput. Syst. 41(3), 479–492 (2007)
16. Dom, M., Lokshtanov, D., Saurabh, S.: Incompressibility through Colors and IDs. In: Albers, S., Marchetti-Spaccamela, A., Matias, Y., Nikoletseas, S., Thomas, W. (eds.) ICALP 2009. LNCS, vol. 5555, pp. 378–389. Springer, Heidelberg (2009)
17. Downey, R.G., Fellows, M.R.: Parameterized complexity. Monographs in Computer Science. Springer, Heidelberg (1999)
18. Guo, J., Gramm, J., Huffner, F., Niedermeier, R., Wernicke, S.: Compression-based fixed-parameter algorithms for feedback vertex set and edge bipartization. J. Comput. Syst. Sci. 72(8), 1386–1396 (2006)
19. Heggernes, P., van 't Hof, P., Jansen, B., Kratsch, S., Villanger, Y.: Parameterized Complexity of Vertex Deletion into Perfect Graph Classes. In: Owe, O., Steffen, M., Telle, J.A. (eds.) FCT 2011. LNCS, vol. 6914, pp. 240–251. Springer, Heidelberg (2011)
20. Heggernes, P., van 't Hof, P., Lévêque, B., Lokshtanov, D., Paul, C.: Contracting graphs to paths and trees. In: Marx, D., Rossmanith, P. (eds.) IPEC 2011. LNCS, vol. 7112, pp. 55–66. Springer, Heidelberg (2012)
21. Kawarabayashi, K., Reed, B.A.: An (almost) linear time algorithm for odd cycles transversal. In: Proc. SODA 2010, pp. 365–378. ACM-SIAM (2010)
22. Marx, D.: Chordal deletion is fixed-parameter tractable. Algorithmica 57(4), 747–768 (2010)
23. Marx, D., Schlotter, I.: Obtaining a Planar Graph by Vertex Deletion. In: Brandstädt, A., Kratsch, D., Müller, H. (eds.) WG 2007. LNCS, vol. 4769, pp. 292–303. Springer, Heidelberg (2007)
24. Naor, M., Schulman, L.J., Srinivasan, A.: Splitters and near-optimal derandomization. In: Proc. FOCS 1995, pp. 182–191. IEEE (1995)
25. Natanzon, A., Shamir, R., Sharan, R.: Complexity classification of some edge modification problems. Disc. Appl. Math. 113(1), 109–128 (2001)
26. Philip, G., Raman, V., Villanger, Y.: A Quartic Kernel for Pathwidth-One Vertex Deletion. In: Thilikos, D.M. (ed.) WG 2010. LNCS, vol. 6410, pp. 196–207. Springer, Heidelberg (2010)
27. Thomassé, S.: A $4k^2$ vertex kernel for feedback vertex set. ACM Transactions on Algorithms 6(2) (2010)
28. Villanger, Y.: Proper Interval Vertex Deletion. In: Raman, V., Saurabh, S. (eds.) IPEC 2010. LNCS, vol. 6478, pp. 228–238. Springer, Heidelberg (2010)
29. Watanabe, T., Ae, T., Nakamura, A.: On the removal of forbidden graphs by edge-deletion or edge-contraction. Disc. Appl. Math. 3, 151–153 (1981)
30. Watanabe, T., Ae, T., Nakamura, A.: On the NP-hardness of edge-deletion and edge-contraction problems. Disc. Appl. Math. 6, 63–78 (1983)
31. Yannakakis, M.: Node and edge-deletion NP-complete problems. In: Proc. STOC 1978, pp. 253–264. ACM (1978)
32. Yannakakis, M.: The effect of a connectivity requirement on the complexity of maximum subgraph problems. J. ACM 26(4), 618–630 (1979)
33. Yannakakis, M.: Edge-deletion problems. SIAM J. Comput. 10(2), 297–309 (1981)

Increasing the Minimum Degree of a Graph by Contractions⋆

Petr A. Golovach[1], Marcin Kamiński[2],
Daniël Paulusma[1], and Dimitrios M. Thilikos[3]

[1] School of Engineering and Computing Sciences, Durham University,
Science Laboratories, South Road, Durham DH1 3LE, United Kingdom
{petr.golovach,daniel.paulusma}@durham.ac.uk
[2] Département d'Informatique, Université Libre de Bruxelles, Belgium
marcin.kaminski@ulb.ac.be
[3] Department of Mathematics, National and Kapodistrian University of Athens,
Panepistimioupolis, GR15784 Athens, Greece
sedthilk@math.uoa.gr

Abstract. The DEGREE CONTRACTIBILITY problem is to test whether a given graph G can be modified to a graph of minimum degree at least d by using at most k contractions. We prove the following three results. First, DEGREE CONTRACTIBILITY is NP-complete even when $d = 14$. Second, it is fixed-parameter tractable when parameterized by k and d. Third, it is W[1]-hard when parameterized by k. We also study its variant where the input graph is weighted, i.e., has some edge weighting and the contractions preserve these weights. The WEIGHTED DEGREE CONTRACTIBILITY problem is to test if a weighted graph G can be contracted to a weighted graph of minimum weighted degree at least d by using at most k weighted contractions. We show that this problem is NP-complete and that it is fixed-parameter tractable when parameterized by k.

1 Introduction

We consider undirected finite graphs that have no loops and no multiple edges. A graph modification problem has as input a graph G and an integer k. The question is whether G can be modified to belong to some specified graph class that satisfies further properties by using at most k operations of a certain specified type such as deleting a vertex or deleting an edge. In our paper the permitted operation is the *contraction* of an edge, which removes both end-vertices of the edge and replaces them by a new vertex adjacent to precisely those vertices to which the two end-vertices were adjacent.

We continue a very recent study [14,15,16] of the following graph modification problem called Π-CONTRACTIBILITY, where Π is some prespecified graph class.

Π-CONTRACTIBILITY
Instance: a graph G and an integer k.
Question: Can G be modified to a graph in Π by at most k contractions?

⋆ This work is supported by EPSRC (EP/G043434/1).

D. Marx and P. Rossmanith (Eds.): IPEC 2011, LNCS 7112, pp. 67–79, 2012.

This research was started by Watanabe, Ae and Nakamura [24,25] who showed that Π-CONTRACTIBILITY is NP-complete whenever Π is finitely characterizable by 3-connected graphs. Their result was generalized by Asano and Hirata [2] who showed that Π-CONTRACTIBILITY is NP-complete whenever Π is a graph class that fulfills the following conditions. First, Π must be closed under contractions. Second, Π must be described by a property that is satisfied by infinitely many connected graphs and violated by infinitely many other connected graphs. Third, a graph belongs to Π if and only if each of its biconnected components belong to Π. Examples [2] of such graph classes Π include planar graphs, outerplanar graphs, series-parallel graphs, and also forests, chordal graphs, or more generally, graphs with no cycles of length at least ℓ for some fixed integer $\ell \geq 3$.

The problem Π-CONTRACTIBILITY is closely related to the problem H-CONTRACTIBILITY, which is to test whether a given graph G can be contracted to a fixed graph H (i.e., which is not part of the input). Brouwer and Veldman [7] showed that the H-CONTRACTIBILITY problem is NP-complete whenever H is a triangle-free graph that contains no vertex adjacent to all the other vertices. Their work has been extended by a series of other papers [17,21,22] showing both polynomial-time solvable and NP-complete cases. Determining a full complexity classification for H-CONTRACTIBILITY is open, although such results restricting the input graph G to be in a special graph class have been obtained [3,4,19].

If Π is the class of paths or cycles, then Π-CONTRACTIBILITY is polynomially equivalent to the problems of determining the length of a longest path and a longest cycle, respectively, to which a given graph can be contracted. The first problem has been shown to be NP-complete by van 't Hof, Paulusma and Woeginger [18] even for graphs with no induced path on 6 vertices; they use the aforementioned NP-completeness result of Brouwer and Veldman [7] for the special case when H is the 4-vertex path. The second problem has been shown to be NP-complete by Hammack [13].

Recently, more papers appeared that study the Π-CONTRACTIBILITY problem, and in particular, its *parameterized* complexity where the parameter is the number k of edges that may be contracted. Heggernes et al. [16] gave an $4^{k+O(\log^2 k)} + n^{O(1)}$ time algorithm for Π-CONTRACTIBILITY if Π is the class of paths. Moreover, they showed that in this case the problem has a linear kernel. When Π is the class of trees, they showed that the problem can be solved in $4.88^k n^{O(1)}$ time and that a polynomial kernel does not exist unless coNP $\subseteq$ NP \ poly. When the input graph is a chordal graph with n vertices and m edges, Heggernes et al. [14] could show that Π-CONTRACTIBILITY can be solved in $O(n + m)$ time when Π is the class of trees and in $O(nm)$ time when Π is the class of paths. When Π is the class of bipartite graphs, Heggernes et al. [15] observed that Π-CONTRACTIBILITY is NP-complete and showed that Π-CONTRACTIBILITY is fixed-parameter tractable when parameterized by k. Finally, due to a close relationship with the problem that is to test whether a given graph contains a so-called disconnected cut set, Martin and Paulusma [20] could show that Π-CONTRACTIBILITY is NP-complete if Π is the class of bicliques $K_{p,q}$ with $p, q \geq 2$.

Bodlaender, Koster and Wolle [6] introduced the related notion of *contraction degeneracy* as a useful tool to improve lower bound heuristics for treewidth. The contraction degeneracy of a graph G is the largest minimum degree of any minor of G. When G is connected, the contraction degeneracy of G is equal to the largest minimum degree of any graph to which G can be contracted [6]. The CONTRACTION DEGENERACY problem is to test whether the contraction degeneracy of a given graph is at least d for some given integer d. Bodlaender, Koster and Wolle [6] proved that this problem is NP-complete, even for bipartite graphs, and that it is is fixed-parameter tractable when parameterized by d. They also evaluated a number of heuristics for computing the contraction degeneracy.

Our Results. We study the Π-CONTRACTIBILITY problem where Π is the class of graphs of minimum degree at least d for some integer d. In this case we call the problem the DEGREE CONTRACTIBILITY (DC) problem. Note that this class does not satisfy the first and third property of Asano and Hirata [2]. Moreover, for connected graphs $G = (V, E)$ and $k \geq |E|$, the DEGREE CONTRACTIBILITY problem is equivalent to the CONTRACTION DEGENERACY problem. In Section 3 we show that DEGREE CONTRACTIBILITY is fixed-parameter tractable when parameterized by k and d. However, as we show in Section 3 as well, when either k or d is part of the input DEGREE CONTRACTIBILITY becomes hard in the following sense. First, DEGREE CONTRACTIBILITY is NP-complete even when d is assumed to be fixed (and k is part of the input). We prove that a value of $d = 14$ already suffices. Second, DEGREE CONTRACTIBILITY is W[1]-hard when parameterized by k. These results complement the result of Amini, Sau and Saurabh [1] who showed that detecting a subgraph with at most k vertices and of minimum degree at least d is W[1]-hard for every fixed $d \geq 4$ when parameterized by k.

In Section 4 we study the weighted version of DEGREE CONTRACTIBILITY. In order to define this variant, let $G = (V, E)$ be a *weighted* graph, i.e., with some *edge weighting* $w: E \to \mathbb{N}$. The *weighted degree* $d_G^w(u)$ of a vertex u is the sum of the weights of the edges incident with u in G, i.e., $d_G^w(u) = \sum_{v \in N(u)} w(uv)$ where $N(u)$ denotes the set of neighbors of u. The *weighted contraction* of an edge $e = uv$ is a contraction of e where the weights on the edges incident with the new vertex x_{uv} are defined as follows:

- $w(x_{uv}y) = w(uy)$ if y is adjacent to u and not adjacent to v;
- $w(x_{uv}y) = w(vy)$ if y is adjacent to v and not adjacent to u;
- $w(x_{uv}y) = w(uy) + w(vy)$ if y is adjacent to u and v.

The WEIGHTED DEGREE CONTRACTIBILITY (WDC) problem is to test if a weighted graph G can be modified to a weighted graph of minimum weighted degree at least d by using at most k weighted contractions. Note that since the weight of an edge $x_{uv}y$ with y adjacent to both u and v is the accumulated weight of the two original edges uy and vy, DEGREE CONTRACTIBILITY is not a special (unweighted) case of WEIGHTED DEGREE CONTRACTIBILITY. We show that WEIGHTED DEGREE CONTRACTIBILITY problem is still NP-complete, even when $k \geq |E|$. However, contrary to the aforementioned W[1]-hardness result for

DEGREE CONTRACTIBILITY when parameterized by k, accumulating the weights after contracting an edge results in the problem not being hard anymore, i.e., we prove that WEIGHTED DEGREE CONTRACTIBILITY is fixed-parameter tractable when parameterized by k. Table 1 summarizes our results. The (only) open case in this table is denoted "?".

Table 1. An overview of our results

input	parameter	DC	WDC
d, k		NP-complete	NP-complete
d	k	W[1]-hard	FPT
k	d	para-NP-complete	?
	d, k	FPT	FPT

2 Preliminaries

We denote the vertex set and edge set of a graph G by V_G and E_G, respectively. If no confusion is possible, we may omit subscripts. Recall that we only consider undirected finite graphs with no loops and no multiple edges. We refer to the text book of Diestel [9] for undefined graph terminology and to the monographs of Downey and Fellows [11] and Niedermeier [23] for more on parameterized complexity.

Let G be a (weighted) graph. A vertex v is a *neighbor* of a vertex u if $uv \in E_G$. We let $N_G(u) = \{uv \mid v \in V_G\}$ denote the *neighborhood* of u. The *degree* of a vertex u is denoted $d_G(u) = |N_G(u)|$. In the case that G is a weighted graph with an edge weighting w, recall that the weighted degree of u is $d_G^w(u) = \sum_{v \in N(u)} w(uv)$. A subset $U \subseteq V$ is a *clique* if there is an edge in G between any two vertices of U, and U is an *independent set* if there is no edge in G between any two vertices of U. We write $G[U]$ to denote the subgraph of G *induced* by $U \subseteq V$, i.e., the graph on vertex set U and an edge between any two vertices whenever there is an edge between them in G. We let G/e denote the (weighted) graph obtained from G by the (weighted) contraction of e. If a (weighted) graph H is obtained from G by a sequence of (weighted) contractions, then H is a *(weighted) contraction* of G.

Let G and H be two graphs. An *H-witness structure* $\mathcal{W}$ is a vertex partition of G into $|V_H|$ (nonempty) sets $W(x)$ called *H-witness bags*, such that

(i) each $W(x)$ induces a connected subgraph of G;
(ii) for all $x, y \in V_H$ with $x \neq y$, bags $W(x)$ and $W(y)$ are adjacent in G if and only if x and y are adjacent in H;

By contracting all bags to singletons we observe that H is a contraction of G if and only if G has an H-witness structure such that conditions (i)-(ii) hold. Note that a graph may have more than one H-witness structure.

3   Contractions

First, we observe that DEGREE CONTRACTIBILITY is FPT when parameterized by k and d.

Proposition 1. DEGREE CONTRACTIBILITY *can be solved in time* $O(d^k(n+m))$ *for graphs with n vertices and m edges.*

Proof. Let G be a graph with n vertices and m edges. We give the following branching algorithm. Let $d_G(u) < d$ for some vertex $u \in V_G$. We consider all edges e incident with u, and call our algorithm recursively for G/e and the parameter $k' = k - 1$. The algorithm returns Yes, if for at least one of the new instances the answer is Yes, and it returns No otherwise. Since for each recursive call of our algorithm, we create at most $d - 1$ instances of the problem, and the depth of the recursion is at most k, the algorithms runs in time $O(d^k(n + m))$. □

If we parameterize only by d or only by k, then DEGREE CONTRACTIBILITY becomes hard. We first prove that the problem is NP-complete even if $d = 14$.

Theorem 1. *For any fixed $d \geq 14$,* DEGREE CONTRACTIBILITY *is* NP-*complete.*

Proof. The inclusion of the problem in NP is obvious. For simplicity, we prove NP-hardness for $d = 14$. We reduce from the NP-complete SET COVER problem [12]. This problem is defined as follows.

Given a set $U = \{u_1, \ldots, u_m\}$, a family of subsets $X_1, \ldots, X_n \subseteq U$ and an integer r, are there at most r subsets that *cover* U, i.e., their union is U?

It is known [12] that this problem remains NP-complete even if

(i) each X_i has cardinality 3, and
(ii) each u_j is included in at least two and at most three subsets of $X_1, \ldots, X_n$.

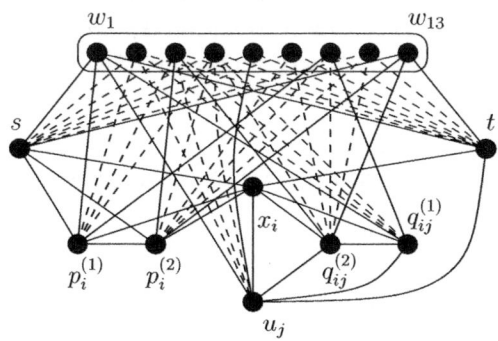

Fig. 1. Construction of G

We consider an instance $(U, X_1, \ldots, X_n)$ of SET COVER with restrictions (i) and (ii). We construct a graph G in the following way; also see Fig. 1. We say that we *connect* a vertex with some other vertex if we add an edge between them.

- Construct a clique with 13 vertices $w_1, \ldots, w_{13}$.
- Add two new vertices s, t and connect each of them with $w_1, \ldots, w_{13}$.
- For $i = 1, \ldots, n$, add a vertex x_i and connect it with s, t.
- For $i = 1, \ldots, n$, add two adjacent vertices $p_i^{(1)}, p_i^{(2)}$, connect $p_i^{(1)}$ with $s, w_1, \ldots, w_{11}, x_i$, and connect $p_i^{(2)}$ with $s, w_3, \ldots, w_{13}, x_i$.
- For $j = 1, \ldots, m$, add a vertex u_j and connect it with t.
- Connect x_i and u_j whenever $u_j \in X_i$. In that case also add two adjacent vertices $q_{ij}^{(1)}, q_{ij}^{(2)}$, connect $q_{ij}^{(1)}$ with $x_i, u_j, w_1, \ldots, w_{11}$ and connect $q_{ij}^{(2)}$ with $x_i, u_j, w_3, \ldots, w_{13}$.
- For $j = 1, \ldots, m$, connect u_j with $w_1, \ldots, w_8$ if u_j occurs in two subsets of $X_1, \ldots, X_n$, and connect u_j with $w_1, \ldots, w_6$ if u_j occurs in three subsets.

We set $k = n + r$ and claim that U can be covered by at most r subsets of $\{X_1, \ldots, X_n\}$ if and only if G can be modified to a graph with minimum degree at least $d = 14$ by at most k contractions.

First suppose that $X_{i_1}, \ldots, X_{i_r}$ is a set cover of U, i.e., $U = X_{i_1} \cup \ldots \cup X_{i_r}$. For $j = 1, \ldots, r$, we contract the edges sx_{i_j} and $p_{i_j}^{(1)} p_{i_j}^{(2)}$. We also contract the edge $x_i t$ for every $i \notin \{i_1, \ldots, i_r\}$. The total number of contractions is $2r + (n - r) = n + r = k$. Moreover, the resulting graph is readily seen to have minimum degree at least 14, as desired.

Now suppose G can be modified to a graph H with minimum degree at least $d = 14$ by at most k contractions. Let $\mathcal{W}$ be an H-witness structure of G. For each bag W of $\mathcal{W}$, we choose an arbitrary spanning tree of $G[W]$. Let $A \subseteq E_G$ denote the union of the sets of edges of these trees. Because H is obtained by contracting the edges of A, we find that $|A| \leq k$.

For each X_i, we define a set of edges $E_i \subseteq E_G$ as follows. The set E_i includes all edges incident with $x_i, p_i^{(1)}, p_i^{(2)}$, and all edges incident with $q_{ij}^{(1)}, q_{ij}^{(2)}$ for every $u_j \in X_i$. Moreover, we choose one vertex $u_j \in X_i$ and also add all (other) edges incident with u_j to E_i. The sets $E_1, \ldots, E_n$ have the following properties.

1. $E_i \cap E_j = \emptyset$ for $1 \leq i < j \leq n$.
2. $E_i \cap A \neq \emptyset$ for $i = 1, \ldots, n$.
3. The number of sets E_i with $|E_i \cap A| \geq 2$ is at most r.

Property 1 is true by definition. Property 2 follows from the fact that $d_G(x_i) = 13 < 14 = d$; therefore, at least one edge incident with x_i must be contracted. Property 3 follows from properties 1 and 2 and the aforementioned observation that $|A| \leq k = n + r$.

Let $I = \{i \mid |E_i \cap A| \geq 2\}$. We claim that $\cup_{i \in I} X_i = U$. In order to obtain a contradiction, assume that there is a vertex $u_j \in U \setminus \cup_{i \in I} X_i$. Then, for each

X_i with $u_j \in X_i$, we find that E_i contains a unique edge $e_i \in E_i \cap A$. Because $d_G(x_i) = 13 < d$, e_i is incident with x_i. If $e_i = sx_i$, then contracting e_i decreases the degree of $p_i^{(1)}$ and $p_i^{(2)}$. Because they both have degree 14, at least one edge incident with them must be contracted as well. Hence, $e_i \neq sx_i$. Similarly, if $e_i = x_i p_i^{(1)}$ then contracting e_i decreases the degree of $p_i^{(2)}$. Hence, $e_i \neq x_i p_i^{(1)}$. We apply the same arguments on the other edges in E_i and conclude that the only possibility is $e_i = x_i t$. Now we consider two cases.

Case 1. u_j is included in exactly two sets X_{i_1}, X_{i_2}. Then edges $x_{i_1} t, x_{i_2} t$ are contracted, whereas all other edges incident with x_{i_1}, x_{i_2} and also edges $p_{i_1}^{(1)} p_{i_1}^{(2)}$, $p_{i_2}^{(1)} p_{i_2}^{(2)}$, $q_{i_1 j}^{(1)} q_{i_1 j}^{(2)}$, $q_{i_2 j}^{(1)} q_{i_2 j}^{(2)}$ are not contracted. Moreover, no edges incident with u_j are contracted, because these belong to $E_{i_1} \cup E_{i_2}$. However, then u_j has degree at most $13 < d$ in H, a contradiction.

Case 2. u_j is included in three sets $X_{i_1}, X_{i_2}, X_{i_3}$. By the same arguments as in Case 1, we find that the degree of u_j in H is at most $13 < d$, a contradiction.

We conclude that $\{X_i \mid i \in I\}$ is a set cover, which contains at most r sets due to Property 3. This completes the proof of Theorem 1. □

While it can be easily seen that for any fixed $d \leq 3$, DEGREE CONTRACTIBILITY can be solved in polynomial time, determining the complexity for $4 \leq d \leq 13$ is an open question.

We observe that DEGREE CONTRACTIBILITY is in XP when parameterized by k; it can be solved in $n^{O(k)}$ time for n-vertex graphs by checking all sequences of at most k contractions. However, it is unlikely to be solvable in FPT-time.

Theorem 2. DEGREE CONTRACTIBILITY *parameterized by k is* W[1]-*hard.*

Proof. The problem MULTICOLORED CLIQUE is to test whether a graph with a proper k-coloring contains a clique of size k with exactly one vertex from each color class. Fellows et al. [10] proved that this problem is W[1]-hard when parameterized by k. Consequently, its dual, the problem MULTICOLORED INDEPENDENT SET, which is to test whether a graph with a partition $X_1, \ldots, X_k$ of the vertex set has an independent set of size k with exactly one vertex from each X_i, is W[1]-hard as well when parameterized by k. This is the problem we reduce from.

Let (G, k) with a partition $X_1, \ldots, X_k$ of V_G be an instance of MULTICOLORED INDEPENDENT SET. Let $X_i = \{x_{i1}, \ldots, x_{in_i}\}$ for $i \in \{1, \ldots, k\}$ where we assume without loss of generality that $n_i \geq 2$. Let $d = n(4k + 3) + 1$. From G we construct a graph G' in the following way. Recall that connecting two vertices means adding an edge between them.

1. Modify each set X_i into a clique.
2. Construct a clique W with vertices $w_1, \ldots, w_{d+1}$.
3. Connect every x_{ij} with $w_1, \ldots, w_{t_{ij}}$ where $t_{ij} = d - d_G(x_{ij}) - n_i - 4k - 2$.
4. Add vertices $y_1, \ldots, y_k$.

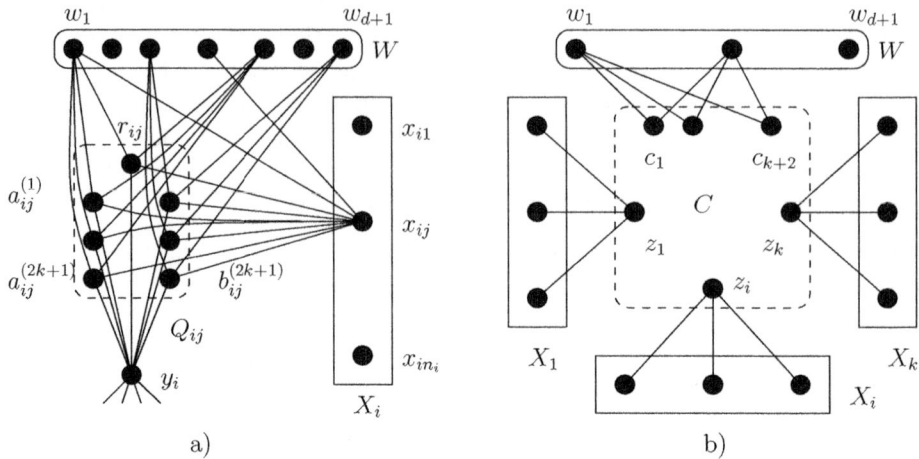

Fig. 2. Construction of G

5. For every x_{ij}, construct a clique Q_{ij} with vertices r_{ij}, $a_{ij}^{(1)}, \ldots, a_{ij}^{(2k+1)}$, $b_{ij}^{(1)}, \ldots, b_{ij}^{(2k+1)}$ and connect every vertex of Q_{ij} with x_{ij} and y_i. Moreover, connect every vertex r_{ij} with $w_1, \ldots, w_{d-(4k+3)}$, every $a_{ij}^{(s)}$ with $w_1, \ldots, w_{d-(4k+3)}$, and every $b_{ij}^{(s)}$ with $w_{4k+5}, \ldots, w_{d+1}$.

6. Construct a clique C with vertices $c_1, \ldots, c_{k+2}, z_1, \ldots, z_k$.

7. For $h = 1, \ldots, k+2$, connect c_h with $w_1, \ldots, w_{d-2k+1}$

8. For $i = 1, \ldots, k$, connect z_i with every vertex of X_i.

Stages 1–5 of the construction are shown in Fig. 2 a), and Stages 6–8 are shown in Fig. 2 b). We let $k' = k(2k+3)$ and claim that G has an independent set with exactly one vertex from each X_i if and only if G' can be modified to a graph H with minimum degree at least d by at most k' contractions.

First suppose that $\{x_{1j_1}, \ldots, x_{kj_k}\}$ is an independent set in G. For $i = 1, \ldots, k$, let $A_i = \{a_{ij_i}^{(1)} b_{ij_i}^{(1)}, \ldots, a_{ij_i}^{(2k+1)} b_{ij_i}^{(2k+1)}, r_{ij_i} y_i, x_{ij_i} z_i\}$ be a set of $2k+3$ edges in G'. We contract every edge in every A_i. Then the total number of contractions is $k(2k+3) = k'$. Moreover, the resulting graph has minimum degree at least d.

Now suppose that G' can be modified to a graph H with minimum degree at least d by at most k' contractions. Let $\mathcal{W}$ be an H-witness structure of G'. For each bag W of $\mathcal{W}$, we choose an arbitrary spanning tree of $G'[W]$. Let $A \subseteq E_{G'}$ denote the union of the sets of edges of these trees. Because we obtain H by contracting the edges of A, we find that $|A| \leq k'$.

Claim 1. $A = A_1 \cup \ldots \cup A_k$, where each $A_i = \{x_{ij} z_i, y_i f, g_1 h_1, \ldots, g_{2k+1} h_{2k+1}\}$ with $\{f, g_1, \ldots, g_{2k+1}, h_1, \ldots, h_{2k+1}\} = Q_{ij}$.

We prove Claim 1 as follows. Let $1 \leq i \leq k$. Because $d_{G'}(y_i) < d$, at least one edge incident with y_i must be included in A_i. Assume that $y_i f \in A_i$ for some

$f \in Q_{ij}$. Note that after contracting $y_i f$, all $4k + 2$ vertices of $Q_{ij} \setminus \{f\}$ have degrees less that d. Hence, at least $2k + 1$ edges incident with these vertices must be contracted. We also note that $d_{G'}(z_i) < d$. Therefore, at least one edge incident with z_i must be in A. Suppose that $z_i t \in A$ for some $t \in C$. Then, after contracting $z_i t$, all other $2k$ vertices of C have degrees less than d. Hence, we must contract at least k edges incident with these vertices. Because the total number of contractions is $k' = k(2k + 3)$ and we also need to contract at least $2k + 3$ edges for every $h \neq i$, this is not possible. We conclude that $z_i x_{ij'} \in A$ for some $1 \leq j' \leq n_i$ and that $A_i = \{x_{ij'} z_i, y_i f, g_1 h_1, \ldots, g_{2k+1} h_{2k+1}\}$ with $\{f, g_1, \ldots, g_{2k+1}, h_1, \ldots, h_{2k+1}\} = Q_{ij}$. We now consider x_{ij} and observe that by contracting the edges $g_1 h_1, \ldots, g_{2k+1} h_{2k+1}$ we decreased the degree of x_{ij} by $2k + 1$. Hence, $j' = j$ and Claim 1 follows.

Due to Claim 1, we can define the set $\{x_{1j_1}, \ldots, x_{kj_k}\}$ with $x_{ij_i} z_i \in A_i$ for $i = 1, \ldots, k$. We prove that this is an independent set in G. In order to obtain a contradiction, assume that there is an edge $x_{ij_i} x_{i'j_{i'}} \in E_G$. Recall that $d_{G'}(x_{i_j j}) = d_G(x_{i_j j}) + (n_i - 1) + (4k + 3) + 1 + t_{ij} = d + 1$. Contracting those edges of A_i that have both end-vertices in Q_{ij} decreases the degree of $x_{i_j j}$ by $2k + 1$. Moreover, after contracting the edges in A_i and $A_{i'}$, the edges $z_i z_{i'}$ and $x_{ij_i} x_{i'j_{i'}}$ have been replaced by one edge. Because z_i is adjacent to all vertices in $C \setminus \{z_i\}$, this means that the degree of the vertex of H obtained by contracting $x_{i_j j} z_i$ is at most $d + 1 - (2k + 1) - 2 + (2k + 1) = d - 1$. This is not possible. Hence, $\{x_{1j_1}, \ldots, x_{kj_k}\}$ is an independent set in G with a vertex, namely $x_{i_j j}$, from each X_i, as desired. This completes the proof of Theorem 2. $\square$

4   Weighted Contractions

We first show that WEIGHTED DEGREE CONTRACTIBILITY is FPT when parameterized by k. Recall that x_{uv} denotes the vertex obtained from u and v after contracting an edge uv in a graph.

Theorem 3. WEIGHTED DEGREE CONTRACTIBILITY *can be solved in time* $O(2^k k^{2k}(n + m))$ *for weighted graphs with n vertices and m edges.*

Proof. Let G be a weighted graph with n vertices and m edges. Let $U = \{u \in V_G \mid d_G^w(u) < d\}$ and let $r = |U|$. Trivially, if $r = 0$, then the answer is Yes. If $r \geq 1$, then we branch according to the following four cases.

Case 1. $r > 2k$.
The algorithm returns No. The reason is that at least one edge incident with each vertex of U must be contracted to get a graph of minimum weighted degree at least d, and every edge is incident with at most two vertices of U.

Case 2. $r \leq 2k$ and there is a vertex $u \in U$ with $d_G(u) \leq k$.
At least one edge incident with u must be contracted to obtain a graph of minimum degree at least d. Hence, for each edge e incident with u, we call our algorithm recursively for G/e and parameter $k' = k - 1$. The algorithm returns Yes if for at least one of the new instances the answer is Yes, and No otherwise.

Case 3. $k < r \leq 2k$ and $d_G(u) \geq k+1$ for all $u \in U$.

If G can be contracted to a graph of minimum weighted degree at least d, then at least one edge with both its end-vertices in U must be contracted. Note that there at most $k(2k-1)$ such edges. If there are no such edges, then the algorithm returns No. Otherwise, for each $e = xy$ with $x, y \in U$, we call our algorithm recursively for G/e and parameter $k' = k - 1$. The algorithm returns Yes if for at least one of the new instances the answer is Yes, and No otherwise.

Case 4. $r \leq k$ and $d_G(u) \geq k+1$ for all $u \in U$.

Let $U = \{u_1, \ldots, u_r\}$. Each u_i is adjacent to at least two vertices in $V_G \setminus U$. For $i = 1, \ldots, r$, we do the following. Let y, z be two neighbors of u in $V_G \setminus U$, where we assume that $w(u_i y) \leq w(u_i z)$. Let $G' = G/u_i y$. Then we deduce that $d_{G'}^w(x_{u_i y}) = d_G^w(u_i) + d_G^w(y) - 2w(u_i y) \geq w(u_i y) + w(u_i z) + d_G^w(y) - 2w(u_i y) = d_G^w(y) - w(u_i y) + w(u_i z) \geq d_G^w(y) \geq d$. Hence, we contract $u_i y$ and recursively proceed with G' and $U' = U \setminus \{u_i\}$. Note that the weighted contraction of $u_i y$ does not change the weighted degrees of the other vertices. Consequently, each vertex in U' is adjacent to at least two vertices of weighted degree at least d in G', and U' is the set of vertices of weighted degree at most $d - 1$ in G'. Then after processing u_r, we obtain a graph of minimum degree at least d by using $r \leq k$ weighted contractions. Hence, our algorithm always returns Yes in this case.

To estimate the running time, observe that for each recursive call of our algorithm, we create at most $k(2k-1)$ instances of the problem, and the depth of the recursion is at most k. Hence, the algorithm runs in time $O(2^k k^{2k}(n+m))$.   □

We call the special case of WEIGHTED DEGREE CONTRACTIBILITY in which there is no upper bound on the number of weighted contractions, i.e., in which $k = |E_G|$, the WEIGHTED CONTRACTION DEGENERACY problem. Our next result shows that already this special case is NP-complete.

Theorem 4. WEIGHTED CONTRACTION DEGENERACY *is* NP-*complete*.

5   Concluding Remarks

Our results are summarized in Table 1. We leave the only missing case in this table, i.e., determining the complexity of WEIGHTED DEGREE CONTRACTIBILITY when parameterized by d, as an open problem. We conclude our paper with some additional results.

Weighted Face Degree Subgraph. First, we introduce a problem on plane (multi)graphs and show how to solve it in FPT time by reducing it to WEIGHTED DEGREE CONTRACTIBILITY; for the definition of a plane graph and other notions, see Diestel [9].

The weighted face degree of a face f of a plane weighted graph G is the sum of all the weights of the edges of G incident with f. The WEIGHTED FACE DEGREE SUBGRAPH problem is to test if a plane weighted graph G can be modified to a plane weighted graph of minimum weighted face degree at least d by using at most k edge removals.

Theorem 5. WEIGHTED FACE DEGREE SUBGRAPH *is* FPT *when parameterized by* k.

Proof. Given an instance of WEIGHTED FACE DEGREE SUBGRAPH with a plane weighted graph G and an integer d we do as follows. Let G^* denote the geometric dual of G. There is a one-to-one correspondence between the edges of G and the edges of G^*. Let e^* be the edge of G that corresponds to the edge e in G. We assign weights to the edges of G^* in the following way: $w_{G^*}(e^*) = w_G(e)$.

An *embedded contraction* of an edge e of a plane graph is a contraction of e that respects the embedding and keeps multiple edges if they appear (that is, if the endpoints of e have common neighbors). We observe that the dual of the graph obtained from a plane graph G by removing an edge e is the graph obtained from G^* by an embedded contraction of e^*.

We apply our algorithm from Theorem 3 for WEIGHTED DEGREE CONTRACTIBILITY for G^* and degree d. Note that there is a one-to-one correspondence between the faces of G and the vertices of G^*. Therefore, weighted contractions can simulate the face degree transformations of a graph with embedded contractions and multiple edges. Due to the equivalence between edge removals in a plane graph and embedded edge contractions in its dual, the sequence of k edges of G^* that is a solution to the WEIGHTED DEGREE CONTRACTIBILITY problem can be transformed into a sequence of k edge removals in G. Hence, WEIGHTED FACE DEGREE SUBGRAPH can be solved in FPT time. $\square$

Weighted Contraction Degeneracy. Recall that the CONTRACTION DEGENERACY problem is NP-complete, even for bipartite graphs, and that it is FPT when parameterized by d [6]. Due to Theorem 4, WEIGHTED CONTRACTION DEGENERACY is NP-complete. We prove the following proposition.

Proposition 2. WEIGHTED CONTRACTION DEGENERACY *is* FPT *when parameterized by* d.

Proof. Let G be a weighted graph. We use Bodlaender's algorithm [5] to check in linear time whether the treewidth of G is at most $2d - 2$.

Suppose that the treewidth of G is at most $2d - 2$. Then, because we can express the problem in monadic second order logic, we may apply the well-known result of Courcelle [8] to solve it in linear time. Suppose that the treewidth of G is at least $2d - 1$. We note that there exists a tree decomposition in which the bags are the sets of vertices incident with some minimal edge cut of G. Hence, G has a minimal edge-cut $C = E(U, \overline{U})$ for some $U \subseteq V_G$ with at least d edges. Contracting the edges of $G[U]$ and $G[\overline{U}]$ yields a graph with two vertices and one edge having weighted degree at least d. So, this case leads to a Yes-answer. $\square$

Acknowledgements. We thank Hans Bodlaender for useful comments.

References

1. Amini, O., Sau, I., Saurabh, S.: Parameterized complexity of finding small degree-constrained subgraphs. Journal of Discrete Algorithms (to appear)
2. Asano, T., Hirata, T.: Edge-contraction problems. J. Comput. Syst. Sci. 26, 197–208 (1983)
3. Belmonte, R., Heggernes, P., van 't Hof, P.: Edge Contractions in Subclasses of Chordal Graphs. In: Ogihara, M., Tarui, J. (eds.) TAMC 2011. LNCS, vol. 6648, pp. 528–539. Springer, Heidelberg (2011)
4. Belmonte, R., Golovach, P.A., Heggernes, P., van 't Hof, P., Kamiński, M., Paulusma, D.: Finding contractions and induced minors in chordal graphs via disjoint paths. In: Proceedings of the 22nd International Symposium on Algorithms and Computation (ISAAC 2011). LNCS, Springer, Heidelberg (to appear, 2011)
5. Bodlaender, H.L.: A linear-time algorithm for finding tree-decompositions of small treewidth. SIAM J. Comput. 25, 1305–1317 (1996)
6. Bodlaender, H.L., Wolle, T., Koster, A.M.C.A.: Contraction and treewidth lower bounds. J. Graph Algorithms Appl. 10, 5–49 (2006)
7. Brouwer, A.E., Veldman, H.J.: Contractibility and NP-completeness. Journal of Graph Theory 11, 71–79 (1987)
8. Courcelle, B.: The Monadic Second-Order Logic of Graphs. I. Recognizable Sets of Finite Graphs. Inf. Comput. 85(1), 12–75 (1990)
9. Diestel, R.: Graph Theory, Electronic edn. Springer, Heidelberg (2005)
10. Fellows, M.R., Hermelin, D., Rosamond, F.A., Vialette, S.: On the parameterized complexity of multiple-interval graph problems. Theor. Comput. Sci. 410, 53–61 (2009)
11. Downey, R.G., Fellows, M.R.: Parameterized Complexity. Springer, Heidelberg (1999)
12. Garey, M.R., Johnson, D.S.: Computers and Intractability: A Guide to the Theory of NP-Completeness. W. H. Freeman (1979)
13. Hammack, R.: A note on the complexity of computing cyclicity. Ars Comb. 63, 89–95 (2002)
14. Heggernes, P., van 't Hof, P., Lévêque, B., Paul, C.: Contracting chordal graphs and bipartite graphs to paths and trees. In: Proceedings of LAGOS 2011. Electronic Notes in Discrete Mathematics, vol. 37, pp. 87–92 (2011)
15. Heggernes, P., van 't Hof, P., Lokshtanov, D., Paul, C.: Obtaining a bipartite graph by contracting few edges, ArXiv:1102.5441 (manuscript)
16. Heggernes, P., van 't Hof, P., Lévêque, B., Lokshtanov, D., Paul, C.: Contracting graphs to paths and trees, ArXiv:1104.3677 (manuscript)
17. van 't Hof, P., Kamiński, M., Paulusma, D., Szeider, S., Thilikos, D.M.: On graph contractions and induced minors. Discrete Appl. Math. (to appear)
18. van 't Hof, P., Paulusma, D., Woeginger, G.J.: Partitioning graphs into connected parts. Theoretical Computer Science 410, 4834–4843 (2009)
19. Kamiński, M., Paulusma, D., Thilikos, D.M.: Contractions of Planar Graphs in Polynomial Time. In: de Berg, M., Meyer, U. (eds.) ESA 2010. LNCS, vol. 6346, pp. 122–133. Springer, Heidelberg (2010)
20. Martin, B., Paulusma, D.: The Computational Complexity of Disconnected Cut and 2K2-Partition. In: Lee, J. (ed.) CP 2011. LNCS, vol. 6876, pp. 561–575. Springer, Heidelberg (2011)

21. Levin, A., Paulusma, D., Woeginger, G.J.: The computational complexity of graph contractions I: polynomially solvable and NP-complete cases. Networks 51, 178–189 (2008)
22. Levin, A., Paulusma, D., Woeginger, G.J.: The computational complexity of graph contractions II: two tough polynomially solvable cases. Networks 52, 32–56 (2008)
23. Niedermeier, R.: Invitation to Fixed-Parameter Algorithms. Oxford University Press (2006)
24. Watanabe, T., Ae, T., Nakamura, A.: On the removal of forbidden graphs by edge-deletion or edge-contraction. Discrete Appl. Math. 3, 151–153 (1981)
25. Watanabe, T., Ae, T., Nakamura, A.: On the NP-hardness of edge-deletion and edge-contraction problems. Discrete Appl. Math. 6, 63–78 (1983)

Planar Disjoint-Paths Completion

Isolde Adler[1], Stavros G. Kolliopoulos[2], and Dimitrios M. Thilikos[3],[*]

[1] Institut für Informatik, Goethe-Universität,
Frankfurt am Main, Robert-Mayer-Str. 11-15,
60325 Frankfurt am Main, Germany
`iadler@informatik.uni-frankfurt.de`
`http://www.tdi.informatik.uni-frankfurt.de/~adler/`
[2] Department of Informatics and Telecommunications,
National and Kapodistrian University of Athens,
Panepistimioupolis,157 84 Athens, Greece
`sgk@di.uoa.gr`
`http://cgi.di.uoa.gr/~sgk/`
[3] Department of Mathematics,
National and Kapodistrian University of Athens,
Panepistimioupolis, 157 84 Athens, Greece
`sedthilk@math.uoa.gr`
`http://www.thilikos.info/`

Abstract. We introduce Planar Disjoint Paths Completion, a completion counterpart of the Disjoint Paths problem, and study its parameterized complexity. The problem can be stated as follows: given a plane
graph G, k pairs of terminals, and a face F of G, find a minimum-size
set of edges, if one exists, to be added inside F so that the embedding
remains planar and the pairs become connected by k disjoint paths in
the augmented network. Our results are twofold: first, we give an explicit
bound on the number of necessary additional edges if a solution exists.
This bound is a function of k, independent of the size of G. Second, we
show that the problem is fixed-parameter tractable, in particular, it can
be solved in time $f(k) \cdot n^2$.

Keywords: Completion Problems, Disjoint Paths, Planar Graphs.

1 Introduction

Suppose we are given a planar road network with n cities and a set of k pairs of
them. An empty area of the network is specified and we wish to add a minimum-
size set of intercity roads in that area so that the augmented network remains
planar and the pairs are connected by k internally disjoint roads. In graph-
theoretic terms, we are looking for a minimum-size edge-completion of a plane
graph so that an infeasible instance of the DISJOINT PATHS problem becomes
feasible without harming planarity. In this paper we give an algorithm that solves

---

[*] Supported by the project "Kapodistrias" (AΠ 02839/28.07.2008) of the National
and Kapodistrian University of Athens.

D. Marx and P. Rossmanith (Eds.): IPEC 2011, LNCS 7112, pp. 80–93, 2012.

this problem in $f(k) \cdot n^2$ steps. Our algorithm uses a combinatorial lemma stating that, whenever such a solution exists, its size depends exclusively on k.

The renowned DISJOINT PATHS PROBLEM (DP) is defined as follows.

$DP(G, s_1, t_1, \ldots, s_k, t_k)$

Input: An undirected graph G and k pairs of *terminals* $s_1, t_1, \ldots, s_k, t_k \in V(G)$.
Question: are there k pairwise internally vertex-disjoint paths $Q_1, \ldots Q_k$ in G such that path Q_i connects s_i to t_i?
(By *pairwise internally vertex-disjoint* we mean that two paths can only intersect at a vertex which is a terminal for both.)

DP is NP-complete even on planar graphs [9] but, when parameterized by k, the problem belongs to the parameterized complexity class FPT, i.e., it can be solved in time $f(k) \cdot n^{O(1)}$, for some function f. More precisely, it can be solved in $O(f(k) \cdot n^3)$ time due to the celebrated algorithm of Robertson and Seymour [11] from the Graph Minors project. For planar graphs, the same problem can be solved in $f(k) \cdot n$ [10]. We write $DP(G, s_1, t_1, \ldots, s_k, t_k)$ for the disjoint paths problem on input G with terminals $s_1, t_1, \ldots, s_k, t_k$.

We introduce a completion counterpart of this problem, PLANAR DISJOINT PATHS COMPLETION (PDPC), which is of interest on infeasible instances of DP, and we study its parameterized complexity, when parameterized by k. We are given an embedding of a, possibly disconnected, planar graph G in the sphere, k pairs of terminals $s_1, t_1, \ldots, s_k, t_k \in V(G)$, a positive integer ℓ, and an open connected subset $\mathbf{F}$ of the surface of the sphere, such that $\mathbf{F}$ and G do not intersect (we stress that the boundary of $\mathbf{F}$ is not necessarily a cycle). We want to determine whether there is a set of at most ℓ edges to add, the so-called *patch*, so that

(i) the new edges lie inside $\mathbf{F}$ and are incident only to vertices of G on the boundary of $\mathbf{F}$,
(ii) the new edges do not cross with each other or with G, and
(iii) in the resulting graph, which consists of G plus the patch, DP has a solution.

PDPC is NP-complete even when ℓ is not a part of the input and G is planar by the following simple reduction from DP: add a triangle T to G and let $\mathbf{F}$ be the interior of T. That way, we force the set of additional edges to be empty and obtain DP as a special case.

Notice that our problem is polynomially equivalent to the minimization problem where we ask for a minimum-size patch: simply solve the problem for all possible values of ℓ. Requiring the size of the patch to be at most ℓ is the primary source of difficulty. In case there is no restriction on the size of the patch and we simply ask whether one exists, the problem is in FPT by a reduction to DP, which is summarized as follows. For simplicity, let $\mathbf{F}$ be an open disk. Let G' be the graph obtained by "sewing" along the boundary of $\mathbf{F}$ an $O(n) \times O(n)$-grid. By standard arguments, PDPC has a solution on G if and only if DP has a solution on G'. A similar, but more involved, construction applies when $\mathbf{F}$ is not an open disk.

Parameterizing completion problems. Completion problems are natural to define: take any graph property, represented by a collection of graphs $\mathcal{P}$, and ask whether it is possible to add edges to a graph so that the new graph is in $\mathcal{P}$. Such problems have been studied for a long time and some of the most prominent are the following: HAMILTONIAN COMPLETION [3, GT34], PATH GRAPH COMPLETION [3, GT36] PROPER INTERVAL GRAPH COMPLETION [4] MINIMUM FILL-IN [12] INTERVAL GRAPH COMPLETION [3, GT35].

Kaplan et al. in their seminal paper [7] initiated the study of the parameterized complexity of completion problems and showed that MINIMUM FILL-IN, PROPER INTERVAL GRAPH COMPLETION and STRONGLY CHORDAL GRAPH COMPLETION are in FPT *when parameterized by the number of edges to add.* Recently, the problem left open by [7], namely INTERVAL GRAPH COMPLETION was also shown to be in FPT [6]. Certainly, for all these problems the testing of the corresponding property is in P, while for problems such as HAMILTONIAN COMPLETION, where $\mathcal{P}$ is the class of Hamiltonian graphs, there is no FPT algorithm, unless P=NP. For the same reason, one cannot expect an FPT-algorithm when $\mathcal{P}$ contains all YES-instances of DP, even on planar graphs. We consider an alternative way to parameterize completion problems, which is appropriate for the hard case, i.e., when testing $\mathcal{P}$ is intractable: we parameterize the property itself. In this paper, we initiate this line of research, by considering the parameterized property $\mathcal{P}_k$ that contains all YES-instances of DP on planar graphs with k pairs of terminals.

Basic concepts. As open sets are not discrete structures, we introduce some formalism that will allow us to move seamlessly from topological to combinatorial arguments. The definitions may look involved at first reading, but this is warranted if one considers, as we do, the problem in its full generality where the input graph is not necessarily connected.

Let G be a graph embedded in the sphere Σ_0. Given an open set $\mathbf{X} \subseteq \Sigma_0$, let $\mathbf{clos}(\mathbf{X})$ and $\partial\mathbf{X}$ denote the closure and the boundary of X, respectively. We define $V(\mathbf{X}) = V(G) \cap \partial\mathbf{X}$. A *noose* is a Jordan curve of Σ_0 that meets G only on vertices. Let $\mathcal{D}$ be a finite collection of mutually non-intersecting open disks of Σ_0 whose boundaries are nooses and such that each point that belongs to at least two such nooses is a vertex of G. We define $\mathbf{I}_\mathcal{D} = \bigcup_{D \in \mathcal{D}} D$ and define $\Gamma_\mathcal{D}$ as the Σ_0-embedded graph whose vertex set is $V(\mathbf{I}_\mathcal{D})$ and whose edge set consists of the connected components of the set $\partial\mathbf{I}_\mathcal{D} \setminus V(\mathbf{I}_\mathcal{D})$. Notice that, in the definition of $\Gamma_\mathcal{D}$, we permit multiple edges, loops, or vertex-less edges.

Let $\mathbf{J}$ be an open subset of Σ_0. $\mathbf{J}$ is a *cactus set* of G if there is a collection $\mathcal{D}$ as above such that $\mathbf{J} = \mathbf{I}_\mathcal{D}$, all biconnected components of the graph $\Gamma_\mathcal{D}$ are cycles, and $G \subseteq \mathbf{clos}(\mathbf{J})$. Given such a $\mathbf{J}$, we define $\Gamma_\mathbf{J} = \Gamma_\mathcal{D}$. Two cactus sets $\mathbf{J}$ and $\mathbf{J}'$ of G are *isomorphic* if $\Gamma_\mathbf{J}$ and $\Gamma_{\mathbf{J}'}$ are topologically isomorphic. Throughout this paper we use the standard notion of topological isomorphism between planar embeddings, see Section 2. Given a cactus set $\mathbf{J}$, we define for each vertex $v \in V(\mathbf{J})$ its *multiplicity* $\mu(v)$ to be equal to the number of connected components of the graph $\Gamma_\mathbf{J} \setminus \{v\}$ minus the number of connected components of $\Gamma_\mathbf{J}$ plus one. We also define $\mu(\mathbf{J}) = \sum_{v \in V(\mathbf{J})} \mu(v)$. Observe that, given a cactus set $\mathbf{J}$ of G, the edges of G lie entirely within the interior of $\mathbf{J}$. The boundary of $\mathbf{J}$ corresponds to a collection of simple closed curves such that (i) no two of them intersect at more than one point and (ii) they intersect with G only at (some of) the vertices in $V(G)$. Cactus sets are useful throughout our paper as "capsule" structures that surround G and thus they abstract the interface of a graph embedding with the rest of the sphere surface.

We say that an open set $\mathbf{F}$ of Σ_0 is an *outer-cactus set* of G if $\Sigma_0 \setminus \mathbf{clos}(\mathbf{F})$ is a cactus set of G. See Fig. 1.(*ii*). For example, if G is planar, any face F of G can be used to define an outer-cactus set, whose boundary meets G only at the vertices incident to F. Our definition of an outer-cactus set is more general: it can be a subset of a face F, meeting the boundary of F only at some of its vertices.

Let G be an input graph to DP, see Fig. 1.(*i*). Given an outer-cactus set $\mathbf{F}$ of G, an $\mathbf{F}$-*patch* of G is a pair $(P, \mathbf{J})$ where (i) $\mathbf{J}$ is a cactus set of G, where $\Sigma_0 \setminus \mathbf{clos}(\mathbf{J}) \subseteq \mathbf{F}$ and (ii) P is a graph embedded in Σ_0 *without crossings* such that $E(P) \subseteq \Sigma_0 \setminus \mathbf{clos}(\mathbf{J})$, $V(P) = V(\mathbf{J})$ (see Figures 1.(*iii*) and 1.(*iv*)). Observe that the edges of P will not cross any edge in $E(G)$. In the definition of the $\mathbf{F}$-patch, the graph P corresponds to the new edges we add. The vertices in $V(\mathbf{F})$ define the vertices of G which we are allowed to include in P. $V(\mathbf{J})$ is meant to contain those vertices of $V(\mathbf{F})$ that eventually become incident with a new edge. In terms of data structures, we assume that a cactus set $\mathbf{J}$ is represented by the (embedded) graph $\Gamma_{\mathbf{J}}$. Similarly an outer-cactus set $\mathbf{F}$ is represented by the (embedded) graph $\Gamma_{\Sigma_0 \setminus \mathbf{clos}(\mathbf{F})}$.

We restate now the definition of the PLANAR DISJOINT PATHS COMPLETION problem as follows:

$\text{PDPC}(G, s_1, t_1, \ldots, s_k, t_k, \ell, \mathbf{F})$

Input: A graph G embedded in Σ_0 without crossings, terminals $s_1, t_1, \ldots, s_k, t_k \in V(G)$, a positive integer ℓ, and an outer-cactus set $\mathbf{F}$ of G.
Parameter: k
Question: Is there an $\mathbf{F}$-patch $(P, \mathbf{J})$ of G, such that $|P| \leq \ell$ and $\text{DP}(G \cup P, s_1, t_1, \ldots, s_k, t_k)$ has a solution? Compute such an $\mathbf{F}$-patch if it exists.

If such an $\mathbf{F}$-patch exists, we call it a *solution* for PDPC. In the corresponding optimization problem, denoted by MIN-PDPC, one asks for the minimum ℓ for which PDPC has a solution, if one exists. See Fig. 1 for an example input of PDPC and a solution to it.

Our results. Notice that in the definition of PDPC the size of the patch does not depend on the parameter k. Thus, it is not even obvious that PDPC belongs to the parameterized complexity class XP, i.e., it has an algorithm of time $n^{f(k)}$ for some function f. Our first contribution, Theorem 2, is a combinatorial one: we prove that if a patch exists, then its size is bounded by k^{2^k}. Therefore, we can always assume that ℓ is bounded by a function of k. This bound is a departure point for the proof of the main algorithmic result of this paper:

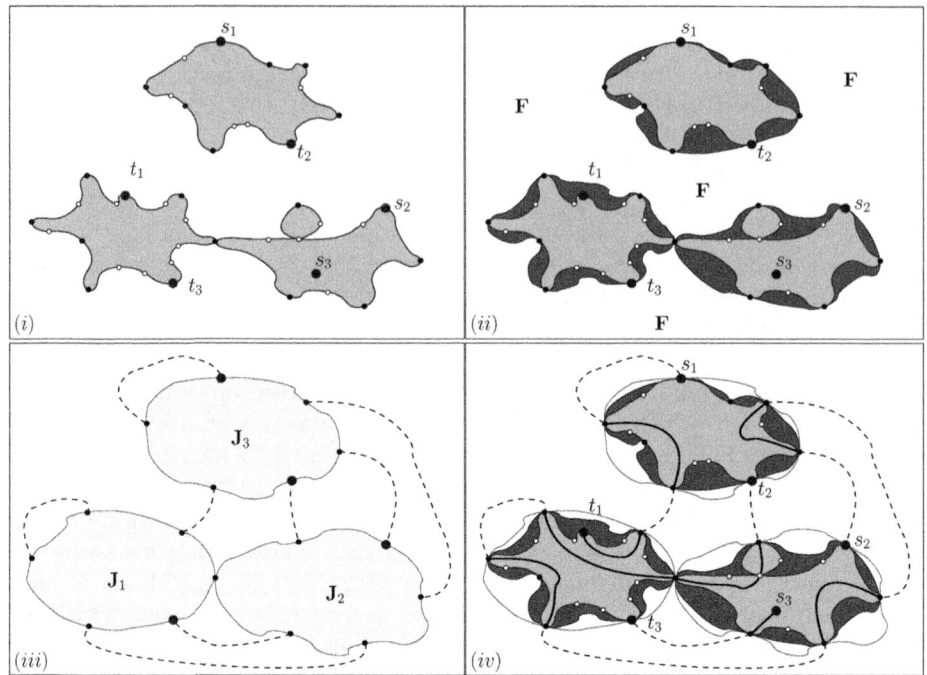

Fig. 1. An example input of the PDPC problem and a solution to it when $\ell = 8$: (i) The graph embedding in the input and the terminals $s_1, t_1, s_2, t_2, s_3, t_3$. The closure of the grey area contains the graph G and the big vertices are the terminals. The white area is a face of G. (ii) The input of the problem, consisting of G, the terminals and the outer-cactus set $\mathbf{F}$. The solid black vertices are the vertices of G that are also vertices of $V(\mathbf{F})$. (iii) The solution of the problem consisting of the $\mathbf{F}$-patch $(P, \mathbf{J})$ where the edges of P are the dashed lines and $\mathbf{J} = \mathbf{J}_1 \cup \mathbf{J}_2 \cup \mathbf{J}_3$. (iv) The input and the solution together where the validity of the patch is certified by 3 disjoint paths.

Theorem 1. PDPC $\in$ FPT. *In particular,* PDPC *can be solved in* $f(k) \cdot n^2$ *steps, where* f *is a function that depends only on* k. *Therefore,* MIN-PDPC *can be solved in* $g(k) \cdot n^2$ *steps.*

We present now the proof strategy and the ideas underlying our results.

1.1 Proof Strategy

Combinatorial Theorem. In Theorem 2, we prove that every patch whose size is larger than k^{2^k}, can be replaced by another one of strictly smaller size. In particular, we identify a region $\mathbf{B}$ of $\mathbf{F}$ that is traversed by a large number of segments of different paths of the DP solution. Within that region, we apply a global topological transformation that replaces the old patch by a new, strictly smaller one, while preserving its embeddability. The planarity of the new patch is based on the fact that the new segments are reflections in $\mathbf{B}$ of a set of segments of the feasible DP solution that previously lied outside **B**. This combinatorial result allows us to reduce the search space of the problem to one whose size is bounded by $\min\{\ell, k^{2^k}\}$. The construction of the corresponding collection of "candidate solutions" can be done in advance, for each given k, without requiring *any a priori knowledge* of the input graph G.

We note that the proof of our combinatorial result could be of independent interest. A simpler variant of it, has been subsequently used in [1].

The algorithm for PDPC. As the number of patches is bounded by a function of k, we need to determine whether there is a correct way to glue one of them on vertices of the boundary of the open set **F** so that the resulting graph is a YES-instance of the DP problem. For each candidate patch $\tilde{P}$, together with its corresponding candidate cactus set $\tilde{\mathbf{J}}$, we define the *set of compatible* graphs embedded in $\tilde{\mathbf{J}}$. Each compatible graph $\tilde{H}$ consists of unit-length paths and has the property that $\tilde{P} \cup \tilde{H}$ contains k disjoint paths. Intuitively, each $\tilde{H}$ is a certificate of the part of the DP solution that lies within G when the patch in **F** is isomorphic to $\tilde{P}$. It therefore remains to check for each $\tilde{H}$ whether it can be realized by a collection of actual paths within G. For this, we set up a collection $\mathcal{H}$ of all such certificates. Checking for a suitable realization of a member of $\mathcal{H}$ in G is still a topological problem that depends on the embedding of G: graphs that are isomorphic, but not *topologically isomorphic,* may certify different completions. For this reason, our next step is to *enhance* the structure of the members of $\mathcal{H}$ so that their realization in G reduces to a purely combinatorial check. (Cf. Section 4.1 for the definition of the enhancement operation). We show in Lemma 5 that for the enhanced certificates, this check can be implemented by rooted topological minor testing. For this check, we can apply the recent algorithm of [5] that runs in $h_1(k) \cdot n^3$ steps and obtain an algorithm of overall complexity $h_2(k) \cdot n^3$.

We note that the use of the complicated machinery of the algorithm in [8] can be bypassed towards obtaining a simpler and faster $f(k) \cdot n^2$ algorithm. This is possible because the generated instances of the rooted topological minor problem satisfy certain structural properties. This allows the direct application of the *Irrelevant Vertex Technique* introduced in [11] for solving, among others, the DISJOINT PATHS Problem. The details of this improvement will appear in the full version of a paper.

2   Preliminaries

We consider finite graphs. For a graph G we denote the vertex set by $V(G)$ and the edge set by $E(G)$. If G is Σ_0-embedded we also refer to the edges of G and the graph G as the corresponding sets of points in Σ_0. Clearly the edges of G correspond to open sets and G itself is a closed set. We denote by $F(G)$ the set of all the faces of G, i.e. all connected components of $\Sigma_0 \setminus G$. Given a set $S \subseteq V(G)$ we say that the pair (G, S) is a graph *rooted* at S. We also denote as $\mathcal{P}(G)$ the set of all paths in G with at least one edge. Given a path $P \in \mathcal{P}(G)$, we denote by $I(P)$ the set of internal vertices of P. Given a vertex $v \in V(G)$, and a positive integer r we denote by $N_G^r(v)$ the set of all vertices in G that are within distance at most r from v.

Rooted Topological Minors. Let H and G be graphs, S_H be a subset of vertices in $V(H)$, S_G be a subset of vertices in $V(G)$, and ρ be a bijection from S_H to S_G. We say that (H, S_H) is a ρ-rooted topological minor of (G, S_G), if there exist injections $\psi_0 \colon V(H) \to V(G)$ and $\psi_1 \colon E(H) \to \mathcal{P}(G)$ such that

   1. $\rho \subseteq \psi_0$,
   2. for every edge $e = \{x, y\} \in E(H)$, $\psi_1(e)$ is a $(\psi_0(x), \psi_0(y))$-path in $\mathcal{P}(G)$, and
   3. all $e_1, e_2 \in E(H)$ with $e_1 \neq e_2$ satisfy $I(\psi_1(e_1)) \cap V(\psi_1(e_2)) = \emptyset$.

In words, when H is a topological minor of G, G contains a subgraph which is isomorphic to a subdivision of H.

Contractions. Let G and H be graphs and let $\sigma \colon V(G) \to V(H)$ be a surjective mapping such that

   1. for every vertex $v \in V(H)$, the graph $G[\sigma^{-1}(v)]$ is connected;
   2. for every edge $\{v, u\} \in E(H)$, the graph $G[\sigma^{-1}(v) \cup \sigma^{-1}(u)]$ is connected;
   3. for every $\{v, u\} \in E(G)$, either $\sigma(v) = \sigma(u)$, or $\{\sigma(v), \sigma(u)\} \in E(H)$.

We say that H is a σ-contraction of G or simply that H is a contraction of G if such a σ exists.

Observation 1. *Let H and G be graphs such that H is a σ-contraction of G. If $x, y \in V(G)$, then the distance in G between x and y is at least the distance in H of $\sigma(x)$ and $\sigma(y)$.*

We also need the following topological lemma.

Lemma 1. *Let G be a Σ_0-embeddable graph and let $\mathbf{J}$ be a cactus set of it. Let also M be a Σ_0-embedded graph such that $M \cap \mathbf{J} = \emptyset$ and $V(M) \subseteq V(\mathbf{J})$. Then there is a closed curve K in $\Sigma \setminus \mathbf{clos}(\mathbf{J})$ meeting each edge of M twice.*

Topological Isomorphism. Given a graph G embedded in Σ_0, let $\mathbf{f}$ be a face in $F(G)$ whose boundary has ξ connected components $A_1, \dots, A_\xi$. We define the set $\pi(\mathbf{f}) = \{\pi_1, \dots, \pi_\xi\}$ such that each π_i is the cyclic ordering of $V(A_i)$, possible with repetitions, defined by the way vertices are met while walking along A_i in a way that the face $\mathbf{f}$ is always on our left side. Clearly, repeated vertices in this walk are cut-vertices of G.

Let G and H be graphs embedded in Σ_0. We say that G and H are *topologically isomorphic* if there exist bijections $\phi \colon V(G) \to V(H)$ and $\theta \colon F(G) \to F(H)$ such that

   1. ϕ is an isomorphism from G to H, i.e. for every pair $\{x, y\}$ of distinct vertices in $V(G)$, $\{x, y\} \in E(G)$ iff $\{\phi(x), \phi(y)\} \in E(H)$.
   2. For every face $\mathbf{f} \in F(G)$, $\phi(\pi(\mathbf{f})) = \pi(\theta(\mathbf{f}))$.

In the definition above, by $\phi(\pi(\mathbf{f}))$ we mean $\{\phi(\pi_1), \dots, \phi(\pi_\xi)\}$, where, if $\pi_i = (x_1, \dots, x_{\zeta_i}, x_1)$, then by $\phi(\pi_i)$, we mean $(\phi(x_1), \dots, \phi(x_{\zeta_i}), \phi(x_1))$. Notice that it is possible for two isomorphic planar graphs to have embeddings that are not topologically isomorphic (see [2, page 93] for such an example and further discussion on this topic).

Treewidth. A *tree decomposition* of a graph G is a pair $(\mathcal{X}, T)$ where T is a tree with nodes $\{1, \ldots, m\}$ and $\mathcal{X} = \{X_i \mid i \in V(T)\}$ is a collection of subsets of $V(G)$ (called *bags*) such that:

1. $\bigcup_{i \in V(T)} X_i = V(G)$,
2. for each edge $\{x, y\} \in E(G)$, $\{x, y\} \subseteq X_i$ for some $\quad i \in V(T)$, and
3. for each $x \in V(G)$ the set $\{i \mid x \in X_i\}$ induces a connected subtree of T.

The *width* of a tree decomposition $(\{X_i \mid i \in V(T)\}, T)$ is $\max_{i \in V(T)} \{|X_i| - 1\}$. The *treewidth* of a graph G denoted $\mathbf{tw}(G)$ is the minimum width over all tree decompositions of G.

3 Bounding the Size of the Completion

In this section we show:

Theorem 2. *If there is a solution for* $\mathrm{PDPC}(G, s_1, t_1, \ldots, s_k, t_k, \ell, \mathbf{F})$, *then there is a solution* $(P, \mathbf{J})$ *with* $|E(P)| \leq k^{2^k}$.

For the proof, we use the following combinatorial lemma.

Lemma 2. *Let* Σ *be an alphabet of size* $|\Sigma| = k$. *Let* $w \in \Sigma^*$ *be a word over* Σ. *If* $|w| > 2^k$, *then* w *contains an infix* y *with* $|y| \geq 2$, *such that every letter occurring in* y *occurs an even number of times in* y.

Proof sketch of Theorem 2. Let $(P, \mathbf{J})$ be a solution for $\mathrm{PDPC}(G, s_1, t_1, \ldots, s_k, t_k, \ell, \mathbf{F})$ with $|E(P)|$ minimal. Consider the embedding of $G \cup P$ in the sphere Σ_0, and let $Q_1, \ldots, Q_k$ be the paths of a DP solution in $G \cup P$. By the minimality of $|E(P)|$ we can assume that the edges of P are exactly the edges of $\bigcup_{i \in \{1, \ldots, k\}} Q_i$ that are not in G. For the same reason, two edges in P have a common endpoint x that is not a terminal *only if* x is a cut-vertex of $\Gamma_{\mathbf{J}}$.

Let P^* denote the graph obtained by the dual of $P \cup \Gamma_{\mathbf{J}}$, after removing the vertices corresponding to the faces of $\Gamma_{\mathbf{J}}$. We show that the maximum degree of P^* is bounded by k and the diameter of P^* is bounded by 2^k. Then $|E(P^*)| = |E(P)| \leq k^{2^k}$ and we are done. Note that every edge in $E(P^*)$ corresponds to an edge in exactly one path of $Q_1, \ldots, Q_k$. Hence, every path $R = r_0, \ldots, r_\zeta$ in P^* corresponds to a word $w \in \{Q_1, \ldots, Q_k\}^*$ in a natural way. It is enough to prove the following claim.

Claim. *The word* w *contains no infix* y *with* $|y| \geq 2$, *such that every letter occurring in* y *occurs an even number of times in* y.

Proof of Claim. Towards a contradiction, suppose that w contains such an infix y. We may assume that $w = y$. Let $E_R \subseteq E(P)$ be the set of edges corresponding to the edges of path $R \subseteq P^*$. Then $|E_R| \geq 2$ because u (and hence R) has length at least 2. Let $\mathbf{B} \subseteq \Sigma_0$ be the open set defined by the union of all edges in E_R and all faces of the graph $P \cup \Gamma_{\mathbf{J}}$ that are incident to them. Clearly, $\mathbf{B}$ is a connected subset of $\mathbf{F}$ with the following properties:

(a) $\mathbf{B}$ contains all edges in E_R and no other edges of P,

(b) the ends of every edge in E_R lie on the boundary $\partial\mathbf{B}$, and

(c) every edge in $E(P) \setminus E_R$ has empty intersection with $\mathbf{B}$.

We consider an 'up-and-down' partition $(U = \{u_1, \ldots, u_r\}, D = \{d_1, \ldots, d_r\})$ of the endpoints of the edges in E_R as follows: traverse the path R in P^* in some arbitrary direction and when the ith edge $e_i \in E_R$ is met, the endpoint u_i of e on the left of this direction is added to U and the right endpoint d_i is added to D. Notice that U and D may be multisets because some vertices in P may have bigger degree. Indeed, if $x \in V(P)$ has degree larger than one, then either x is a terminal and has degree at most k or x is a cutpoint of $\Gamma_{\mathbf{J}}$ and has degree exactly 2. For each $i \in \{1, \ldots, r\}$ we say that u_i is the *counterpart* of d_i and vice versa.

By assumption, every path Q_i crosses R an even, say $2n_i$, number of times. Now for every path Q_i satisfying $E(Q_i) \cap E_R \neq \emptyset$, we number the edges in $E(Q_i) \cap E_R$ by $e_1^i, \ldots, e_{2n_i}^i$ in the order of their appearance when traversing Q_i from s_i to t_i and we orient them from s_i to t_i. We introduce shortcuts for Q_i as follows: for every odd number $j \in \{1, \ldots, 2n_i\}$, we replace the subpath of Q_i from $tail(e_j^i)$ to $head(e_{j+1}^i)$ by a new edge f_j^i in D.

After having done this for all odd numbers $j \in \{1, \ldots, 2n_i\}$, we obtain a new path Q_i' from s_i to t_i that uses strictly less edges in $\mathbf{B}$ than Q_i. Having replaced all paths Q_i with $E(Q_i) \cap E_R \neq \emptyset$ in this way by a new path Q_i', we obtain from P a new graph P' by replacing every pair of edges $e_j^i, e_{j+1}^i \in E(P)$ by f_j^i for all $i \in \{1, \ldots, k\}$ with $E(Q_i) \cap E_R \neq \emptyset$, and for all $j \in \{1, \ldots, 2n_i\}$, j odd. We denote by E_R' the set of all replacement edges f_j^i. We also remove every vertex that becomes isolated in P' during this operation. Then it is easy to verify that: 1) None of the edges of E_R survives in $E(P')$, 2) $|E(P')| < |E(P)|$, and 3) $\mathrm{DP}(G \cup P', s_1, t_1, \ldots, s_k, t_k)$ has a solution.

If we show that, for some suitable cactus set $\mathbf{J}'$ of G, $(P', \mathbf{J}')$ is an $\mathbf{F}$-patch, then we are done, because $|E(P')| < |E(P)|$. In what follows, we prove that P' can also be embedded without crossings in $\mathbf{clos}(\mathbf{F})$ such that $E(P') \subseteq \Sigma_0 \setminus \partial(\mathbf{F})$. For this it suffices to prove that the edges in E_R' can be embedded in $\mathbf{B}$ without crossings.

For every path Q_i with $E(Q_i) \cap E_R \neq \emptyset$ let F_j^i denote the subpath of Q_i from $head(e_j^i)$ to $tail(e_{j+1}^i)$, for $j \in \{1, \ldots, 2n_i\}$, j odd (this path may be edgeless only in the case where $head(e_j^i) = tail(e_{j+1}^i)$ is a cut-vertex of $\Gamma_{\mathbf{J}}$). We replace F_j^i by a single edge c_j^i (when the corresponding path is edgeless, the edge c_j^i is a loop outside B). We consider the graph C with vertex set $V(P)$ and edge set $\{c_j^i \mid i \in \{1, \ldots, k\}, E(Q_i) \cap E_R \neq \emptyset, j \in \{1, \ldots, 2n_i\}, j \text{ odd}\}$.

Our strategy consists of a two-step transformation of this embedding. The first step creates an embedding of C inside $\mathbf{clos}(\mathbf{B})$ without moving the vertices. Indeed, notice that C is embedded in $\Sigma_0 \setminus \mathbf{B}$ without crossings such that all the endpoints of the edges in C lie on the boundary of $\mathbf{B}$. Moreover, because none of the endpoints of F_j^i can be a terminal, no two edges of C have a common endpoint. By standard topological arguments, we consider a new non-crossing embedding of C where all of its edges lie inside $\mathbf{B}$. (Recall that all edges of E_R have been deleted.) This transformation maps every edge c_j^i to a new edge inside $\mathbf{B}$ with the same endpoints.

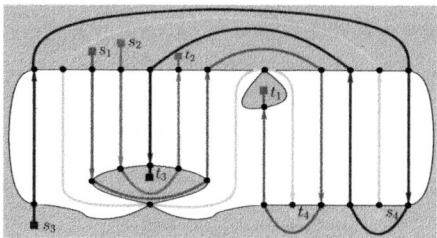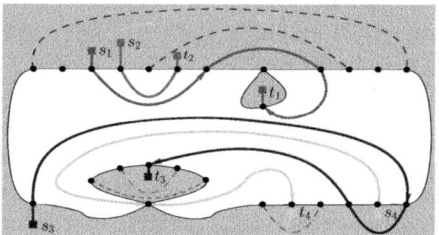

Fig. 2. Example of the transformation in the proof of the Claim; P is on the left and P' is shown on the right. The dashed lines represent the edges of C.

The second step "reflects" the resulting embedding along the axis defined by the path R such that each vertex is exchanged with its counterpart. Now define $(c^i_j)'$ so that it connects $tail(e^i_j)$ and $head(e^i_{j+1})$ – these are exactly the counterparts of $head(e^i_j)$ and $tail(e^i_{j+1})$. Due to symmetry, the $(c^i_j)'$ are pairwise non-crossing, and none of them crosses a drawing of an edge in $E(P) \setminus E_R$. Hence the $(c^i_j)'$ together with the drawing of edges in $E(P) \setminus E_R$ provide a planar drawing of P' (where $(c^i_j)'$ is the drawing of f^i_j). We finally define (up to isomorphism) the cactus set $\mathbf{J}'$ of G such that $\Gamma_{\mathbf{J}'}$ is obtained from $\Gamma_{\mathbf{J}}$ after dissolving the vertices that became isolated during the construction of P'. It is easy to verify that $(P', \mathbf{J})$ is an $\mathbf{F}$-patch of G and this concludes the proof of the Claim.

Indeed, if the above claim holds, then, by Lemma 2, it follows that $n \leq 2^k$, and hence the diameter of P^* is bounded by 2^k. Notice also that the degree of any $v \in V(P^*)$ is bounded by k. Otherwise, $v \in V(P^*)$ is incident to two edges $e_1, e_2 \in E(P^*)$ that correspond to the same letter $Q_i \in \{Q_1 \ldots, Q_k\}$ and the path e_1, e_2 contradicts the claim.   □

Let $\mathcal{L}$ be a list of all simple planar graphs with at most $\min\{\ell, k^{2^k}\}$ edges and no isolated vertices. We call a graph in $\mathcal{L}$ a *completion*. As a first step, our algorithm for PDPC computes the list $\mathcal{L}$. Obviously, the running time of this process is bounded by a function depending only on k.

4 The Algorithm for PLANAR-DPC

The fact that the size of $\mathcal{L}$ is bounded by a function of k implies that PDPC is in XP. Indeed, given the list $\mathcal{L}$, for each completion $\tilde{P} \in \mathcal{L}$ we define the graph $Q_{\tilde{P}} = (V(\tilde{P}), \emptyset)$ and we consider all cactus sets $\tilde{\mathbf{J}}$ of $Q_{\tilde{P}}$ where $(\tilde{P}, \tilde{\mathbf{J}})$ is a $(\Sigma_0 \setminus \mathbf{clos}(\tilde{\mathbf{J}}))$-patch of $Q_{\tilde{P}}$ and $V(\tilde{\mathbf{J}}) = V(\tilde{P})$. We denote the set of all such pairs $(\tilde{P}, \tilde{\mathbf{J}})$ by $\mathcal{J}$ and observe that the number of its elements (up to topological isomorphism of the graph $\tilde{P} \cup \Gamma_{\tilde{\mathbf{J}}}$) is bounded by a function of k.

For each pair $(\tilde{P}, \tilde{\mathbf{J}}) \in \mathcal{J}$, we check whether there exists an $\mathbf{F}$-patch $(P, \mathbf{J})$ of G such that $\tilde{P} \cup \Gamma_{\tilde{\mathbf{J}}}$ and $P \cup \Gamma_{\mathbf{J}}$ are topologically isomorphic and DP has a solution in the graph $G \cup P$. As there are $n^{z(k)}$ ways to choose $(P, \mathbf{J})$ and each check can be done in $O(z_1(k) \cdot n^3)$ steps, we conclude that PDPC can be solved in $n^{z_2(k)}$ steps. In the remainder of the paper, we will prove that the problem is actually in FPT.

The main bottleneck is that there are too many ways to identify $V(\tilde{\mathbf{J}})$ with vertices of $V(\mathbf{F})$, because we cannot bound $|V(\mathbf{F})|$ by a function of k. To overcome this, we characterize the positive instances of PDPC by a rooted topological minor $(\tilde{H}, \tilde{T})$ of the original graph G, that witnesses the fact that $(\tilde{P}, \tilde{\mathbf{J}})$ corresponds to the desired $\mathbf{F}$-patch of G.

Given a pair $(\tilde{P}, \tilde{\mathbf{J}}) \in \mathcal{J}$, we say that a rooted simple graph $(\tilde{H}, \tilde{T} = \{a_1, b_1, \ldots, a_k, b_k\})$ embedded in Σ_0, is *compatible* with $(\tilde{P}, \tilde{\mathbf{J}})$ when

1. for every $e \in E(\tilde{H})$, $e \subseteq \tilde{\mathbf{J}}$,
2. $\tilde{H}$ has at most $2(|E(\tilde{P})| + k)$ vertices,
3. $V(\tilde{H}) \setminus \tilde{T} \subseteq V(\tilde{\mathbf{J}}) \subseteq V(\tilde{H})$,
4. $\mathrm{DP}(\tilde{P} \cup \tilde{H}, a_1, b_1, \ldots, a_k, b_k)$ has a solution.

We define $\mathcal{H} = \{(\tilde{\mathbf{J}}, \tilde{H}, \tilde{T}) \mid$ there exists a $(\tilde{P}, \tilde{\mathbf{J}}) \in \mathcal{J}$ such that $(\tilde{H}, \tilde{T})$ is compatible with $(\tilde{P}, \tilde{\mathbf{J}})\}$.

And notice that $|\mathcal{H}|$ is bounded by some function of k. Assuming that $(P, \mathbf{J})$ is a solution for $\mathrm{PDPC}(G, s_1, t_1, \ldots, s_k, t_k, \ell, \mathbf{F})$, consider the parts of the corresponding disjoint paths that lie within G. The intuition behind the definition above is that $\tilde{H}$ is a certificate of these "partial paths" in G. Clearly, the number of these certificates is bounded by $|\mathcal{H}|$ and they can be enumerated in $f_0(k)$ steps, for some suitable function f_0. For example, for the solution depicted in Fig. 1.(iv), $\tilde{H}$ consists of 7 disjoint edges, one for each subpath within G. Our task is to find an FPT-algorithm that for every such certificate checks whether the corresponding partial paths exist in G.

Given an open set $\mathbf{O}$, a *weakly connected component* of $\mathbf{O}$ is the interior of some connected component of the set $\mathbf{clos}(\mathbf{O})$. Notice that a weakly connected component is not necessarily a connected set.

Let $\bar{\mathbf{F}}^1, \ldots, \bar{\mathbf{F}}^\lambda$ be the weakly connected components of the set $\Sigma_0 \setminus \mathbf{clos}(\mathbf{F})$. We call such a component $\bar{\mathbf{F}}^i$ *active* if $\mathbf{clos}(\bar{\mathbf{F}}^i) \cap T \neq \emptyset$. We denote the collection of all active components by $\mathcal{F}_\mathbf{F}$. A crucial observation is that if an $\mathbf{F}$-patch exists we can always replace it by one that bypasses the inactive components.

Lemma 3. *Let* $(G, s_1, t_1, \ldots, s_k, t_k, \mathbf{F})$ *be an instance for the* PDPC *problem and let* $G' = G[\bigcup_{\bar{\mathbf{F}}^i \in \mathcal{F}_\mathbf{F}} \mathbf{clos}(\bar{\mathbf{F}}^i) \cap V(G)]$ *and* $\mathbf{F}' = \Sigma_0 \setminus \bigcup_{\bar{\mathbf{F}}^i \in \mathcal{F}_\mathbf{F}} \mathbf{clos}(\bar{\mathbf{F}}^i)$. *Then* $(G', s_1, t_1, \ldots, s_k, t_k, \mathbf{F}')$ *is an equivalent instance.*

By Lemma 3, we can assume from now on that $\lambda \leq 2k$. Also we *restrict* $\mathcal{H}$ so that it contains only triples $(\tilde{\mathbf{J}}, \tilde{H}, \tilde{T})$ such that the weakly connected components of the set $\tilde{\mathbf{J}}$ are exactly λ.

4.1 The Enhancement Operation

Consider the triple $\tau = (\tilde{\mathbf{J}}, \tilde{H}, \tilde{T}) \in \mathcal{H}$. Let $\tilde{\mathbf{J}}^1, \ldots, \tilde{\mathbf{J}}^\lambda$ be the weakly connected components of the set $\tilde{\mathbf{J}}$. Then we define $\tilde{C}^i = \Gamma_{\tilde{\mathbf{J}}^i} \cup (\mathbf{clos}(\tilde{\mathbf{J}}^i) \cap \tilde{H})$ for $i \in$

$\{1, \ldots, \lambda\}$ and we call them *parts* of τ. Also we set $\tilde{T}^i = \tilde{T} \cap V(\tilde{C}_i)$, $1 \le i \le \lambda$. We now apply the following *enhancement* operation on each part of τ: For $i = 1, \ldots, \lambda$, we consider the sequence $\mathcal{R}_\tau = (R_\tau^1, \ldots, R_\tau^\lambda)$ where R_τ^i is the rooted graph $(R_\tau'^i, \tilde{T}^i \cup \{x_{\text{new}}^i\})$ such that $R_\tau'^i$ is defined as follows: Take the disjoint union of the graph $\tilde{C}_i$ and a copy of the the wheel $W_{\mu(\tilde{J}^i)}$ with center x_{new}^i and add $\mu(\tilde{J}^i)$ edges, called *i-external* between the vertices of $V(\tilde{J}^i)$ and the peripheral vertices of $W_{\mu(\tilde{J}^i)}$ such that the resulting graph remains Σ_0-embedded and each vertex $v \in V(\tilde{J}^i)$ is incident to $\mu(v)$ non-homotopic edges not in $\tilde{J}$. As the graph $\Gamma_{\tilde{J}^i}$ is connected and planar, the construction of $R_\tau'^i$ is possible. Observe also that $R_\tau'^i \setminus \tilde{J}^i$ is *unique up to topological isomorphism*. To see this, it is enough to verify that for every two vertices in $R_\tau'^i \setminus \tilde{J}^i$ of degree ≥ 3 there are always 3 disjoint paths connecting them.

We define $\mathcal{R} = \{\mathcal{R}_\tau \mid \tau \in \mathcal{H}\}$ and observe that $|\mathcal{R}|$ is bounded by a function

We now define $(C^1, \ldots, C^\lambda)$ such that $C^i = \Gamma_{\bar{F}^i} \cup (\mathbf{clos}(\bar{F}^i) \cap G)$, $i \in \{1, \ldots, \lambda\}$. We call the graphs in $(C^1, \ldots, C^\lambda)$ *parts* of G and let $T^i = T \cap V(C^i)$, $1 \le i \le \lambda$, where $T = \{s_1, t_1, \ldots, s_k, t_k\}$. As above we define the *enhancement* of the parts of G as follows. For each $i = 1, \ldots, \lambda$ we define the rooted graph $G^{*i} = (G'^i, T^i \cup \{x_{\text{new}}^{*i}\})$ where G'^i is defined as follows: take the disjoint union of C^i and the wheel $W_{\mu(\bar{F}^i)}^*$ with center x_{new}^{*i} and add $\mu(\bar{F}^i)$ edges, called *$*i$-external,* between the vertices of $V(\bar{F}^i)$ and the peripheral vertices of $W_{\mu(\bar{F}^i)}^*$ such that the resulting graph remains Σ_0-embedded and each vertex $v \in V(\bar{F}^i)$ is incident to $\mu(v)$ non-homotopic edges. As above, each G'^i is possible to construct and $G'^i \setminus \bar{F}^i$ is unique up to topological isomorphism.

The purpose of the above definitions is twofold. First, they help us to treat separately each of the parts of G and try to match them with the correct parts of τ. Second, the addition of the wheels to each part gives rise to a single, uniquely embeddable interface, between the part and its "exterior" and this helps us to treat embeddings as abstract graphs. Therefore, to check whether a part of τ is realizable within the corresponding part of G, we can use the rooted version of the topological minor relation on graphs as defined in Section 2.

4.2   The Stretching Lemma

A bijection ρ from $\tilde{T} = \{a_1, b_1, \ldots, a_k, b_k\}$ to $T = \{s_1, t_1, \ldots, s_k, t_k\}$ is *legal* if for every $i \in \{1, \ldots, k\}$, there exists some $j \in \{1, \ldots, k\}$ such that $\rho((a_i, b_i)) = (s_j, t_j)$.

Let $\tau \in \mathcal{H}$ and let ρ be a legal bijection from $\tilde{T}$ to T and let ρ_i be the restriction of ρ in $\tilde{T}_i$. We say that $\mathcal{R}_\tau = (R_\tau^1, \ldots, R_\tau^\lambda)$ is *ρ-realizable* in G if there exists a bijection $\phi: \{1, \ldots, \lambda\} \to \{1, \ldots, \lambda\}$ such that for $i = 1, \ldots, \lambda$, R_τ^i is a $\hat{\rho}_i$-rooted topological minor of $G^{*\phi(i)}$ were $\hat{\rho}_i = \rho_i \cup \{(x_{\text{new}}^i, x_{\text{new}}^{*\phi(i)})\}$.

By enumerating all possible bijections ϕ, we enumerate all possible correspondences between the parts of of G and the parts of τ. In order to simplify notation, we assume in the remainder of this section that ϕ is the identity function.

The following lemma is crucial. It shows that when R_τ^i is a topological minor of G^{*i} we can always assume that all vertices and edges of $\tilde{C}^i$ are mapped via

ψ_0 and ψ_1 to vertices and paths in $\mathbf{clos}(\bar{\mathbf{F}}^i)$; the wheel $W_{\mu(\tilde{\mathbf{J}}^i)}$ is mapped to a "sub-wheel" of $W_{\mu(\bar{\mathbf{F}}^i)}$ while i-external edges are mapped to $*i$-external edges. This proves useful in the proof of Lemma 5 as the i-external edges represent the interface of the completion $\tilde{P}$ with $\tilde{C}^i$. The topological minor relation certifies that the same interface is feasible between the corresponding part C^i of G and its "exterior". Lemma 4 establishes also that the image of $\Gamma_{\tilde{\mathbf{J}}^i}$ can be "stretched" so that it falls on $\Gamma_{\bar{\mathbf{F}}^i}$. As all the vertices in $V(\bar{\mathbf{F}}^i)$ are within distance 2 from the artificial terminal x_{new}^{*i} in G^{*i}, this allows us to locate within G^{*i} the possible images of $V(R_\tau'^i)$ in a neighborhood of the terminals. It is then safe to look for an irrelevant vertex far away from this neighborhood.

Lemma 4. *Let R_τ^i be a $\hat{\rho}_i$-rooted topological minor of G^{*i} were $\hat{\rho}_i = \rho_i \cup \{(x_{\text{new}}^i, x_{\text{new}}^{*i})\}$, for $i = 1, \dots, \lambda$. Let also ψ_0^i and ψ_1^i be the functions (cf. Section 2) certifying this topological minor relation. Then ψ_0^i and ψ_1^i can be modified so that the following properties are satisfied.*

1. *if $\tilde{e}$ is an edge of the wheel $W_{\mu(\tilde{\mathbf{J}}^i)}$ incident to x_{new}^i, then $\psi_1^i(\tilde{e})$ is an edge incident to x_{new}^{*i}.*
2. *if $\tilde{e}$ is an edge of the wheel $W_{\mu(\tilde{\mathbf{J}}^i)}$ not incident to x_{new}^i, then $\psi_1^i(\tilde{e})$ is an x_{new}^{*i}-avoiding path of $W_{\mu(\bar{\mathbf{F}}^i)}^*$.*
3. *if $\tilde{e}$ is an i-external edge between $V(\tilde{\mathbf{J}}^i)$ and $V(W_{\mu(\tilde{\mathbf{J}}^i)}) \setminus \{x_{\text{new}}^i\}$, then $\psi_1(e)$ is a path consisting of an $*i$-external edge between $V(\bar{\mathbf{F}}^i)$ and $V(W_{\mu(\bar{\mathbf{F}}^i)}) \setminus \{x_{\text{new}}^{*i}\}$.*
4. *$\psi_0^i(V(\tilde{\mathbf{J}}^i)) \subseteq V(\bar{\mathbf{F}}^i)$.*

4.3   Reducing PDPC to Topological Minor Testing

Lemma 5. *PDPC$(G, t_1, s_1, \dots, t_k, s_k, \ell, \mathbf{F})$ has a solution if and only if there exists a $\tau = (\tilde{\mathbf{J}}, \tilde{H}, \tilde{T}) \in \mathcal{H}$ and a legal bijection $\rho\colon \tilde{T} \to T$ such that $\mathcal{R}_\tau$ is ρ-realizable in G.*

In [8], Grohe, Kawarabayashi, Marx, and Wollan gave an $h_1(k) \cdot n^3$ algorithm for checking rooted topological minor testing. Combining their algorithm with Lemma 5, we obtain an $h_2(k) \cdot n^3$ algorithm for PDPC. Therefore, PDPC $\in$ FPT.

5   Further Extensions and Open Problems

We chose to tackle the disjoint-paths completion problem with the topological restriction of having non-crossing patch edges. A natural extension of this problem is to allow a fixed number $\xi > 0$ of crossings in the patch. Using the same techniques, we can devise an $f(k) \cdot n^2$ algorithm for this problem as well. The only substantial difference is a generalization of our combinatorial result (Theorem 2) under the presence of crossings. The proof is omitted in this extended abstract.

An interesting topic for future work is to define and solve the disjoint-paths completion problem for graphs embedded in surfaces of higher genus. A necessary step in this direction is to extend Theorem 2 for the case where the face to be patched contains handles.

Another issue is to extend the whole approach for the case where the patched faces are more than one. This aim can be achieved without significant deviation from our methodology, in case the number of these faces is bounded. However, when this restriction does not apply, the problem seems challenging and, in our opinion, it is not even clear whether it belongs to FPT.

Acknowledgement. We wish to thank the anonymous reviewers of an earlier version of this paper for valuable comments and suggestions.

References

1. Adler, I., Kolliopoulos, S.G., Krause, P.K., Lokshtanov, D., Saurabh, S., Thilikos, D.: Tight Bounds for Linkages in Planar Graphs. In: Aceto, L., Henzinger, M., Sgall, J. (eds.) ICALP 2011. LNCS, vol. 6755, pp. 110–121. Springer, Heidelberg (2011)
2. Diestel, R.: Graph theory, 3rd edn. Graduate Texts in Mathematics, vol. 173. Springer, Berlin (2005)
3. Garey, M.R., Johnson, D.S.: Computers and Intractability, A Guide to the Theory of NP-Completeness. W.H. Freeman and Company, New York (1979)
4. Golumbic, M.C., Kaplan, H., Shamir, R.: On the complexity of DNA physical mapping. Adv. in Appl. Math. 15, 251–261 (1994)
5. Grohe, M., Kawarabayashi, K., Marx, D., Wollan, P.: Finding topological subgraphs is fixed-parameter tractable. In: 43rd ACM Symposium on Theory of Computing (STOC 2011), San Jose, California, June 6-8 (2011)
6. Heggernes, P., Paul, C., Telle, J.A., Villanger, Y.: Interval completion with few edges. In: Proceedings of the 39th Annual ACM Symposium on the Theory of Computing (STOC 2007), pp. 374–381. American Mathematical Society, San Diego (2007)
7. Kaplan, H., Shamir, R., Tarjan, R.E.: Tractability of parameterized completion problems on chordal, strongly chordal, and proper interval graphs. SIAM Journal on Computing 28, 1906–1922 (1999)
8. Kawarabayashi, K., Wollan, P.: A shorter proof of the graph minor algorithm - the unique linkage theorem. In: Proc. of the 42nd annual ACM Symposium on Theory of Computing, STOC 2010 (2010)
9. Kramer, M.R., van Leeuven, J.: The complexity of wire-routing and finding minimum area layouts for arbitrary VLSI circuits. Advances in Comp. Research 2, 129–146 (1984)
10. Reed, B.A., Robertson, N., Schrijver, A., Seymour, P.D.: Finding disjoint trees in planar graphs in linear time. In: Graph Structure Theory, Seattle, WA. Contemp. Math., vol. 147, pp. 295–301. Amer. Math. Soc., Providence (1993)
11. Robertson, N., Seymour, P.D.: Graph minors. XIII. The disjoint paths problem. J. Combin. Theory Ser. B 63, 65–110 (1995)
12. Yannakakis, M.: Computing the minimum fill-in is NP-complete. SIAM J. Algebraic Discrete Methods 2, 77–79 (1981)

Sparse Solutions of Sparse Linear Systems: Fixed-Parameter Tractability and an Application of Complex Group Testing

Peter Damaschke

Department of Computer Science and Engineering
Chalmers University, 41296 Göteborg, Sweden
ptr@chalmers.se

Abstract. A vector with at most k nonzeros is called k-sparse. We show
that enumerating the support vectors of k-sparse solutions to a system
$Ax = b$ of r-sparse linear equations (i.e., where the rows of A are r-
sparse) is fixed-parameter tractable (FPT) in the combined parameter
r, k. For $r = 2$ the problem is simple. For $0, 1$-matrices A we can also
compute an $O(rk^r)$ kernel. For systems of linear inequalities we get an
FPT result in the combined parameter d, k, where d is the total number
of minimal solutions. This is achieved by interpeting the problem as a
case of group testing in the complex model. The problems stem from the
reconstruction of chemical mixtures by observable reaction products.

1   Introduction

Let A be an $m \times n$ matrix with entries $a_{ij} \geq 0$ ($1 \leq i \leq m$, $1 \leq j \leq n$), and let
b be a vector of length m with entries $b_i \geq 0$. A vector with at most k nonzero
entries is k-sparse. Given a number k, usually much smaller than n, we want to
determine the k-sparse nonnegative solutions x of $Ax = b$, where the rows of A
are r-sparse. We pose the same problem for systems of linear inequalities, where
both relations $\leq$ and $\geq$ may appear mixed in the different rows.

This is certainly a fundamental problem, appearing in machine learning and
related areas like inference or reconstruction problems in computational biology,
see [12] for an example. A particular application we have in mind is the quantification of proteins in an unknown mixture. (For some background information
on protein inference see [7,15].) There the columns of A correspond to candidate proteins, and the rows correspond to peptides, i.e., products of enzymatic
digestion, or just masses of peptides. Entry a_{ij} is the number of occurrences of
peptide i in protein j. The a_{ij} are, of course, nonnegative integers, moreover
they are mostly 0, and the nonzeros are typically just 1. The real-valued vector
b indicates the measured amounts of peptides, obtained by mass spectroscopy.
We want to infer which proteins are in the mixture, and their amounts.

A is a matrix of simulated digestion results with several 100,000 rows and
colums. However, some other input parameters are small, which suggests the
question of parameterized complexity: After separation procedures some small

D. Marx and P. Rossmanith (Eds.): IPEC 2011, LNCS 7112, pp. 94–105, 2012.
© Springer-Verlag Berlin Heidelberg 2012

number k of proteins are present, and a peptide typically appears in a small number r of candidate proteins, thus the rows of A are r-sparse. Due to these facts, only some hundreds of entries of b are nonzero. Of course, we can ignore rows i with $b_i = 0$, and we immediately know $x_j = 0$ if some i exists with $b_i = 0$ but $a_{ij} > 0$. After deletion of these trivial rows and columns there remains a submatrix of manageable size. We still denote the resulting system $Ax = b$, where b is now strictly positive. Simulations with protein data suggest that many rows are only 2-sparse. By solving $Ax = b$ we work under the idealized assumption that b has been accurately measured. Under experimental conditions with much noise it is more realistic to consider inequalities, in the simplest case just with a reliable lower und upper bound for each b_i (so that every peptide gives rise to two inequalities).

Formal Notation and Results. Let R be any set of rows, and let C be any set of columns of matrix A. We denote by $b[R]$ and $x[C]$ the vector b and x restricted to its entries corresponding to R and C, respectively. $A[R]$ and $A[C]$ denotes the submatrix of A restricted to R and C, respectively, and $A[R, C]$ denotes the submatrix of A being the intersection of $A[R]$ and $A[C]$. Sometimes we identify row and column sets with the sets of their indices, without risk of confusion. For any vector y, as usual, y_i is the entry at index i. The *support* of a vector is the set of indices with nonzero entries. A column set $C \neq \emptyset$ is called *feasible* if the system $A[C] \cdot x[C] = b$ has some solution where all entries of $x[C]$ are positive. A column set C is *minimal feasible* if C is feasible but no $C' \subset C$ is. The definition applies similarly to a system of linear inequalities.

A problem with input size n and some other parameter k is fixed-parameter tractable (FPT) if it is solvable in $f(k) \cdot p(n)$ time, where f is any computable function but p is a polynomial. When the polynomial factor is not in the focus, the time bound is often expressed as $O^*(f(k))$.

Finding a sparsest solution to a linear system is NP-hard in general [10,14]. In Section 2 we show that enumerating all minimal feasible sets of at most k columns, for systems of linear equations with r-sparse rows, is an FPT problem in the combined parameter r, k. For $r = 2$ the problem is polynomial, even very simple, and this observation can be extended to a heuristic to speed up the branching in cases where we have $r > 2$ but many 2-sparse rows appear in the matrix. In Section 3 we compute an $O(rk^r)$ size full kernel, i.e., set of columns that includes all minimal feasible sets, by adaptation of an earlier result for hitting sets. In Section 4 we show that the same problem for systems of linear inequalities is in FPT, in the combined parameter d, k, where d is the number of minimal feasible sets. This result is an application of a strategy for the more abstract *group testing problem in the complex model* (also known as searching for defective hyperedges; we shall give the necessary definitions later). Section 5 concludes the paper with some discussion. While the algorithmic techniques are standard and some proofs elaborate on earlier ones for related problems, they are not just straightforward extensions. Some combination of linear algebra and FPT techniques is used. We also remark that d may be exponential in k, but due to known uniqueness results for sparse solutions there is hope for small d in real data sets, and even if a system has many solutions, one is not forced to list them all. Anyway, the algorithms need exponential (FPT) time.

Related Literature. Our problem is close to hitting set enumeration in hypergraphs. This problem asks to enumerate the hitting sets of size at most k, that is, vertex sets that intersect all hyperedges, in a hypergraph of rank r, where the rank is the maximum size of the hyperedges. (The connection between the two problems will be explained in Section 2.) Trivially, hitting set enumeration is in FPT in the combined parameter r, k. However it is not trivial to improve the time bound $O^*(r^k)$, and quite some work has been devoted to this, see [8,9]. While our branching approach is superficially similar to that for hitting set enumeration, it is not an immediate generalization. Despite some connections to hypergraphs and graphs we hope to bring here some fresh contribution to the "not about graphs" direction of parameterized algorithms research that seems to be somewhat neglected. For certain classes of matrices, such as those used in compressive sensing/sampling [2], sparsest solutions are unique and can be computed surprisingly simply, by a linear program that minimizes the sum of entries of x [1,6,13,17]. However, in our case the matrices A are part of the inpput and cannot be chosen, thus we cannot assume special structural properties of A, instead we have to consider the worst case. In general, the sparsest solution is not unique, as the vector b may be in the convex hulls of various small sets of columns of A. (Still we may first test for a given matrix A whether the linear program already yields some sparse solution.)

2   Row-Sparse Linear Systems of Equations

In this section we address the problem of enumerating all minimal feasible sets of at most k columns, thus also determining all k-sparse nonnegative solutions x, to a system $Ax = b$ where every row of A is r-sparse.

Lemma 1. *Let C denote any set of columns in A. If C is minimal feasible then C is linearly independent.*

We omit the simple proof. Geometrically the lemma says that a vector being in the convex hull of other vectors is already in the convex hull of a linearly independent subset of them.

Proposition 1. *Every feasible set of columns is the union of some minimal feasible sets.*

Proof. Consider any feasible set C. If C itself is minimal, there is nothing to prove. Otherwise let $D \subset C$ be minimal feasible. Clearly, there exist positive solutions to $A[C] \cdot x[C] = b$ and $A[D] \cdot y[D] = b$. With a slight abuse of notation, let $y[C]$ be the vector obtained from $y[D]$ by filling all entries in $C \setminus D$ with zeros. Note that still $A[C] \cdot y[C] = b$. Since the two matrix-vector products above are equal to b, all numbers are nonnegative, and $D \subset C$, there must exist some index $i \in D$ where $y_i > x_i$. Let t be the largest number such that

$x[C] - t \cdot y[C]$ is still nonnegative. Due to the previous observation we have $t < 1$, hence $A[C] \cdot (x[C] - t \cdot y[C]) = (1 - t)b$, and multiplication with $1/(1-t)$ yields a nonegative solution z to $Az = b$ whose support C' fulfills $C' \subset C$, $C' \not\supseteq D$ (since $z_i = 0$), and $C \setminus C' \subseteq D$ (since $z_j > 0$ for all $j \in C \setminus D$). In words: All columns that we removed from the C belong to some minimal feasible set. We repeat this procedure with C' in the role of C, and so on. By an inductive argument, eventually every column of C is in some minimal feasible set. □

Proposition 1 and Lemma 1 imply that the minimal feasible sets have the role of vertices of the (convex) space of nonnegative solutions to $Ax = b$. For each minimal feasible set C, the solution vector x with support C is unique (since the columns of C are linearly independent), and trivially, any convex linear combination of any nonnegative solutions is a nonnegative solution, too. In this sense we have characterized all nonnegative solutions once we know the minimal feasible sets. This motivates the problem of enumerating these sets.

A tempting idea of a branching algorithm for this task is the following. Pick a row i and decide exactly which of the x_j, $a_{ij} > 0$, shall be positive. At least one of them must be positive, and since the rows are r-sparse, we get an $O(r)$ branching number. When all rows are treated, check whether the columns j of all positive x_j are linearly independent (cf. Lemma 1), and if so, compute the unique nonnegative solution, or find that there is none. But the catch is that, in rows i where some x_j, $a_{ij} > 0$, are already deemed positive, there is an option not to select further positive variables, and then this branch does not reduce the parameter k. It may happen that the above branching rule has to stop, but the obtained set C of columns is not yet feasible. At this point we have to identify a "small" set of candidates to be added to C. The following lemma is the key.

Lemma 2. *Let C be a set of linearly independent columns such that $A[C] \cdot x[C] = b$ lacks a nonnegative solution. Then there exists a set R of at most $|C| + 1$ rows such that $A[R, C] \cdot x[C] = b[R]$ lacks a nonnegative solution, too. Moreover, we can find such R in polynomial time.*

Proof. Since the columns in C are linearly independent, there is a set R' of $|C|$ rows such that $A[R', C]$ has full rank, and R' can be computed in polynomial time, e.g., by Gauss elimination. Note that $A[R', C] \cdot x[C] = b[R']$ has at most one solution, which can be computed in polynomial time. If $x[C]$ does not exist, or if $x[C]$ exists but has some negative entry, we set $R := R'$. Suppose that $x[C]$ does exist and is nonnegative. Due to the assumption on C, this $x[C]$ does not solve $A[C] \cdot x[C] = b$. Hence there is a row index i with $A[i, C] \cdot x[C] \neq b_i$, and we can trivially find such i. Finally set $R := R' \cup \{i\}$. Since already the solution to $A[R', C] \cdot x[C] = b[R']$ was unique, $A[R, C] \cdot x[C] = b[R]$ has no alternative solution either. □

Now we get the first main result of this section.

Theorem 1. *For systems $Ax = b$ where all rows of A are r-sparse, we can enumerate all minimal feasible sets of size at most k in $O^*(r^k k!)$ time. In particular, this problem is in FPT, in the combined parameter r, k.*

Proof. Starting from a family with one member $C = \emptyset$ we generate a family of sets C of linearly independent columns. It evolves as follows. We pick any C from this family. For $C \neq \emptyset$ we check in polynomial time, by linear programming, whether $A[C] \cdot x[C] = b$ has a nonnegative solution. If so, we remove C from the family and put it aside. (We know that C contains some feasible set, hence we also know that extending C by further columns cannot generate new minimal feasible sets.) If not, or if $C = \emptyset$, it is clear that C does not include any feasible subset. Then we find a small family E of columns, with the property that every feasible set containing C as a subset must also contain some of the columns from E. (This step will be detailed below.) Then we check linear independence of every such set $C \cup \{j\}$, $j \in E$, and we replace C in our family with all $C \cup \{j\}$ that pass this test.

Since we are only interested in minimal feasible sets of size at most k, we also throw away sets that exceed the size limit. It is easy to see that we cannot miss any solution: By Lemma 1, only linearly independent column sets need to be considered, and their subsets are linearly independent as well. We keep all column sets that are candidates for being extendible to a minimal feasible set.

In order to find a set E as specified above, we apply Lemma 2. With $c := |C|$, we determine a set R of at most $c + 1$ rows such that $A[R, C] \cdot x[C] = b[R]$ lacks a nonnegative solution. Since the rows of A are r-sparse, $A[R]$ has nonzeros in a set E of at most $(c + 1)r$ columns. We extend C with any one column from E, that is, we generate at most $(c + 1)r$ new column sets. Note that at least some column of E must be inserted in C to make $A[R, C] \cdot x[C] = b[R]$ solvable, which is a necessary condition for feasibility.

On every path of the search tree generating our sets C, their cardinalities c grow from 0 to at most $k - 1$. Since the outdegrees of search tree nodes are at most $(c + 1)r$, the number of leaves of the search tree is bounded by $(r)(2r)(3r) \cdot \ldots \cdot (kr) = r^k k!$ This bounds the number of column sets C we have put aside, and all minimal feasible sets of size at most k are among them. It remains to check the minimality of every candidate C. Recall that C is linearly independent, hence it comes with a unique solution to $A[C] \cdot x[C] = b$. Thus C is minimal if and only if $x[C]$ has only positive entries.   □

So far we have silently assumed a model of computation with precise real numbers. To turn the algorithm into a practical method, note that a vector b being in the convex hull of some set C of columns is, after a small perturbation, still close to a point in the convex hull. Instead of looking for an exact solution to $A[C] \cdot x[C] = b$ we append all unit vectors to the matrix and compute, still by a linear program, a solution that minimizes the coefficients of these extra vectors, and we accept solutions within some tolerance. Similarly we relax the equality tests by some tolerance. This way we can still recover minimal feasible sets after small perturbations of b. Solutions are changed only if, roughly speaking, the noise is comparable to the distance to the next candidate solutions.

The $k!$ term in the time bound is somewhat unsatisfactory. Before we give an alternative branching strategy avoiding that, we introduce the following notion.

Definition 1. *We define the hypergraph H associated with the system $Ax = b$ to be the hypergraph whose vertices and hyperedges are the columns and rows of A, respectively, and vertex j belongs to edge i iff $a_{ij} \neq 0$.*

Observe that every minimal feasible set of columns of A *contains* some minimal hitting set of H (but is not necessarily *equal* to some minimal hitting set of H). Therefore we may first enumerate the minimal hitting sets C of H, which can be trivially done with branching number r, and then start the procedure of Theorem 1 from these sets C rather than from the empty set. The associated hypergraph is not only useful in this heuristic. We also use it in:

Theorem 2. *For systems $Ax = b$ where all rows of A are r-sparse, we can enumerate all minimal feasible sets of size at most k in $O^*(r^{2k})$ time.*

Proof. For the ease of presentation we first describe a simpler algorithm with a worse time bound of $O^*((r+1)^{2k})$, and then we refine it to achieve $O^*(r^{2k})$.

We generate a family of records, where every record consists of a set C of columns and a system of linear equations $Qx = s$, such that the rows of Q are linearly independent, and Q has nonzeros in $Q[C]$ only. Note that, consequently, Q has at most $|C|$ rows. We start from a family with one record where C and the row set of Q are empty; then the above condition is vacuously true. The family evolves as follows. We pick any record $(C, Qx = s)$. If $C \neq \emptyset$, we check in polynomial time, by linear programming, whether $A[C] \cdot x[C] = b$, $Q[C] \cdot x[C] = s$ has a nonnegative solution. If so, we remove C from the family and put it aside. (As before, we know that C contains some feasible set and needs no further extension.) If not, or if $C = \emptyset$, clearly C does not include any feasible subset that also satisfies the extra system $Qx = s$. Then we find a small family E of columns and a new linear equation, with the property that every feasible set of the compound system $Ax = b$, $Qx = s$ containing C as a subset must also contain some of the columns from E or must have a solution that fulfills the new equation. (This step will be detailed below.) Then we extend our record in all possible ways, that is, we either replace C with some $C \cup \{j\}$, $j \in E$, or we keep C but insert the new linear equation. Again we abandon records where C is not linearly independet or exceeds the size limit k, and again, the "exhaustive" branching ensures that we cannot miss any solution.

Specifically, in order to do the branching we fix some row i of A that has some nonzero outside C. (Such a row exists, as we can w.l.o.g. assume that no column of A is the zero vector, and we can trivially stop if C is already the full set of columns.) Let E be the set of columns $j \notin C$ where $a_{ij} > 0$ in this fixed row i. If we decide to append none of the $j \in E$ to C, the equation in row i must be fulfilled already by nonzero variables in C, formally $A[i, C] \cdot x[C] = b_i$. We append this equation to the extra system $Qx = s$, provided that it is linearly independent of those already being in $Qx = s$. In the opposite case it follows $x_j = 0$ for all $j \in E$ (since the equation is already enforced, and everything is nonnegative). Then we fix a another row i and repeat the procedure until we either find an i and E for branching, or all variables outside C are fixed to 0, and thus C is a dead end. Since the rows of A are r-sparse, obviously the branching number is at most $r + 1$, where the worst case is that the selected row i has all its r nonzeros outside C.

The minimality of every candidate C is checked as before. For the time analysis we use an auxiliary parameter that is initially $2k$. We deduct 1 from this parameter, for every column inserted in C and for every equation inserted in $Qx = s$. Since Q has at most $|C| \leq k$ rows, in fact we deduct never more than $2k$. We also remark that the extra equations have been introduced only for the sake of a simple analysis. Of course, it is equivalent to fix the variables x_j, $j \in E$ to 0 in the affected branches.

Now we give the improvement to $O^*(r^{2k})$. Instead of starting from scratch, we first take the hypergraph associated with $Ax = b$ and enumerate all minimal hitting sets C of size at most k. Every feasible set must contain some of them, and the branching number is trivially r. For all these C, the extra system $Qx = s$ is still empty. After that we continue as above. Since now every row has at most $r - 1$ nonzeros outside C, the branching number is r.   □

Obviously $O^*(r^{2k})$ beats $O^*(r^k k!)$ when k grows, in relation to r. Still the former branching strategy could be faster for k close to r, as one can see from Stirling's formula. A direct comparison in theory is difficult, as the hidden polynomial factors depend on implementation details. Moreover, additional heuristics may be applied that sometimes allow cheaper branchings also in Theorem 1. We discuss some observation below.

We may look for pairs of rows i, i' where $b_i > b_{i'}$ but $a_{ij} \leq a_{i'j}$ for all $j \in C$. Then we add a column j with $a_{ij} > a_{i'j}$, clearly some of them must be put in C. Note that these are at most r columns. When all these conditions for pairs of rows are fulfilled, we similarly look for branchings based on triples of rows, etc. If we are lucky, we can grow our sets C by moderate branchings. Another improvement comes from the special case $r = 2$ which is interesting in itself.

Theorem 3. *For systems $Ax = b$ where all rows of A are 2-sparse, we get an implicit enumeration of all minimal feasible sets in polynomial time.*

Proof. Every 1-sparse row is a linear equation with only one variable, and its unique solution value is obtained instantly. Hence we can remove all 1-sparse rows from A, as well as all columns where these rows have their nonzero entries. That is, these columns must appear in every minimal feasible set. Now we have exactly two positive entries in each row.

We construct a graph with every column being a vertex, and every row being an edge that joins the vertices representing the columns of the two nonzeros. (The graph formulation is not really needed algorithmically, but we use it in order to have the notion of connectivity; see below.) Solving $Ax = b$ is now equivalent to a graph problem: Given a graph where the edges e are labeled with real numbers b_e and the vertex-edge pairs (v, e), $v \in e$, are weighted with real numbers a_{ev}, the task is to label the vertices with real numbers $x_v \geq 0$, such that $a_{eu}x_u + a_{ev}x_v = b_e$ holds on every edge $e = uv$.

But this graph problem is rather simple, too, as we discuss now. First assume that the graph is connected. Then any label x_v determines, by propagation, the x_u labels of all vertices u. Hence it suffices to try every vertex v and set its label zero, and check whether the unique solution forced by that choice is nonnegative. This yields all minimal feasible sets, unless the test fails for every v. In the latter case, the entire vertex set is the only candidate feasible solution. If so, by Lemma 1, the columns are linearly independent, hence we can determine the unique solution by Gauss elimination and check for nonnegativity.

Finally, if the graph is not connected, the reasoning applies to every connected component independently, and all combinations of minimal feasible sets of the components are exactly the minimal feasible sets in the whole instance. In this way we can implicitly describe all solutions in polynomial time, although their number is, of course, not polynomial in general.   □

The proof also reveals the structure of the solution space: In every connected component Y of the graph, there exist pairwise disjoint subsets $Z \subset Y$ so that exactly the sets $Y \setminus Z$ appear as intersections of Y with minimal feasible sets.

When matrix A is r-sparse for some $r > 2$, we can still begin and apply the method in Theorem 3 to the 2-sparse rows, temporarily ignoring the other rows, to find all possible solutions on the affected entries of x. Then we may branch on these solutions, remove the settled rows and columns, and apply the method iteratively to the remaining systems, as long as new 2-sparse rows are obtained. From the aforementioned structure of the solution space of 2-sparse systems one can derive that the branching number in this phase is only 2, or better. (The worst case is graphs consisting of isolated edges.) We have to skip details here.

3   A Problem Kernel for Binary Matrices

In [4] we introduced the notion of a *full kernel* of an FPT enumeration problem, which is a set that contains all minimal solutions. For the minimal hitting sets of size at most k in hypergraphs with hyperedges of size r there is an $(r-1)k^r + k$ size full kernel [4]. In order to establish a similar bound for the present problem we have to adapt the *proof* in [4], i.e., the next theorem is not a consequence of the old *result*. The following result holds when all entries in A are 0 or 1.

Theorem 4. *For systems $Ax = b$ with 0-1-matrices A where all rows are r-sparse, all minimal feasible sets of at most k columns are contained in a set of $(r-1)k^r + k$ columns.*

Proof. First we need some notation: The *hyperedges* are those of the associated hypergraph defined above. With respect to a solution vector x, we call a vertex *positive* if the corresponding variable x_j is positive. The sum $s(C)$ of a set C of vertices is defined by $s(C) := \sum_{j \in C} x_j$.

As an inductive hypothesis we suppose that every set C of $r - i$ vertices (i.e., columns) is contained in at most k^i hyperedges, or there is no feasible set at all. To establish the induction base $i = 0$, note that a set of r vertices can be in only one hyperedge, due to r-sparsity. (Otherwise the system $Ax = b$ has identical rows, and all copies but one can be deleted.)

For the induction step, with a fixed i, assume that some set C of $r - i$ vertices belongs to $k^i + 1$ or more hyperedges. By the inductive hypothesis, every set $C \cup \{v\}$ is in at most k^{i-1} hyperedges. It follows that k vertices outside C would not be enough to hit all $R \setminus C$, for the hyperedges $R \supset C$. In a k-sparse nonnegative solution we cannot have $s(C) < s(R)$ for every hyperedge $R \supset C$, since then we need a positive vertex in every $R \setminus C$, but these would be more than k positive vertices, due to the previous observation. Furthermore, in a nonnegative solution, $s(C) > s(R)$ is not possible either, for any $R \supset C$. Thus, in every k-sparse nonnegative solution, $s(C)$ must be equal to the (given) smallest $s(R)$, $R \supset C$. This establishes a new equation for a new hyperedge C, whereas all hyperedges $R \supset C$ can be replaced with $R \setminus C$, where the required sums are adjusted in the obvious way. In summary, if some set C of $r - i$ vertices belongs to $k^i + 1$ or more hyperedges, we get a system with the same k-sparse nonnegative solutions, where C is no longer a subset of other hyperedges, and the total size of all hyperedges, i.e., the number of 1 entries in A, has strictly decreased. Hence this transformation can be done only finitely many times, and eventually the inductive hypothesis holds for i, in the transformed system.

For $i = r - 1$ the inductive hypothesis says that every vertex is contained in at most k^{r-1} hyperedges. Since every feasible solution is also a hitting set, and at most k positive vertices are permitted, at most k^r hyperedges remain, or no feasible set can exist. Now the size bound follows obviously.   □

Note that the proof also describes an efficient method to obtain a full kernel of the claimed size, and that only subsets C of the given hyperedges need to be examined; these are at most 2^r per hyperedge. But the proof does not apply to matrices A with arbitrary positive coefficients, because then the step where new equations on smaller hyperedges are enforced does not work.

4   Sparse Nonnegative Solutions of Systems of Linear Inequalities via the Complex Model of Group Testing

Our FPT result for systems of linear equations do not immediately generalize to inequalities. In this section we give a completely different approach for systems of linear inequalities. Recall the notion of a minimal feasible column set, and the basic fact that, given the support of x, some solution x can be computed by linear programming. Hence we have a test that tells us, for any given set C of columns, whether C contains some minimal feasible set. The goal is to find all minimal feasible sets. This is a problem known elsewhere, as the *complex model of group testing* or *searching for defective edges*: In a set, an unknown family of subsets are *defective*, and a *group test* (for brevity: *test*) on a subset C answers positively if C contains (entirely) some defective set, otherwise it answers negatively. Accordingly, a tested set C is also called a positive or negative *pool*. Instead of *defective set* we also speak of a *complex*.

It follows from [3] that all complexes can be found using $k d \log_2 n + k^{k/2} d^k + o(d^k)$ tests, where n is the number of elements, k the maximum size of a complex, and d the number of complexes. No previous knowledge of d is assumed. However, it is assumed that all complexes have at most a prescribed number k of elements. In our application the problem is slightly more general: For a given k we wish to enumerate all complexes of size at most k, but there may exist larger complexes as well, and clearly they can affect the test results. Therefore the algorithm from [3] does not carry over to our problem, moreover, the algorithm and its analysis are intricate. Below we give an algorithm that has a somewhat worse dependency on parameters k and d, but it also works in our case and is conceptually simpler. We start with an adaptation of the Triesch-Johann procedure [16,11] used in [3]. In the following, a *k-complex* is either a complex with at most k elements, or a k-element subset of a larger complex.

Lemma 3. *Some k-complex can be found by $k \log_2 n$ tests.*

Proof. First consider the problem of finding a complex, rather than a k-complex. We index the elements arbitrarily by $v_1, \ldots, v_n$. By binary search using $\log_2 n$ tests we determine the largest j such that $\{v_j, \ldots, v_n\}$ is a positive pool. Note that $j = n$ if $\{v_n\}$ is still a positive pool, and then this is also a complex. In the following consider $j < n$. Clearly $\{v_j, \ldots, v_n\}$ contains a complex while $\{v_{j+1}, \ldots, v_n\}$ does not. Hence all complexes in $\{v_j, \ldots, v_n\}$ necessarily contain v_j, and such a complex does exist. In order to find a complex of this form, we fix v_j, that is, we henceforth add v_j to every pool, and we search for a complex in $\{v_{j+1}, \ldots, v_n\}$ relative to that. (In other words, we search for a positive pool such that removal of any element v_l, $l > j$, yields a negative pool. We know already that removal of v_j yields a negative pool.)

Iterating this reasoning, we successively fix elements, each by at most $\log_2 n$ tests, that together form a subset of a complex. After each round we test the current subset C. If C is a negative pool, we continue the search on the suffix after the last inserted element. If C is a positive pool, we have finished a complex.

When searching for a k-complex we proceed in the same way, we just stop after k rounds if the complex is not yet completed.   □

Now we are using this routine in our enumeration algorithm.

Theorem 5. *Given an integer k we can determine all complexes of size at most k using $kd \log_2 n + \min(k^k d^{k+1}, dk^d)$ tests, if d complexes exist. No previous knowledge of d is assumed.*

Proof. Let V denote the set of all elements. We maintain a family $\mathcal{C}$ of k-complexes and their union U. Initially, $\mathcal{C}$ and U are empty.

In each round of the algorithm we probe all sets S that have the following properties: (1) $S \subseteq U$, (2) $|S| \le k$, (3) S contains none of the $C \in \mathcal{C}$ as a subset. To probe S means to test $S \cup (V \setminus U)$. If this pool is positive, it contains a complex which is not already in $\mathcal{C}$ nor extends any k-complex from $\mathcal{C}$, due to (3). Thus, by applying Lemma 3 we find either another complex (of size at most k) or another k-complex C'. In the latter case we have $C' \subset C''$ for some complex C'' not already "covered" by $\mathcal{C}$, as said before. This fact is important, as it implies that every round deals with some new complex, hence the routine of Lemma 3 is called at most d times. If all probes give a negative answer, then every complex C of size at most k appears already in $\mathcal{C}$: If not, then $S := C \cap U$ fulfills (1)–(3), hence S would be probed and answer positively, a contradiction.

All calls of Lemma 3 need at most $kd \log_2 n$ tests. Since $|C| \leq k$ for all $C \in \mathcal{C}$, we always have $|U| \leq kd$. Due to (1) and (2) we probe fewer than $(kd)^k$ sets S in every round, furthermore each S is probed at most d times. This yields the first bound $kd \log_2 n + k^k d^{k+1}$.

An extension of this strategy gives the second part of the bound. In case $d \leq k$ we can bound the number of probes in a round by k^d, which is smaller than $(kd)^k$: Property (3) requires that some element from each $C \in \mathcal{C}$ be excluded from S. In the worst case we have k^d different choices, and then we take S as U minus the excluded elements. Now S can be larger than k, however, property (2) above was only used to bound the number of probes.

Thus, we finally proceed as follows: In the first k rounds we apply the probing strategy that excludes a hitting set of $\mathcal{C}$, and then we switch to the former strategy that guesses the intersection of a new complex with U. □

Corollary 1. *For systems of linear inequalities with d minimal feasible sets, the problem of enumerating all minimal feasible sets of size at most k is in FPT, in the combined parameter d, k.* □

A concern is that the parameter d could be too large to be practical. But the famous results about unique sparsest solutions for some natural classes of random matrices [1,6,13,17] give hope that one would actually encounter small d in real data. This question needs experimental research.

5   Conclusions

It would be interesting to prove even better FPT time bounds and a kernel bound for the general case, and to extend the polynomial result for 2-sparse equations. We also remark that we improved in [5] the full kernel size bound for hitting set enumeration to essentially k^r, using more sophisticated counting arguments. A natural question is whether these techniques can be applied also to r-sparse linear systems. For the case of inequalities we used complex group testing where the tests correspond to linear programs. Other than that, our current algorithm does not further use the mere fact that we are working on linear systems. Whereas this modularity of our algorithm might be appealing, some clever use of polyhedral combinatorics instead of group testing might lead to more efficient algorithms. One could also think of other meaningful problem versions that bridge between strict equations, arbitrary inequalities, and the Boolean case (hitting set problem). Finally, implementation of the methods, taking the numerical issues into account, and experiments on protein mixture data would give important insights.

Acknowledgments. This work has been supported by the Swedish Research Council (Vetenskapsrådet), grant no. 2010-4661 "Generalized and fast search strategies for parameterized problems", and the questions were inspired by discussions with Leonid Molokov (PhD student in Computational Systems Biology). Remarks of the reviewers at IPEC 2011 led to further thinking about the heuristics and to substantial improvements upon the original submission.

References

1. Bruckstein, A.M., Elad, M., Zibulevsky, M.: On the Uniqueness of Nonnegative Sparse Solutions to Underdetermined Systems of Equations. IEEE Trans. on Info. Theory 54, 4813–4820 (2008)
2. Candés, E.J., Wakin, M.B.: An Introduction to Compressive Sampling. IEEE Signal Proc. Magazine, 21–30 (March 2008)
3. Chen, T., Hwang, F.K.: A Competitive Algorithm in Searching for Many Edges in a Hypergraph. Discr. Appl. Math. 155, 566–571 (2007)
4. Damaschke, P.: Parameterized Enumeration, Transversals, and Imperfect Phylogeny Reconstruction. Theor. Comput. Sci. 351, 337–350 (2006)
5. Damaschke, P., Molokov, L.: The Union of Minimal Hitting Sets: Parameterized Combinatorial Bounds and Counting. J. Discr. Algor. 7, 391–401 (2009)
6. Donoho, D.L., Tanner, J.: Sparse Nonnegative Solution of Underdetermined Linear Equations by Linear Programming. Proc. Nat. Acad. of Sciences 102, 9446–9451 (2005)
7. Dost, B., Bandeira, N., Li, X., Shen, Z., Briggs, S., Bafna, V.: Shared Peptides in Mass Spectrometry Based Protein Quantification. In: Batzoglou, S. (ed.) RECOMB 2009. LNCS, vol. 5541, pp. 356–371. Springer, Heidelberg (2009)
8. Fernau, H.: Parameterized Algorithms for d-Hitting Set: The Weighted Case. Theor. Comput. Sci. 411, 1698–1713 (2010)
9. Fernau, H.: A Top-Down Approach to Search-Trees: Improved Algorithmics for 3-Hitting Set. Algorithmica 57, 97–118 (2010)
10. Garey, M.R., Johnson, D.S.: Computers and Intractability. A Guide to the Theory of NP-Completeness. Freeman and Company, New York (1979)
11. Johann, P.: A Group Testing Problem for Graphs with Several Defective Edges. Discr. Appl. Math. 117, 99–108 (2002)
12. Lacroix, V., Sammeth, M., Guigo, R., Bergeron, A.: Exact Transcriptome Reconstruction from Short Sequence Reads. In: Crandall, K.A., Lagergren, J. (eds.) WABI 2008. LNCS (LNBI), vol. 5251, pp. 50–63. Springer, Heidelberg (2008)
13. Lai, M.J.: On Sparse Solutions of Underdetermined Linear Systems. J. Concr. Applic. Math. 8, 296–327 (2010)
14. Natarajan, B.K.: Sparse Approximate Solutions to Linear Systems. SIAM J. Comput. 24, 227–234 (1995)
15. Nesvizhskii, A.I., Aebersold, R.: Interpretation of Shotgun Proteomic Data: The Protein Inference Problem. Mol. Cellular Proteomics 4, 1419–1440 (2005)
16. Triesch, E.: A Group Testing Problem for Hypergraphs of Bounded Rank. Discr. Appl. Math. 66, 185–188 (1996)
17. Wang, M., Xu, W., Tang, A.: A Unique "Nonnegative" Solution to an Underdetermined System: From Vectors to Matrices. arXiv:1003.4778v1 (2010) (manuscript)

New Upper Bounds for MAX-2-SAT and MAX-2-CSP w.r.t. the Average Variable Degree[*]

Alexander Golovnev

St. Petersburg University of the Russian Academy of Sciences, St. Petersburg, Russia
alex.golovnev@gmail.com

Abstract. MAX-2-SAT and MAX-2-CSP are important NP-hard optimization problems generalizing many graph problems. Despite many efforts, the only known algorithm (due to Williams) solving them in less than 2^n steps uses exponential space. Scott and Sorkin give an algorithm with $2^{n(1-\frac{2}{d+1})}$ time and polynomial space for these problems, where d is the average variable degree. We improve this bound to $O^*(2^{n(1-\frac{10/3}{d+1})})$ for MAX-2-SAT and $O^*(2^{n(1-\frac{3}{d+1})})$ for MAX-2-CSP. We also prove stronger upper bounds for d bounded from below. E.g., for $d \geq 10$ the bounds improve to $O^*(2^{n(1-\frac{3.469}{d+1})})$ and $O^*(2^{n(1-\frac{3.221}{d+1})})$, respectively. As a byproduct we get a simple proof of an $O^*(2^{\frac{m}{5.263}})$ upper bound for MAX-2-CSP, where m is the number of constraints. This matches the best known upper bound w.r.t. m due to Gaspers and Sorkin.

Keywords: algorithm, satisfiability, maximum satisfiability, constraint satisfaction, maximum constraint satisfaction.

1 Introduction

1.1 Problem Statement

The maximum satisfiability problem (MAX-SAT) is, given a boolean formula in conjunctive normal form (CNF), to find a maximum number of simultaneously satisfiable clauses of this formula. MAX-2-SAT is restricted MAX-SAT, where each clause contains at most two literals. MAX-SAT and MAX-2-SAT are NP-hard problems. Moreover, it is still not known whether MAX-2-SAT can be solved in less than $O^*(2^n)$[1] with polynomial memory.

MAX-2-SAT is a special case of the maximum 2-constraint satisfaction problem (MAX-2-CSP). In MAX-2-CSP problem one is given a graph $G = (V, E)$ along with sets of functions $S_v : \{0, 1\} \to \mathbb{Z}$ for each vertex v and $S_e : \{0, 1\}^2 \to \mathbb{Z}$ for each edge e. The goal is to find an assignment $\phi : V \to \{0, 1\}$ maximizing the sum

$$\sum_{e=(v_1,v_2)\in E} S_e(\phi(v_1), \phi(v_2)) + \sum_{v\in V} S_v(\phi(v)). \tag{1}$$

---

[*] Research is partially supported by Yandex, Parallels and JetBrains.
[1] As usual, $O^*(\cdot)$ suppresses polynomial factors.

D. Marx and P. Rossmanith (Eds.): IPEC 2011, LNCS 7112, pp. 106–117, 2012.

It is easy to see that MAX-2-SAT corresponds to the case when all functions from S_e are disjunctions (of variables and their negations). MAX-2-SAT and MAX-2-CSP are important NP-hard optimization problems generalizing many graph problems.

1.2   The Main Definitions

Let F be an instance of MAX-2-SAT or MAX-2-CSP. By $n(F), m(F)$ we denote, respectively, the number of vertices (variables) and the number of edges (clauses) of the formula F. By the degree $deg(x)$ of a vertex x we mean the number of edges incident to x. We say variable y is the neighbor of variable x if there is an edge (x, y) in the graph (i.e. there is a 2-clause with these variables in F). By $\Delta(F)$ we denote the maximum vertex degree. $d(F) = 2m/n$ is the average vertex degree. We omit F if it is clear from the context. By the length $|F|$ of a formula F we mean its number of clauses.

Note that in case of MAX-2-CSP one can assume without loss of generality that the corresponding graph does not contain multiple edges (as any two parallel edges can be replaced by their "sum"). At the same time one cannot exclude multiple edges from a MAX-2-SAT graph by the same argument (e.g., the graph of a formula $(x \vee y)(\neg x \vee y)(y \vee z)$ has two edges between x and y).

By (n, Δ)-MAX-2-SAT and (n, Δ)-MAX-2-CSP we denote, respectively, MAX-2-SAT and MAX-2-CSP problems restricted to instances in which each variable appears in at most Δ 2-clauses. By $Opt(F)$ we denote the maximal value of (1) for F over all possible assignments (for MAX-2-SAT, this is the maximal number of simultaneously satisfiable clauses of the formula F).

Let F be an instance of MAX-2-SAT or MAX-2-CSP, l be a literal of F. By $F[l]$ we denote a formula resulting from F by replacing l by 1 and $\neg l$ by 0. Under this assignment, all 2-clauses containing l or $\neg l$ become 1-clauses.

1.3   Known Results

In this subsection, we review some known results for the considered problems. Williams [9] proved that MAX-2-CSP can be solved in time $O^*(2^{\frac{\omega n}{3}})$, where $\omega \approx 2.376$ is the matrix multiplication exponent. Williams' algorithm beats the 2^n barrier at the cost of requiring exponential space. It is a big challenge of the field to solve MAX-2-CSP in less than 2^n steps with only polynomial space. However the trivial 2^n upper bound was improved for several special cases of the considered problems. Dantsin and Wolpert [3] showed that MAX-SAT for formulas with constant clause density can be solved faster than in $O^*(2^n)$ time with exponential space. Kulikov and Kutzkov [7] developed an algorithm for MAX-SAT with polynomial space (and all the algorithms mentioned below use polynomial space) and running time c^n for formulas with constant clause density, where $c < 2$ is a constant.

Fürer and Kasiviswanathan [4] developed an algorithm for MAX-2-SAT with the running time $O^*(2^{n(1-\frac{1}{d-1})})$. Scott and Sorkin [8] improved this bound to $O^*(2^{n(1-\frac{2}{d+1})})$. For $(n, 3)$-MAX-2-SAT, Kojevnikov and Kulikov [6] proved $O^*(2^{\frac{n}{6}})$ bound. This was later improved to $O^*(2^{\frac{n}{6.7}})$ by Kulikov and Kutzkov [7].

Concerning the number of clauses m, the best known upper bound $O^*(2^{\frac{m}{2.465}})$ for MAX-SAT was given by Chen and Kanj [1]. For MAX-2-SAT and MAX-2-CSP, Gaspers and Sorkin [5] proved $O^*(2^{\frac{m}{6.321}})$ and $O^*(2^{\frac{m}{5.263}})$ bounds, respectively.

MAX-CUT is a special case of MAX-2-CSP. Della Croce, Kaminski and Paschos [2] developed an algorithm for MAX-CUT with the running time $O^*(2^{n(1-\frac{2}{\Delta})})$.

1.4 New Upper Bounds

In this paper, we present an elementary algorithm solving MAX-2-SAT and MAX-2-CSP in time $O^*(2^{n(1-\frac{10/3}{d+1})})$ and $O^*(2^{n(1-\frac{3}{d+1})})$, respectively. We show also how to improve these bounds for d bounded from below. E.g., for $d \geq 5$ we get upper bounds $O^*(2^{n(1-\frac{3.40}{d+1})})$ and $O^*(2^{n(1-\frac{3.15}{d+1})})$ and for $d \geq 10$ we get $O^*(2^{n(1-\frac{3.469}{d+1})})$ and $O^*(2^{n(1-\frac{3.221}{d+1})})$. The key point of our algorithm is branching on a vertex of maximal degree that has at least one neighbor with smaller degree.

From these improved upper bounds w.r.t. the average degree d we can derive an upper bound $O^*(2^{\frac{m}{5.263}})$ w.r.t. the number of clauses for MAX-2-CSP. This bound matches the best known upper bound by Gaspers and Sorkin [5]. We also show that any improvement of upper bound for (n, Δ)-MAX-2-CSP for any $\Delta \leq 5$ would improve this record bound.

Since MAX-CUT is s special case of MAX-2-CSP, we also get an improved upper bound $O^*(2^{n(1-\frac{3}{d+1})})$ for MAX-CUT (again, the bound decreases when d increases).

1.5 Organization of the Paper

In Section 2, we construct a simple algorithm for MAX-2-SAT and MAX-2-CSP. The main idea of the algorithm is branching on a vertex of maximal degree. We prove $O^*(2^{n(1-\frac{10/3}{\Delta+1})})$ and $O^*(2^{n(1-\frac{3}{\Delta+1})})$ upper bounds for this algorithm. Section 3 generalizes upper bounds w.r.t. Δ to upper bounds w.r.t. d. Also, we apply this theorem to the algorithm from Section 2. In Section 4, we slightly change the algorithm and get stronger upper bounds for it.

2 A Simple Algorithm for MAX-2-SAT and MAX-2-CSP

In this section, we present a simple algorithm for (n, Δ)-MAX-2-SAT and (n, Δ)-MAX-2-CSP with upper bounds $O^*(2^{n(1-\frac{10/3}{\Delta+1})})$ and $O^*(2^{n(1-\frac{3}{\Delta+1})})$, respectively. To solve an instance of the maximal degree Δ, our algorithm branches on a variable of maximal degree until it gets an instance of maximal degree $\Delta - 1$.

2.1   Removing Variables of Degree 2

Lemma 1. *Let F be an instance of MAX-2-SAT or MAX-2-CSP containing a vertex u of degree at most 2. Then F can be transformed in polynomial time into a formula F' s.t.*

1. *$deg_{F'}(u) = 0$,*
2. *for all v, $deg_{F'}(v) \leq deg_F(v)$,*
3. *$Opt(F)$ can be computed from $Opt(F')$ in polynomial time.*

This lemma is proved for MAX-2-SAT in [6, Lemma 3.1], and for MAX-2-CSP in [5, Section 5.9]. It allows us to assume that a simplified formula contains variables of degree at least 3 only.

2.2   An Algorithm

The algorithm branches on a vertex of maximal degree Δ until it gets a graph of maximal degree 3. It then calls a known algorithm for $(n, 3)$-MAX-2-SAT or $(n, 3)$-MAX-2-CSP, respectively.

Denote by n_i the number of vertices of degree i for $i \in \{3, \ldots, \Delta\}$. Consider the problem (n, Δ)-MAX-2-SAT $((n, \Delta)$-MAX-2-CSP). We use the following formula complexity measure:

$$\mu = \alpha_3 n_3 + \ldots + \alpha_\Delta n_\Delta,$$

where α_i denotes the weight of a variable of degree i. The values of α_i's will be determined later. We would like to find α_i's such that for any formula F the algorithm has the running time $poly(|F|) \cdot 2^{\mu(F)}$.

Assume that an algorithm A solves $(n, \Delta - 1)$-MAX-2-SAT $((n, \Delta - 1)$-MAX-2-CSP) in time $2^{\alpha_3 n_3 + \ldots + \alpha_{\Delta-1} n_{\Delta-1}}$. Consider the following algorithm for (n, Δ)-MAX-2-SAT $((n, \Delta)$-MAX-2-CSP).

> METAALG
>
> *Parameter:* Algorithm A for $(n, \Delta - 1)$-MAX-2-SAT $((n, \Delta - 1)$-MAX-2-CSP).
> *Input:* F – instance of MAX-2-SAT or MAX-2-CSP.
> *Output:* $Opt(F)$.
> *Method.*
>
> 1. Remove all vertices of degree < 3 (using Lemma 1).
> 2. If F does not contain 2-clauses, then return the result.
> 3. If the maximal vertex degree of F is less than Δ, then return $A(F)$.
> 4. Choose a vertex x of maximal degree Δ.
> 5. Return $\max(\text{METAALG}(A, F[x]), \text{METAALG}(A, F[\neg x]))$.

Lemma 2. *Let $\Delta > 3$, $\alpha_i < 1$, for all i. If*

$$\delta = \min(\alpha_\Delta - \alpha_{\Delta-1}, \alpha_{\Delta-1} - \alpha_{\Delta-2}, \ldots, \alpha_4 - \alpha_3, \alpha_3) \geq \frac{1 - \alpha_\Delta}{\Delta}, \qquad (2)$$

then the running time of the algorithm METAALG for (n, Δ)-MAX-2-SAT $((n, \Delta)$-MAX-2-CSP) is $2^{\alpha_3 n_3 + \ldots + \alpha_\Delta n_\Delta}$.

Proof. Denote by $T(n_3, \ldots, n_\Delta)$ the running time of the algorithm on a formula that has n_i vertices of degree i, for all $3 \leq i \leq \Delta$. If there are no vertices of degree Δ (i.e., $n_\Delta = 0$), then METAALG just calls A. Then, clearly,

$$T(n_3, \ldots, n_\Delta) \leq 2^{\alpha_3 n_3 + \ldots + \alpha_{\Delta-1} n_{\Delta-1}} = 2^{\alpha_3 n_3 + \ldots + \alpha_\Delta n_\Delta}.$$

Assume now that there exists a vertex x of degree Δ. Then METAALG at step 4 branches on a vertex of degree Δ. We show that in both branches $F[x]$ and $F[\neg x]$, μ is reduced at least by 1.

Indeed, the measure decreases by α_Δ, because the algorithm branches on a vertex of degree Δ. The degree of each neighbor of x is reduced, so the complexity is decreased at least by δ (as δ is the minimal amount by which μ is decreased when the degree of a vertex is reduced). This causes a complexity decrease of $\Delta \cdot \delta$. Lemma 1 guarantees that removing variables of degree 2 does not increase μ. It follows from (2) that $\Delta \cdot \delta + \alpha_\Delta \geq 1$. Therefore μ decreases at least by 1. Then

$$T(n_3, \ldots, n_\Delta) \leq 2 \cdot 2^{\alpha_3 n_3 + \ldots + \alpha_\Delta n_\Delta - 1} + poly(|F|) \leq 2^{\alpha_3 n_3 + \ldots + \alpha_\Delta n_\Delta} + poly(|F|).$$

Thus, the running time of the algorithm METAALG is $O^*(2^{\alpha_3 n_3 + \ldots + \alpha_\Delta n_\Delta})$.   □

As easy consequence of the just proved lemma is an upper bound $O^*(2^{\alpha n})$, where $\alpha = \max(\alpha_\Delta, \ldots \alpha_3)$. From $\alpha_i < 1$ and (2) we conclude that α_i's increase with i, which means that $\alpha = \alpha_\Delta$.

It is known [7] that $(n, 3)$-MAX-2-SAT can be solved in time $O^*(2^{n/6.7})$. Also, the fact that vertices of degree at most 2 can be removed implies that $(n, 3)$-MAX-2-CSP can be solved in $O^*(2^{n/4})$. Indeed, when branching on a vertex of degree 3 we can remove all its neighbors in both branches (so, the number of vertices is decreased at least by 4).

Corollary 1. *The following algorithm solves MAX-2-SAT (MAX-2-CSP) in* $O^*(2^{n(1-\frac{10/3}{\Delta+1})})$ *($O^*(2^{n(1-\frac{3}{\Delta+1})})$) time.*

SIMPLEALG

Input: F – an instance of MAX-2-SAT or MAX-2-CSP.
Output: $Opt(F)$.
Method.

1. Remove all vertices of degree < 3 (using Lemma 1).
2. If F does not contain 2-clauses, then return the result.
3. If the maximal vertex degree of F is 3, then call the known algorithm for $(n, 3)$-MAX-2-SAT or $(n, 3)$-MAX-2-CSP, respectively.
4. Choose a vertex x of maximal degree Δ.
5. Return $\max(\text{SIMPLEALG}(F[x]), \text{SIMPLEALG}(F[\neg x]))$.

Proof. SIMPLEALG is obtained from METAALG. As a parameter A SIMPLEALG takes himself if $i > 3$ and described algorithms if $i = 3$. We will choose α_i satisfying (2).

As mentioned above, $(n, 3)$-MAX-2-SAT $((n, 3)$-MAX-2-CSP) can be solved by SimpleAlg in $O^*(2^{n/6.7})$ $(O^*(2^{n/4}))$ time. Hence, to minimize $\max(\alpha_3, \alpha_4)$, according to (2), we can choose α_3 and α_4 as follows:

$$\text{MAX-2-SAT:} \quad \alpha_3 = \frac{1}{6}, \alpha_4 = \frac{1}{3}$$

$$\text{MAX-2-CSP:} \quad \alpha_3 = \frac{1}{4}, \alpha_4 = \frac{2}{5}.$$

Thus, we get upper bounds $O^*(2^{n/3})$ and $O^*(2^{2n/5})$ for $(n, 4)$-MAX-2-SAT and $(n, 4)$-MAX-2-CSP, respectively. Now let $\Delta > 4$. To (2) to hold we can set α_i as follows:

$$\alpha_i = \frac{1 + i \cdot \alpha_{i-1}}{i+1}. \tag{3}$$

Below we state several simple properties of α_i's that will be needed in further analysis.

- $\alpha_i = 1 - 4\frac{1-\alpha_3}{i+1}$.
  By expanding (3), one gets:

$$\alpha_i = \frac{1 + i \cdot \alpha_{i-1}}{i+1} = \frac{1 + i \cdot \frac{1+(i-1)\alpha_{i-2}}{i}}{i+1} =$$
$$\frac{2 + (i-1)\alpha_{i-2}}{i+1} = \ldots = \frac{i - 3 + 4\alpha_3}{i+1} = 1 - 4\frac{1-\alpha_3}{i+1}. \tag{4}$$

- $\alpha_i < 1$.
  This follows immediately from the previous property and the fact that $\alpha_3 < 1$.
- α_i increases with i.
  This follows immediately from $\alpha_i = 1 - 4\frac{1-\alpha_3}{i+1}$.
- $\alpha_i - \alpha_{i-1} = \frac{1-\alpha_i}{i}$.
  This follows from (3).

For MAX-2-SAT $(\alpha_3 = \frac{1}{6})$ we get: $\alpha_\Delta = 1 - \frac{10/3}{\Delta+1}$. For MAX-2-CSP $(\alpha_3 = \frac{1}{4})$: $\alpha_\Delta = 1 - \frac{3}{\Delta+1}$.

Show that these α_i satisfy the condition of the lemma. First, SimpleAlg solves $(n, \Delta - 1)$-MAX-2-SAT $((n, \Delta - 1)$-MAX-2-CSP) in $2^{\alpha_3 n_3 + \ldots + \alpha_{\Delta-1} n_{\Delta-1}}$ time by induction.

It remains to show that (2) holds. First, show that $\alpha_3 \geq \frac{1-\alpha_\Delta}{\Delta}$, for $\Delta \geq 4$.

$$\frac{1-\alpha_\Delta}{\Delta} \leq \frac{1 - 1 + \frac{10/3}{\Delta+1}}{\Delta} \leq \frac{10}{3\Delta(\Delta+1)} \leq \frac{10}{3 \cdot 4 \cdot (4+1)} \leq \frac{1}{6} \leq \alpha_3.$$

Now show that $\alpha_i - \alpha_{i-1} \geq \frac{1-\alpha_\Delta}{\Delta}$ for $i \leq \Delta$.

From properties of α_i we know that $\alpha_i - \alpha_{i-1} = \frac{1-\alpha_{i-1}}{i+1}$. α_i increases, so for $i < \Delta$, $\frac{1-\alpha_{i-1}}{i+1} > \frac{1-\alpha_\Delta}{\Delta}$. Hence, $\alpha_i - \alpha_{i-1} = \frac{1-\alpha_{i-1}}{i+1} > \frac{1-\alpha_\Delta}{\Delta}$ for $i < \Delta$.
For $i = \Delta$, $\alpha_i - \alpha_{i-1} = \frac{1-\alpha_\Delta}{\Delta}$.   □

The described algorithm has the running time $O^*(2^{n(1-\frac{10/3}{\Delta+1})})$ and $O^*(2^{n(1-\frac{3}{\Delta+1})})$ for MAX-2-SAT and MAX-2-CSP, respectively. Note that our algorithm is based on upper bound $2^{n/6}$ [6], while a stronger bound $2^{n/6.7}$ [7] is known. The latter bound would not improve our algorithm as for smaller α_3 one needs a larger α_4 to satisfy $\alpha_3 \geq \frac{1-\alpha_4}{4}$ from (2).

The presented algorithm already improves the known bounds $O^*(2^{n(1-\frac{1}{\Delta-1})})$ [4] and $O^*(2^{n(1-\frac{2}{\Delta+1})})$ [8] w.r.t. Δ. In Section 4, we further improve these bounds by changing the algorithm slightly.

3   Going from the Maximal Degree to the Average Degree

In this section, we generalize the results of the previous section from the maximal degree to the average degree. Informally, we show that the worst case of the considered algorithm is achieved in case when all the vertices have the same degree (and so $d = \Delta$). In this section we consider only simplified formulas (i.e. formulas without vertices of degree less than 3).

Theorem 1. *If an algorithm X solves MAX-2-SAT (MAX-2-CSP) in time $O^*(2^{\alpha_3 n_3 + \ldots + \alpha_\Delta n_\Delta})$ and for all $i > 3$,*

$$2\alpha_i \geq \alpha_{i+1} + \alpha_{i-1},\tag{5}$$

then   the   algorithm X solves   MAX-2-SAT   (MAX-2-CSP)   in   time $O^*(2^{n(\alpha_D + \epsilon(\alpha_{D+1} - \alpha_D))})$, where $D = \lfloor d \rfloor$, $\epsilon = d - D$ and $d = \frac{2m}{n}$ is the average vertex degree.

Proof. As d is the average degree of the simplified graph,

$$3n_3 + 4n_4 + \ldots \Delta n_\Delta = nd.\tag{6}$$

Subtract from (6) the equation $n_3 + n_4 + \ldots + n_\Delta = n$ multiplied by d:

$$\sum_{i=3}^{\Delta}(i - d)n_i = 0.$$

By substitution d by $D + \epsilon$ in this equality, we get

$$\sum_{i=3}^{\Delta}(i - D)n_i = \epsilon n.\tag{7}$$

Let $\sigma = \alpha_{D+1} - \alpha_D$. Show that

$$n_i\alpha_i \leq n_i\alpha_D + n_i(i - D)\sigma.\tag{8}$$

Condition (5) can be written as follows:

$$\alpha_i - \alpha_{i-1} \geq \alpha_{i+1} - \alpha_i.$$

Then, for all $i \leq D$,

$$\alpha_{i+1} - \alpha_i \geq \alpha_{i+2} - \alpha_{i+1} \geq \ldots \geq \alpha_{D+1} - \alpha_D = \sigma.$$

Hence

$$\alpha_i \leq \alpha_{i+1} - \sigma \leq \alpha_{i+2} - 2\sigma \leq \ldots \leq \alpha_D + (i - D)\sigma.$$

Therefore $\alpha_i n_i \leq \alpha_D n_i + \sigma(i - D)n_i$, for all $i \leq D$. By the same argument, $\alpha_i n_i \leq \alpha_D n_i + \sigma(i - D)n_i$, for all $i \geq D$.

Then the exponent of the running time of the algorithm is

$$\alpha_3 n_3 + \ldots + \alpha_\Delta n_\Delta \leq \alpha_D n_3 + \sigma(3 - D)n_3 + \ldots + \alpha_D n_\Delta + \sigma(\Delta - D)n_\Delta =$$

$$\alpha_D(n_3 + n_4 + \ldots + n_\Delta) + \sigma \sum_{i=3}^{\Delta}(i - D)n_i = \alpha_D n + \sigma \sum_{i=3}^{\Delta}(i - D)n_i \overset{(by(7))}{=}$$

$$\alpha_D n + \sigma \epsilon n = n(\alpha_D + \epsilon(\alpha_{D+1} - \alpha_D)).$$

□

It can be shown that this bound holds also for formulas containing variables of degree less than 3 for MAX-2-CSP with average degree $d \geq 3$ and for MAX-2-SAT with average degree $d \geq 4$.

Corollary 2. *The algorithm* SIMPLEALG *solves MAX-2-SAT (MAX-2-CSP) in time* $O^*(2^{n(1-\frac{10/3}{d+1})})$ $(O^*(2^{n(1-\frac{3}{d+1})}))$.

Proof. From (3):

$$\alpha_{i+1} = \frac{1 + \alpha_i(i + 1)}{i + 2}, \qquad \alpha_{i-1} = \frac{\alpha_i(i + 1) - 1}{i}.$$

Then

$$\alpha_{i+1} + \alpha_{i-1} = \frac{\alpha_i(i + 1) + 1}{i + 2} + \frac{\alpha_i(i + 1) - 1}{i} =$$

$$2\alpha_i + \frac{1 - \alpha_i}{i + 2} + \frac{\alpha_i - 1}{i} = 2\alpha_i - (1 - \alpha_i)(\frac{1}{i} - \frac{1}{i + 2}) < 2\alpha_i.$$

Therefore (5) holds. From Theorem 1 it follows that the exponent of the running time is $\alpha_D + \epsilon(\alpha_{D+1} - \alpha_D)$.

To prove $O^*(2^{n(1-\frac{10/3}{d+1})})$ and $O^*(2^{n(1-\frac{3}{d+1})})$ upper bounds for MAX-2-SAT and MAX-2-CSP it remains to show that $\alpha_D + \epsilon(\alpha_{D+1} - \alpha_D) \leq 1 - 4\frac{1-\alpha_3}{d+1}$.

We know that $\alpha_i = 1 - 4\frac{1-\alpha_3}{i+1}$.

$$\alpha_D + \epsilon(\alpha_{D+1} - \alpha_D) = (1 - \epsilon)\alpha_D + \epsilon\alpha_{D+1} =$$

$$1 - \epsilon - 4\frac{(1 - \epsilon)(1 - \alpha_3)}{D + 1} + \epsilon - 4\frac{\epsilon(1 - \alpha_3)}{D + 2} = 1 - 4\frac{(1 - \alpha_3)(D + 2 - \epsilon)}{(D + 1)(D + 2)} =$$

$$1 - 4\frac{(1 - \alpha_3)(D + 2 - \epsilon)(D + 1 + \epsilon)}{(D + 1)(D + 2)(D + 1 + \epsilon)} = 1 - 4\frac{(1 - \alpha_3)((D + 1)(D + 2) + \epsilon - \epsilon^2)}{(D + 1)(D + 2)(D + 1 + \epsilon)} <$$

$$1 - 4\frac{1 - \alpha_3}{D + 1 + \epsilon} = 1 - 4\frac{1 - \alpha_3}{d + 1}.$$

□

4   An Algorithm for MAX-2-SAT and MAX-2-CSP

Recall that $\delta = \min(\alpha_\Delta - \alpha_{\Delta-1}, \alpha_{\Delta-1} - \alpha_{\Delta-2}, \ldots, \alpha_4 - \alpha_3, \alpha_3)$ is the minimal amount by which μ is decreased when the degree of a vertex is decreased. The algorithm from Section 2 at each iteration branches on a vertex of maximal degree Δ. Therefore the complexity measure decreases at least by $\alpha_\Delta + \Delta \cdot \delta$. In Corollary 1 we showed that $\delta = \alpha_\Delta - \alpha_{\Delta-1}$, and in accordance with this we choose α_Δ such that

$$\alpha_\Delta + \Delta \cdot \delta = \alpha_\Delta + \Delta(\alpha_\Delta - \alpha_{\Delta-1}) = 1.$$

Therefore we get $\alpha_\Delta = \frac{1 + \Delta\alpha_{\Delta-1}}{\Delta+1}$.

We now improve these bounds. To do this, at each iteration we choose a branching vertex such that it has a neighbor with degree less than Δ. Then we can choose α_Δ based on the following equation:

$$\alpha_\Delta + (\Delta - 1)(\alpha_\Delta - \alpha_{\Delta-1}) + (\alpha_{\Delta-1} - \alpha_{\Delta-2}) = 1.$$

Then

$$\alpha_\Delta = \frac{1 + \alpha_{\Delta-2} + (\Delta - 2)\alpha_{\Delta-1}}{\Delta}. \tag{9}$$

We present an algorithm for which the recurrence (9) holds.

MAX2ALG
Input: F – instance of MAX-2-SAT or MAX-2-CSP.
Output: $Opt(F)$.
Method.

1. Remove all vertices of degree < 3 (using Lemma 1).
2. If F does not contain 2-clauses, then return the result.
3. **If the formula F has connected components F_1 and F_2, then return** MAX2ALG(F_1) **+** MAX2ALG(F_2).
4. If the maximal vertex degree of F is 3, then call the known algorithm for $(n, 3)$-MAX-2-SAT or $(n, 3)$-MAX-2-CSP, respectively.
5. **If F contains a vertex of maximal degree Δ that has at least one neighbor whose degree is not maximal, then let x be this vertex.** Otherwise, let x be any vertex of maximal degree.
6. Return $\max($MAX2ALG$(F[x]),$ MAX2ALG$(F[\neg x]))$.

The algorithm MAX2ALG is SIMPLEALG extended by two steps (given in bold).

Lemma 3. *If*

1. *Algorithm called at step 4 solves $(n, 3)$-MAX-2-SAT ($(n, 3)$-MAX-2-CSP) in time $O^*(2^{\alpha_3 n})$,*
2. $\delta_i = \min(\alpha_i - \alpha_{i-1}, \ldots, \alpha_4 - \alpha_3, \alpha_3)$,
   for each i:

$$\alpha_i + (i - 1)\delta_i + \delta_{i-1} \geq 1, \tag{10}$$

then the algorithm MAX2ALG *solves MAX-2-SAT (MAX-2-CSP) in time*
$O^*(2^{\alpha_3 n_3 + \ldots + \alpha_\Delta n_\Delta + \tau})$, *where* $\tau = \sum_i (1 - \tau_i)$,

$$\tau_i = \alpha_i + i\delta_i.$$

Proof. We prove this by induction on the number of vertices. Again we use the formula complexity measure $\mu = \alpha_3 n_3 + \ldots + \alpha_\Delta n_\Delta$. By $T(n_3, \ldots, n_\Delta)$ we denote the running time of the algorithm on formula with n_i vertices of degree i, $3 \leq i \leq \Delta$.

It is clear that connected components of an input formula can be handled independently. This is done at the step 4 of the algorithm. So, below we assume that the graph of the formula is connected.

If there is both a vertex of degree Δ and a vertex of degree less than Δ, then recursive calls decrease μ at least by 1. Indeed, we branch on a vertex x of degree Δ and we decrease μ by α_Δ. We choose x such that x has a neighbor of degree less than Δ. This neighbor causes a complexity decrease of at least $\delta_{\Delta-1}$. Each of the remaining $\Delta - 1$ neighbors decrease μ at least by δ_Δ. It follows from (10), that $\alpha_\Delta + (i - 1)\delta_\Delta + \delta_{\Delta-1} \geq 1$. Therefore,

$$T(n_3, \ldots, n_\Delta) \leq 2 \cdot 2^{\alpha_3 n_3 + \ldots + \alpha_\Delta n_\Delta + \tau - 1} + poly(|F|) \leq 2^{\alpha_3 n_3 + \ldots + \alpha_\Delta n_\Delta + \tau} + poly(|F|),$$

Now assume that the graph contains degree Δ variables only. Since during the work of the algorithm the degrees of variables can only decrease such a graph cannot appear in this branch again. So, at this iteration the algorithm makes two recursive calls for formulas whose complexity measure is less than the complexity of the initial formula at least by τ_Δ.

$$T(n_3, \ldots, n_\Delta) \leq 2 \cdot 2^k \cdot 2^{\alpha_3 n_3 + \ldots + \alpha_\Delta n_\Delta - k - \tau_\Delta + \sum_{i=3}^{\Delta-1}(1-\tau_i)} + poly(|F|)$$
$$= 2^{\alpha_3 n_3 + \ldots + \alpha_\Delta n_\Delta + \sum_{i=3}^{\Delta}(1-\tau_i)} + poly(|F|),$$

where k is the number of iterations of the algorithm while maximal degree is Δ. □

Corollary 3. *If for all $i > 3$,*

$$2\alpha_i \geq \alpha_{i+1} + \alpha_{i-1}, \quad \alpha_i - \alpha_{i-1} \leq \alpha_3,$$

then the running time of MAX2ALG *is*

$$O^*(2^{n(\alpha_D + \epsilon(\alpha_{D+1} - \alpha_D))}),$$

where $D = \lfloor d \rfloor$, $d = D + \epsilon$, $d = \frac{2m}{n}$ *is the average degree of the vertices.*

Proof. It can be shown by induction on i, that $\alpha_i \geq 1 - \frac{4}{i+1} = \frac{i-3}{i+1}$ From the corollary condition it follows that $\alpha_i - \alpha_{i-1}$ decreases with increasing i. Therefore $\delta_i = \alpha_i - \alpha_{i-1}$. Then

$$\tau = \sum_{i=3}^{\Delta}(1 - \tau_i) = \sum_{i=3}^{\Delta}(1 - \alpha_i - i\delta_i) = \sum_{i=3}^{\Delta}(1 - \alpha_i - i\alpha_i + i\alpha_{i-1}) =$$

$$\Delta - 2 - (\Delta + 1)\alpha_\Delta = \Delta(1 - \alpha_\Delta) + O(1) = \Delta(1 - \frac{\Delta - 3}{\Delta + 1}) + O(1) = O(1).$$

We use Theorem 1 to complete the proof.   □

It is easy to show that α_i's, chosen by (9), satisfy the condition of Corollary 3, so we have the bounds w.r.t. d. We get the following sequence for MAX-2-SAT: $\alpha_3 = 1/6$, $\alpha_4 = 1/3$, $\alpha_5 = 13/30$, $\alpha_6 = 23/45$, and the following sequence for MAX-2-CSP: $\alpha_3 = 1/4$, $\alpha_4 = 3/8$, $\alpha_5 = 19/40$, $\alpha_6 = 131/240$. At each step we can continue computing α_i's with weaker equality (4). This gives us an explicit formula for α_i for all i, if α_k is already computed from a stronger equality (9):

$$\alpha_i = \frac{i - k + (k+1)\alpha_k}{i+1} =$$
$$1 - \frac{k + 1 - (k+1)\alpha_k}{i+1} = 1 - \frac{k+1}{i+1}(1 - \alpha_k). \tag{11}$$

The values of the first α_k for MAX-2-SAT and MAX-2-CSP are shown in Table 1. According to the table we can calculate the running time of MAX2ALG for graphs with the average degree i. The running time is $O^*(2^{n\alpha_i})$.

Table 1. The values of α_k for $3 \leq k \leq 10$

d	MAX-2-SAT		MAX-2-CSP	
3	1/6	≈ 0.1666	1/4	≈ 0.2500
4	1/3	≈ 0.3333	3/8	≈ 0.3750
5	13/30	≈ 0.4333	19/40	≈ 0.4750
6	23/45	≈ 0.5111	131/240	≈ 0.5458
7	359/630	≈ 0.5698	1009/1680	≈ 0.6005
8	1553/2520	≈ 0.6162	8651/13440	≈ 0.6436
9	14827/22680	≈ 0.6537	82069/120960	≈ 0.6784
10	155273/226800	≈ 0.6846	855371/1209600	≈ 0.7071

Consider, e.g., the case $k = 5$. From (11) for MAX-2-SAT $\alpha_i = 1 - \frac{3.4}{d+1}$, for MAX-2-CSP $\alpha_i = 1 - \frac{3.15}{d+1}$, if $d \geq 5$. For $k = 8$, $\alpha_i = 1 - \frac{3.45}{d+1}$ for MAX-2-SAT, $\alpha_i = 1 - \frac{3.20}{d+1}$ for MAX-2-CSP, if $d \geq 8$. So, the running time of MAX2ALG for MAX-2-SAT (MAX-2-CSP) is $O^*(2^{n(1-\frac{3.45}{d+1})})$ $(O^*(2^{n(1-\frac{3.20}{d+1})}))$, if $d \geq 8$.

The upper bounds w.r.t. d imply upper bounds w.r.t. m. Indeed, $n = \frac{2m}{d}$, so the running time is $O^*(2^{\frac{2m(\alpha_D + \epsilon(\alpha_{D+1} - \alpha_D))}{d}})$. For MAX-2-CSP, the minimum of this function is at $d = 5$ and is equal to $O^*(2^{\frac{m}{5.263}})$. This matches the best known upper bound for MAX-2-CSP w.r.t. m [5].

5   Further Directions

To improve the upper bounds for MAX-2-SAT it is enough to improve any α_i, $i \geq 4$. All the subsequent α_i's will also be improved recursively. We can improve bounds from Section 4 for (n, Δ)-MAX-2-SAT by using item 5 of Lemma 4.1

in [7], saying that a neighbor of a variable x of degree 3 in at least one of two branches $F[x]$ and $F[\neg x]$ is not just eliminated by simplification rules, but is assigned a constant. Using this lemma and Lemma 3 we can get $\alpha_4 = 1/3.43$. Also for small i, α_i can be chosen using the algorithm from [5].

Also, as shown in Section 4, improving an upper bound for either $(n, 3)$-, $(n, 4)$-, or $(n, 5)$-MAX-2-CSP w.r.t. n, gives an improved upper bound for MAX-2-CSP w.r.t. m (the number of clauses).

Acknowledgments. I would like to thank my supervisor Alexander S. Kulikov for help in writing this paper and valuable comments.

References

1. Chen, J., Kanj, I.: Improved exact algorithms for Max-Sat. Discrete Applied Mathematics 142(1-3), 17–27 (2004)
2. Croce, F.D., Kaminski, M., Paschos, V.: An exact algorithm for MAX-CUT in sparse graphs. Operations Research Letters 35(3), 403–408 (2007)
3. Dantsin, E., Wolpert, A.: MAX-SAT for formulas with Constant Clause Density can be Solved Faster than in $O(2^n)$ Time. In: Biere, A., Gomes, C.P. (eds.) SAT 2006. LNCS, vol. 4121, pp. 266–276. Springer, Heidelberg (2006)
4. Fürer, M., Kasiviswanathan, S.P.: Exact Max 2-Sat: Easier and Faster. In: van Leeuwen, J., Italiano, G.F., van der Hoek, W., Meinel, C., Sack, H., Plášil, F. (eds.) SOFSEM 2007. LNCS, vol. 4362, pp. 272–283. Springer, Heidelberg (2007)
5. Gaspers, S., Sorkin, G.B.: A universally fastest algorithm for Max 2-Sat, Max 2-CSP, and everything in between, pp. 606–615 (2009)
6. Kojevnikov, A., Kulikov, A.S.: A new approach to proving upper bounds for MAX-2-SAT. In: Proceedings of the Seventeenth Annual ACM-SIAM Symposium on Discrete Algorithm, SODA 2006, pp. 11–17 (2006)
7. Kulikov, A., Kutzkov, K.: New upper bounds for the problem of maximal satisfiability. Discrete Mathematics and Applications 19, 155–172 (2009)
8. Scott, A.D., Sorkin, G.B.: Linear-programming design and analysis of fast algorithms for Max 2-CSP. Discrete Optimization 4(3-4), 260–287 (2007)
9. Williams, R.: A new algorithm for optimal 2-constraint satisfaction and its implications. Theoretical Computer Science 348(2-3), 357–365 (2005)

Improved Parameterized Algorithms for above Average Constraint Satisfaction

Eun Jung Kim[1,*] and Ryan Williams[2,**]

[1] LAMSADE-CNRS, Université de Paris-Dauphine, 75775 Paris, France
eunjungkim78@gmail.com
[2] Stanford University, Stanford, CA, USA
rrwilliams@gmail.com

Abstract. For many constraint satisfaction problems, the algorithm which chooses a random assignment achieves the best possible approximation ratio. For instance, a simple random assignment for MAX-E3-SAT allows 7/8-approximation and for every $\varepsilon > 0$ there is no polynomial-time $(7/8+\varepsilon)$-approximation unless P=NP. Another example is the PERMUTATION CSP of bounded arity. Given the expected fraction ρ of the constraints satisfied by a random assignment (i.e. permutation), there is no $(\rho+\varepsilon)$-approximation algorithm for every $\varepsilon > 0$, assuming the Unique Games Conjecture (UGC).

In this work, we consider the following parameterization of constraint satisfaction problems. Given a set of m constraints of constant arity, can we satisfy at least $\rho m + k$ constraint, where ρ is the expected fraction of constraints satisfied by a random assignment? CONSTRAINT SATISFACTION PROBLEMS ABOVE AVERAGE have been posed in different forms in the literature [18,17]. We present a faster parameterized algorithm for deciding whether $m/2 + k/2$ equations can be simultaneously satisfied over $\mathbb{F}_2$. As a consequence, we obtain $O(k)$-variable bikernels for BOOLEAN CSPs of arity c for every fixed c, and for PERMUTATION CSPs of arity 3. This implies linear bikernels for many problems under the "above average" parameterization, such as MAX-c-SAT, SET-SPLITTING, BETWEENNESS and MAX ACYCLIC SUBGRAPH. As a result, all the parameterized problems we consider in this paper admit $2^{O(k)}$-time algorithms.

We also obtain non-trivial hybrid algorithms for every Max c-CSP: for every instance I, we can either approximate I beyond the random assignment threshold in polynomial time, or we can find an optimal solution to I in subexponential time.

1 Introduction

The constraint satisfaction problem (CSP) is a general language to express many combinatorial problems such as graph coloring, satisfiability and various permutation problems. An instance of a CSP is a set V of variables, a domain D for

---

* This work was performed while the author was at LIRMM-CNRS, supported by the ANR-project AGAPE.

** This work was performed while the author was at IBM Research–Almaden, supported by the Josef Raviv Memorial Fellowship.

D. Marx and P. Rossmanith (Eds.): IPEC 2011, LNCS 7112, pp. 118–131, 2012.
© Springer-Verlag Berlin Heidelberg 2012

the variables and C a set of constraints. The objective is to assign a value from D to each variable of V so as to maximize the number of satisfied constraints. For example, 3-COLORING can be seen as a CSP over a three-element domain, and the constraints correspond to edges (thus the arity of each constraint is 2), indicating that values assigned to the endpoints of an edge must differ. In this work, we are interested in two types of CSPs. In a boolean CSP, the domain D is $\{-1, +1\}$. In a permutation CSP, the size of the domain equals $|V|$ and we request that the assignment is a bijection.

As solving CSPs is NP-hard in general, the next question is whether they allow efficient approximation algorithms. Interestingly, many constraint satisfaction problems exhibit a *hardness threshold*, where it is relatively easy to obtain a feasible solution that satisfies a certain fraction of the optimum number of constraints, yet it is difficult to find a solution that is even slightly better. Perhaps the best known example is MAX-E3-SAT, where we are given a CNF formula in which all clauses have exactly three literals, and wish to find a truth assignment satisfying as many clauses as possible. Although a uniform random assignment satisfies 7/8 of the clauses, Håstad [14] proved that it is NP-hard to satisfy $7/8 + \varepsilon$ for every $\varepsilon > 0$. The list of problems exhibiting a hardness threshold contains MAX-EC-SAT for $c \geq 3$, MAX-EC-LIN-2 for $c \geq 3$, and EC-SET SPLITTING for $c \geq 4$. For all these problems, a uniform random assignment achieves the best approximation ratio based on $P \neq NP$. Furthermore, for these problems the lower bounds on the optimum are also *tight* in the sense that they are optimal for an infinite sequence of instances.

For permutation CSPs, similar results have been identified, conditioned on the Unique Games Conjecture (UGC) of Khot [15]. In the BETWEENNESS problem, we have a set of *betweenness constraints* of the form "v_i is between v_j and v_k" for distinct variables $v_i, v_j, v_k \in V$ and the task is to find a permutation of the variables that satisfies the maximum number of constraints. More formally, constraints have the form

$$(\pi(v_j) < \pi(v_i) < \pi(v_k)) \vee (\pi(v_k) < \pi(v_i) < \pi(v_j))$$

and the task is to find a bijection $\pi : V \rightarrow \{1, \ldots, |V|\}$ that satisfies the maximum. We can satisfy one-third of the constraints in expectation by choosing a uniform random permutation. Moreover, it is hard to achieve a better approximation ratio, assuming UGC [5]. Hardness thresholds under UGC were known for permutation CSPs of arity 2 and 3 [9,5]. Recently these results were generalized to arbitrary fixed arity [8]. Here, the lower bounds on the optimum obtained by a random assignment is tight as well: consider, for example, an instance of BETWEENNESS in which we have all possible three constraints for triplets of variables.

These threshold phenomena are fascinating in that they provide a sharp boundary between feasibility and infeasibility: while the average is easy to obtain, satisfying an "above average" fraction is intractable. While the identification of such thresholds is extremely interesting, it does not end the story but rather initiates another one. As it is likely that practitioners will require feasible solutions that exceed these easy thresholds, it is important to understand how much computational effort is required to solve a problem beyond its threshold. One way of cleanly formalizing this question uses parameterized complexity.

A *parameterized problem* is a subset $L \subseteq \Sigma^* \times \mathbb{N}$ over a finite alphabet Σ. L is *fixed-parameter tractable* if the membership of (x, k) in $\Sigma^* \times \mathbb{N}$ can be decided in time $|x|^{O(1)} \cdot f(k)$ where f is a computable function of the parameter [7]. Given a pair of parameterized problems L and L', a *bikernelization* is a polynomial-time pre-processing algorithm that maps an instance (x, k) to an instance (x', k') (the *bikernel*) such that (i) $(x, k) \in L$ if and only if $(x', k') \in L'$, (ii) $k' \leq f(k)$, and (iii) $|x'| \leq g(k)$ for some functions f and g. The function $g(k)$ is called the *size* of the bikernel. A parameterized problem is fixed-parameter tractable if and only if it is decidable and admits a bikernelization [7]. A *kernelization* of a parameterized problem is a bikernelization to itself. For an overview of kernelization, see the recent survey [3].

Motivated by the discussion above, our work focuses on the following question:

(PERMUTATION) MAX-c-CSP ABOVE AVERAGE: We are given a parameter k and a set of constraints with at most c variables per constraint. Each constraint has a positive integer weight. Determine if there is a variable assignment (or permutation) that satisfies a subset of constraints with total weight at least $\rho \cdot W + k$, where W is the total weight of all constraints and ρ is the expected fraction of weighted constraints satisfied by a uniform random assignment.

Previous Work. Parameterizations *above a guaranteed value* were first considered by Mahajan and Raman [16] for the problems MAX-SAT and MAX-CUT. In a recent paper [17], Mahajan, Raman and Sikdar argue, in detail, that a practical (and challenging) parameter for a maximization problem is the number of clauses satisfied above a *tight* lower bound, which is $(1 - 2^{-c})m$ for MAX-SAT if each clause contains exactly c different variables. In the monograph by Neidermeier [18], an open problem attributed to Benny Chor [25, p.43] asks whether BETWEENNESS ABOVE AVERAGE is fixed parameter tractable.

A way for systematic investigation of above-average parameterization was recently presented by Gutin et al. [12]. They presented reductions to quadratic (bi)kernels (i.e., (bi)kernels with $O(k^2)$ variables) for the above-average versions of problems such as MAXIMUM ACYCLIC SUBGRAPH and MAX-c-LIN-2. Alon et al. [1] pushed forward the idea of representing a CSP instance algebraically, presenting a quadratic kernel for MAX-c-CSP using a similar method. This method was used to give a fixed-parameter algorithm for BETWEENNESS ABOVE AVERAGE [11], in which the idea of a *coarse* ordering is used. This result was later generalized in [13] to obtain a quadratic bikernel for PERMUTATION MAX-3-CSP.

Our Contribution. We show that every MAX-c-CSP and PERMUTATION MAX-3-CSP admits a problem bikernel with only $O(k)$ variables in the above-average parameterization. More precisely, we prove that essentially any *hard* instance of MAX-c-CSP must have less than $c(c + 1)k/2$ variables; instances with more variables are yes-instances for which good assignments can be generated in polynomial time. This improves over the main results of Alon *et al.* [1] who gave kernels of $O(k^2)$ variables, Crowston *et al.* [6] for MAX-c-CSP with $O(k \log k)$ variables and [13] for PERMUTATION MAX-3-CSP of $O(k^2)$ variables. This implies linear variable bikernels for the above-average versions of many different problems, such as MAX-c-LIN-2, MAX-c-SAT, SET SPLITTING when the sets are of size ≥ 4, MAXIMUM ACYCLIC SUBGRAPH, BETWEENNESS, CIRCULAR ORDERING, and 3-LINEAR ORDERING. As a result, both MAX-c-CSP and PERMUTATION MAX-3-CSP admit $2^{O(k)}$-time algorithms.

The key to our results is a fixed-parameter algorithm for the following problem:

> MAX-c-LIN-2 ABOVE AVERAGE: *We are given a parameter k and a system of linear equations over $\mathbb{F}_2$ with at most c variables per equation. Each equation e has a positive integer weight, and the weight of an assignment in the system is defined to be the total sum of weights of equations satisfied by the assignment. Determine if there is an assignment of weight at least $W/2 + k/2$, where W is the total weight of all equations.*

Note the random assignment algorithm yields $W/2$ weight, and it is famously NP-hard to attain $W/2 + \varepsilon W$ for every $\varepsilon > 0$ [14]. Our proofs imply the stronger result that *every* constraint satisfaction problem admits a hybrid algorithm, in the following sense:

Theorem 1. *For every Boolean MAX-c-CSP, there is an algorithm with the property that, for every $\varepsilon > 0$, on any instance I, the algorithm outputs either:*

- *an optimal solution to I within $O^\star(2^{c(c+1)\varepsilon m/2})$ time, or*
- *a $(\rho + \varepsilon/2^c)$-approximation to I within polynomial time, where ρ is the expected fraction of weighted constraints satisfied by a uniform random assignment to the CSP.*

This resolves an open problem of Vassilevska, Williams, and Woo [20], who asked if MAX-3-SAT had an algorithm of this form.

2   Preliminaries

We define the BOOLEAN MAX-c-CSP and MAX-c-PERMUTATION CSP. A boolean constraint satisfaction problem is specified by the domain $\{-1, +1\}$ and a set of predicates $\mathbb{P}$, called *payoff functions* as well. A predicate $P \in \mathbb{P}$ is a function from $\{-1, +1\}^{c'}$ to $\{0, 1\}$ for $c' \leq c$. The maximum number c of inputs to the predicates in $\mathbb{P}$ is the arity of the problem. We interpret -1 as the value True and $+1$ as False.

An instance of boolean CSP is specified as a set of variables V along with a collection of triples $\mathcal{C} = \{(f_1, S_1, w_1), \ldots, (f_m, S_m, w_m)\}$, where $f_i \in \mathbb{P}$, every S_i is an ordered tuple from V of size at most c, every w_i is a positive integer. A variable assignment $\phi : V \to D$ *satisfies* a constraint $(f_i, (s_1, \ldots, s_{c'}), w_i)$ provided that $f(\phi(s_1), \ldots, \phi(s_{c'})) = 1$. Our goal is to find an assignment ϕ:

$V \rightarrow \{-1, +1\}$ of maximum weight. Here the *weight of an assignment* ϕ is defined to be the total sum of weights w_i of constraints (f_i, S_i, w_i) satisfied by ϕ. For example, MAX-E3-SAT is specified by a single predicate $P : \{-1, +1\} \rightarrow \{0, 1\}$, where $P(x, y, z) = 0$ if and only if $x = y = z = +1$.

We consider the following parameterization of boolean CSP.

MAX-c-CSP ABOVE AVERAGE (c-CSP$_{AA}$)
Input: A variable set V and constraints $\mathcal{C} = \{(f_1, S_1, w_1), \ldots, (f_m, S_m, w_m)\}$ with $|S_i| \leq c$ for every i, an integer $k \geq 0$.
Parameter: k
Goal: Determine if there is an assignment with weight at least $\rho \cdot W + k/2^c$, where W is the total weight of all constraints and ρ is the expected fraction of weighted constraints satisfied by a uniform random assignment.

In the permutation CSP problem, the domain D is $[n] = \{1, \ldots, n\}$ and a predicate $P \in \mathbb{P}$ is a function from $\mathcal{S}_{c'}$ to $\{0, 1\}$ for $c' \leq c$, where $\mathbb{S}_{c'}$ is the set of permutations on $\{1, \ldots, c'\}$. Let $\mathbb{P}$ be a set of predicates. An instance of permutation CSP is given as a variable set V and a collection $\mathcal{C} = \{(f_1, S_1, w_1), \ldots, (f_m, S_m, w_m)\}$, where every S_i is an ordered tuple from V of size at most c, every w_i is a positive integer, $f_i \in \mathbb{P}$ is applied to the tuple S_i. In a permutation CSP, a variable assignment $\phi : V \rightarrow D$ is required to be a bijection, or equivalently, a permutation. A permutation ϕ *satisfies* a constraint $(f_i, (s_1, \ldots, s_{c'}), w_i)$ provided that $f(\phi(S_i)) = 1$, viewing the local permutation $\phi(S_i)$ as an element of $\mathbb{S}_{c'}$. Our goal is to find an assignment $\phi : V \rightarrow [n]$ of maximum weight, where the weight of ϕ is the total sum of weights of constraints satisfied by ϕ. For example, BETWEENNESS is specified by a single predicate of arity 3, i.e. $P : \mathbb{S}_3 \rightarrow \{0, 1\}$, where $P(x, y, z) = 1$ if and only if $xyz \in \{123, 321\}$.

In the full version of the paper, we also consider the following parameterization of permutation CSP:

MAX-c-PERMUTATION CSP ABOVE AVERAGE
Input: A variable set V and constraints $\mathcal{C} = \{(f_1, S_1, w_1), \ldots, (f_m, S_m, w_m)\}$, an integer $k \geq 0$.
Parameter: k
Goal: Determine if there is an assignment with weight at least $\rho \cdot W + k/(c!4^c)$, where W is the total weight of all constraints and ρ is the expected fraction of weighted constraints satisfied by a uniform random permutation.

In what follows, we omit ABOVE AVERAGE and simply say MAX-c-LIN-2, MAX-c-CSP and MAX-c-PERMUTATION CSP to refer to the parameterized problems.

For MAX-c-PERMUTATION CSP, let $\mathbb{P}$ be the associated set of predicates. For each predicate $P \in \mathbb{P}$, we can identify the set $\Pi_P = \{\pi \in \mathbb{S}_c : P(\pi) = 1\}$. Notice that ϕ satisfies $(f_i, S_i = (v_1, \ldots, v_{c'}), w_i)$ if and only if there is a $\pi \in \Pi_{f_i}$ s.t. $\phi(v_1)\phi(v_2) \cdots \phi(v_{c'}) \cong \pi(1)\pi(2) \cdots \pi(c')$, and thus if and only if there is $\pi \in \Pi_{f_i}$ s.t. $\phi(v_{\pi^{-1}(1)}) < \phi(v_{\pi^{-1}(2)}) < \cdots < \phi(v_{\pi^{-1}(c')})$. The folklore result below allows us to focus on the case when $\mathbb{P}$ contains a single predicate P such that $\Pi_P = \{12 \cdots c\}$, which we denote c-LINEAR ORDERING. The proof for the case $c = 3$ and $|\mathbb{P}| = 1$ can be found in [13] and its extension for arbitrary fixed c is straightforward. As the proof of [13] considers only the case when $\mathbb{P}$ has a single predicate, we sketch the proof here even though its generalization is immediate.

Proposition 1. *Let $(V, \mathcal{C}, k)$ be an instance of c-OCSP$_{AA}$. There is a polynomial time transformation R from* MAX-c-PERMUTATION CSP *to c-LINEAR ORDERING such that an instance $(V, \mathcal{C}, k)$ of* MAX-c-PERMUTATION CSP *is a yes-instance if and only if $R(V, \mathcal{C}, k)$ is a yes-instance of c-LINEAR ORDERING.*

Proof. From an instance $(V, \mathcal{C}, k)$ of MAX-c-PERMUTATION CSP, we construct an instance $(V, \mathcal{C}_0, k)$ of c-LINEAR ORDERING as follows. We shall express a constraint (f_i, S_i, w_i) by a set of constraints $R(f_i, S_i, w_i)$ such that (f_i, S_i, w_i) is satisfied if and only if exactly one of $R(f_i, S_i, w_i)$ is satisfied. Let S_i be $(v_1, v_2, \ldots, v_{c'})$.

For every element π of $\Pi_{f_i} = \{\pi \in \mathbb{S}_c : f_i(\pi) = 1\}$, we add to the instance of c-LINEAR ORDERING the constraint $(P_{id}, (v_{\pi^{-1}(1)}, v_{\pi^{-1}(2)}, \ldots, v_{\pi^{-1}(c')}), w_i)$. Here P_{id} is a function mapping identity permutation to 1 and other permutations to 0. Notice that (f_i, S_i, w_i) is satisfied if and only if exactly one of the constraints $(P_{id}, (v_{\pi^{-1}(1)}, v_{\pi^{-1}(2)}, \ldots, v_{\pi^{-1}(c')}), w_i)$ for $\pi \in \Pi_{f_i}$ is satisfied. Hence the weight of a linear ordering ϕ remains the same in the original and transformed instances. Moreover, the expected satisfied fraction of the constraint (f_i, S_i, w_i) is $|\Pi_{f_i}|/(c')!$ and the expected satisfied fraction of the new constraints is the same. Hence, the instance of MAX-c-PERMUTATION CSP has a linear ordering of weight $\rho \cdot W + k$ if and only if the constructed instance of c-LINEAR ORDERING a linear ordering of weight $\rho \cdot W + k$. □

It is well-known that for every function $f : \{1, -1\}^n \to \mathbb{R}$ can be uniquely expressed as a multilinear polynomial

$$f(x) = \sum_{S \subseteq [n]} \hat{f}(S) \chi_S(x),$$

where $\hat{f}(S)$ is the fourier coefficient of f on S, defined as

$$\hat{f}(S) := \mathop{\mathbf{E}}_{x \in \{1, -1\}^n} [f(x) \chi_S(x)]$$

and the character function $\chi_S(x)$ is defined as $\chi_S(x) := \prod_{i \in S} x_i$. Given the truth table of f, the fourier coefficients of f can be computed via the inverse fourier transform (one reference is [19]).

3   Max-c-Lin-2 above Average

We now turn to describing improved parameterized algorithms for maximum constraint satisfaction problems with a constant number of variables per constraint, including the problems of satisfying a maximum subset of linear equations and maximum CNF satisfiability. At the heart of our approach is a faster algorithm for MAX-c-LIN-2 ABOVE AVERAGE that can be applied in a general way to solve other CSPs.

Theorem 2. *For every $c \geq 2$,* MAX-c-LIN-2 ABOVE AVERAGE *can be solved in* $O(2^{(c(c+1)/2)k} \cdot m)$ *time.*

In [20], the authors gave a "hybrid algorithm" for the unweighted problem MAX-E3-LIN-2 (where exactly three variables appear in each equation), with the property that, after a polynomial time test of the instance, the algorithm either outputs an assignment satisfying $(1/2 + \varepsilon)m$ equations in polynomial time, or outputs the optimal satisfying assignment in $2^{O(\varepsilon m)}$ time. The algorithm works by finding a maximal subset of equations such that every pair of equations share no variables; based on the size of this set, the hybrid algorithm decides to either approximately solve the instance or solve it exactly. Our algorithm is in a similar spirit, but requires several modifications to yield a parameterized algorithm for the weighted case, to deal with any $c \geq 2$, and to deal with "mixed" equations that can have different numbers of variables.

Let F be a set of equations over $\mathbb{F}_2$, where each equation e contains at most c variables and has a positive integral weight $w(e)$. For a single equation $e \in F$, let $var(e)$ be the set of all variables appearing in e. Let $var(F) = \bigcup_{e \in F} var(e)$. For a set of equations F', the weight $w(F')$ is the sum of weights $w(e)$ over $e \in F'$. The *weight of an assignment* is the total weight of equations that are satisfied by the assignment.

Note that MAX-2-LIN-2 ABOVE AVERAGE is a generalization of MAX CUT ABOVE AVERAGE on weighted graphs: by simulating each edge $\{u, v\}$ of weight w with an equation $x_u + x_v = 1$ of weight w, the MAX-2-LIN-2 problem easily captures MAX CUT.

We assume that the given instance is *reduced* in the sense that there is no pair of equations e, e' with $e \equiv e' + 1 \pmod 2$. (Such an equation e is said to be *degenerate* in [20].) If such a pair exists, one can remove the equation of lesser weight (call it e') and subtract $w(e')$ from $w(e)$. Note the weight of every variable assignment has now been subtracted by $w(e')$.

Proof of Theorem 2. It is convenient to view an equation e as a set $var(e)$. We first find a maximal independent (i.e. disjoint) collection $S_c \subseteq F$ of c-sets. More precisely, we treat each equation as a set of variables, ignore those sets of cardinality less than c, and find a maximal disjoint set over the c-sets using the standard greedy algorithm. All remaining equations in F now have at most $c-1$ variables if we remove all occurrences of variables in $var(S_c)$ from F.

Next, we pick another collection $S_{c-1} \subseteq F$ of sets with the property that, after we remove all variables in $var(S_c)$ from F, S_{c-1} forms a maximally independent collection of $(c-1)$-sets in the remaining set system. In general, for $j = c - 2$ down to 1, once the variables in $var(S_c \cup \cdots \cup S_{j+1})$ have been removed from the remaining equations, a maximal independent set of j-sets is chosen greedily, and we set S_j to be a collection of corresponding original sets in F (with the variables in $var(S_c \cup \cdots \cup S_{j+1})$ added back). We continue until S_1, in which each set in the collection has exactly one variable after those in $var(S_c \cup \cdots \cup S_2)$

have been removed. For convenience, let $S_{c+1} = var(S_{c+1}) = \emptyset$. By properties of maximal disjoint sets, we have:

Observation 1. *For every $1 \leq j \leq c$, eliminating the variables appearing in $var(S_c \cup \cdots \cup S_{j+1})$ leaves at most j variables in every equation of F.*

Now, either **(1)** $w(S_j) < k$ for every $j = 1, \ldots, c$, or **(2)** there is a j such that $w(S_j) \geq k$.

Case **(1)** is easily handled: for every j, each equation in S_j contains j variables which do not appear in $S_c \cup \cdots \cup S_{j+1}$. Hence, $|var(F)| = |var(\bigcup_{i=1}^{c} S_i)| < ck + (c-1)k + \cdots + k < (c(c+1)/2) \cdot k$. By trying all $O(2^{(c(c+1)/2)k})$ assignments to $var(F)$, we can find an optimal assignment for F.

Case **(2)** is more delicate and is handled by the two claims below. We will show that in this case, F is a yes-instance of the problem and we can efficiently recover a solution for it. Recall an equation $e \in F$ is non-degenerate if there is no $e' \in F$ such that $e \equiv e' + 1 \pmod 2$. As mentioned earlier, we may assume without loss of generality that every equation in F is non-degenerate.

Claim. For every $1 \leq j \leq c$, a random assignment satisfying all equations in S_j will satisfy every non-degenerate equation in $F - S_j$ with probability $1/2$. Moreover, we can output such a random assignment in polynomial time.

Proof. To prove the first part of the claim, it suffices to show that no equation $e \in F - S_j$ (or its negation $e + 1$) can be expressed as a linear combination of one or more equations in S_j. Put another way, we will show that every equation in $e \in F - S_j$ is *linearly independent* of the equations in S_j.

Suppose there are equations $e_1, \ldots, e_m$ from S_j such that their summation (modulo 2) results in a variable subset that is equal to the set of variables in another equation $e \in F$. That is, viewing $e_1, \ldots, e_m$ and e as indicator n-bit vectors (one bit for each of the n variables, omitting the constant terms in the equations), we have $e = \sum_{i=1}^{m} e_i \pmod 2$. Recall that every equation in S_j has j variables which do not appear in $var(S_c \cup \cdots \cup S_{j+1})$, and every pair of equations in S_j involves disjoint sets of variables, by construction. Hence, if $m > 1$, then the equation e (composed of variables from $e_1, \ldots, e_m$) has more than j variables which do not appear in $var(S_c \cup \cdots \cup S_{j+1})$, which is impossible by Observation 1. Therefore $m = 1$, and every subset $\{e_1, \ldots, e_m\}$ of equations from S_j whose modulo sum is the same as another equation $e \in F - S_j$ has cardinality 1. But then the equation e is degenerate, which is a contradiction to the non-degeneracy assumption. Therefore no non-degenerate equation in F (or its negation) can be represented as a linear combination of one or more equations from S_j.

Now, given that every non-degenerate equation in $e \in F - S_j$ is linearly independent of the equations in S_j, we claim that a random assignment that is consistent with the equations in S_j will satisfy e with probability $1/2$. This is a simple consequence of linear algebra over $\mathbb{F}_2$. Put the system of equations S_j in the form $Ax = b$, where $A \in \mathbb{F}_2^{|S_j| \times n}$, $x \in \mathbb{F}_2^n$, and $b \in \mathbb{F}_2^{|S_j|}$. Let $e \in F - S_j$. Define $B_e \in \mathbb{F}_2^{(|S_j|+1) \times n}$ to be identical to A in its first $|S_j|$ rows, and in the last row, B_e contains the indicator vector for the variables of e. Define $c_e \in \mathbb{F}_2^{|S_j|+1}$ to be identical to b in its first $|S_j|$ components, and c contains the constant term of e in its last component. Saying that $e \in F - S_j$ is linearly independent of S_j is equivalent to saying $rowrank(B_e) = rowrank(A) + 1$, and the set of solutions to $Ax = b$ contains the set of solutions to $B_e x = c_e$. The number of solutions to a system of rank r is 2^{n-r}. Therefore a uniform random variable assignment that satisfies $Ax = b$ will also satisfy $B_e x = c_e$ with probability $1/2$.

Finally, we describe how to produce a uniform random assignment over all assignments that satisfy the equations in S_j. Produce a random assignment to the variables in $var(S_{c+1} \cup \cdots \cup S_{j+1})$, then produce a random assignment to those variables in the maximal independent collection of j-sets obtained after removing $var(S_{c+1} \cup \cdots \cup S_{j+1})$, in such a way that every equation in S_j is satisfied. (Exactly one variable in each equation of S_j will be "forced" to be a certain value, but note that none of these forced variables appear in more than one equation of S_j, by construction.) The remaining variables are set to 0 or 1 uniformly at random. Note that if $j = 1$ and some equation $e \in S_1$ has $|var(e)| = 1$, the assignment to the variable of e is decided uniquely. □

Claim. If there is a j with $w(S_j) \geq k$, then we can find an assignment with weight at least $W/2 + k/2$ in polynomial time.

Proof. Suppose that $j \geq 1$ is the largest integer with $w(S_j) \geq k$. By Claim 3, a random assignment satisfying all equations in S_j will satisfy every other non-degenerate equation with probability $1/2$. Hence the weight of such an assignment is at least $(W - w(S_j))/2 + w(S_j) \geq W/2 + k/2$ on average. An assignment can also be found deterministically using conditional expectation. □

This completes the proof of Theorem 2. □

The above proof shows that the following stronger statement is also true.

Theorem 3. *For every $c \geq 2$, let I be an instance of* MAX-c-LIN-2 ABOVE AV-ERAGE. *If $|var(F)| \geq (c(c+1)/2)k$, then I is an yes-instance and an assignment satisfying equations with at least $W/2 + k/2$ weight can be found in polynomial time.*

Observe that the running time of our algorithm is optimal up to constant factors in the exponent, assuming the Exponential Time Hypothesis:

Theorem 4. *If* MAX-3-LIN-2 ABOVE AVERAGE *can be solved in $O(2^{\varepsilon k} 2^{\varepsilon m})$ time for every $\varepsilon > 0$, then 3SAT can be solved in $O(2^{\delta n})$ time for every $\delta > 0$, where n is the number of variables.*

Proof. First, by the improved Sparsification Lemma of [4], for every $\delta > 0$ we can reduce 3SAT on n variables and m clauses in $2^{\delta n}$ time to 3SAT on n variables and $m' = (1/\delta)^c n$ clauses, for some fixed constant $c > 1$. This 3SAT instance on n variables and m' clauses can further be reduced to Max-3-Lin-2 on n variables and $O(m')$ clauses using the reduction of Lemma 1 (proved below). Provided that we can determine whether $m'/2 + k/2$ equations can be satisfied in $2^{\varepsilon k} 2^{\varepsilon m'}$ time, then by trying each k in the interval $[1, m']$ we can solve the Max-3-Lin-2 instance exactly in at most $2^{2\varepsilon m'} \leq O(2^{2\varepsilon(1/\delta)^c n})$ time.

This results in an $O(2^{\delta n + 2\varepsilon(1/\delta)^c n})$ algorithm for 3SAT. Setting $\varepsilon = \delta^{c+1}$, we obtain $O(2^{3\delta n})$ time. As this reduction works for every $\delta > 0$, the conclusion follows.   □

4   Boolean MAX-c-CSP above Average

To apply our algorithm to general CSPs, we use the following reduction.

Lemma 1 ([1],[6]). *There is a polynomial time reduction from* MAX-c-CSP ABOVE AVERAGE *with n variables and parameter k to* MAX-c-LIN-2 ABOVE AVERAGE *with n variables and parameter k.*

The proof of the lemma for unweighted c-CSP is sketched in [1], and a full proof is given in [6]. An alternative proof which also covers the weighted case is provided in the full version of this paper.

Theorem 5. *For every $c \geq 2$,* MAX-c-CSP ABOVE AVERAGE *can be solved in* $O(2^{(c(c+1)/2)k} \cdot m)$ *time.*

Proof of Theorem 5. Using the reduction of Lemma 1, reduce an instance of MAX-c-CSP ABOVE AVERAGE with m constraints to MAX-c-LIN-2 ABOVE AVERAGE with $O(2^c \cdot m)$ equations. Using the algorithm of Theorem 2 we solve the obtained instance of MAX-c-LIN-2 ABOVE AVERAGE. Thus we can determine if the given c-CSP has an assignment with weight at least $AVG + k$ in $O(2^{(c(c+1)/2)k} \cdot 2^c \cdot m) = O(2^{(c(c+1)/2)k} \cdot m)$ time. To finding an actual solution for MAX-c-CSP ABOVE AVERAGE, we can simply use the transformation given in the proof of Lemma 1.   □

Theorem 3 and Lemma 1 show in fact that every CSP admits a hybrid algorithm [20].

Reminder of Theorem 1. *For every Boolean* MAX-c-CSP, *there is an algorithm with the property that, for every $\varepsilon > 0$, on any instance I, the algorithm outputs either:*

- *an optimal solution to I within $O^\star(2^{c(c+1)\varepsilon m/2})$ time, or*
- *a $(\rho + \varepsilon/2^{c+1})$-approximation to I within polynomial time, where ρ is the expected fraction of weighted constraints satisfied by a uniform random assignment to the CSP.*

Proof. Given an instance I of MAX-c-CSP with m constraints, Lemma 1 shows that we can reduce I to an weighted instance I' of MAX-c-LIN-2 with $O(2^c m)$ equations, in polynomial time, such that at least $\rho m + \delta k$ constraints can be satisfied in I if and only if at least $W/2 + k/2$ weight of equations can be satisfied in I', where $\delta \geq 1/2^c$ and depends on the underlying constraints. Now set $k = \varepsilon m$ and run the algorithm of Theorem 3. If $|var(I')| \geq (c(c+1)/2)k$, then an assignment satisfying at least $W/2 + k/2 \geq W/2 + \varepsilon m/2$ weight of equations can be found in polynomial time, hence we obtain an assignment for I satisfying at least $\rho m + \delta \varepsilon m/2$ constraints. Otherwise, exhaustive search over the $c(c+1)\varepsilon m/2$ variables of I' will uncover an exact solution to I in $O^\star(2^{c(c+1)\varepsilon m/2})$ time.   □

We close this section with showing how our algorithm can provide linear size kernels for MAX-c-LIN-2 ABOVE AVERAGE and a kernel for MAX-c-CSP ABOVE AVERAGE.

Corollary 1. *For every $c \geq 3$, the problem* MAX-c-LIN-2 ABOVE AVERAGE *can be reduced to a problem kernel with at most $(c(c+1)/2)k$ variables in polynomial time.*

Proof. Consider executing the algorithm of Theorem 2, up to the point before it performs an exhaustive search of assignments. At this point, the algorithm has taken only polynomial time. If there is a S_j with weight at least k, the algorithm outputs an assignment with weight at least $W/2 + k/2$ in polynomial time. Otherwise, for all $j = c, c-1, \ldots, 1$, S_j has weight less than k. It follows (from Case **(1)** in the proof of Theorem 2) that the total number of variables in the instance is at most $(c(c+1)/2)k$.   □

Note that the size of the kernel in Corollary 1 matches the prior work for $c = 2$ [1].

Corollary 2. *For every $c \geq 3$, the problem* MAX-c-CSP ABOVE AVERAGE *can be reduced to a problem kernel with at most $(c(c+1)/2)k$ variables in polynomial time.*

Proof. In the proof of Theorem 1 in [1], a procedure $\mathcal{P}$ is given that reduces any instance of MAX-c-LIN-2 ABOVE AVERAGE with total sum of weights W and parameter k into an instance of MAX-c-CSP ABOVE AVERAGE[1] with (a multiset of) $2^{c-1}W$ constraints and parameter $2^{c-1}k$. More precisely, the procedure $\mathcal{P}$ considers an instance of MAX-c-LIN-2 ABOVE AVERAGE in which each equation has weight 1 and the multiplicity of an equation may be larger than one. $\mathcal{P}$ maps an equation into a set of 2^{c-1} clauses.

Given a MAX-c-CSP ABOVE AVERAGE instance on n variables and m constraints, we first perform the transformation given by Lemma 1 and obtain a MAX-c-LIN-2 ABOVE AVERAGE instance, with $O(2^c m)$ equations and n variables. By applying the kernelization of Theorem 2 we obtain an equivalent instance with at most $(c(c+1)/2)k$ variables and no more than $(c(c+1)k/2)^c$ (weighted) equations. Finally, apply procedure $\mathcal{P}$ to reduce the problem *back* into a MAX-c-CSP ABOVE AVERAGE instance, having $(c(c+1)/2)k$ variables, $O(2^{c-1} \cdot (c(c+1)k/2)^c)$ constraints and parameter $2^{c-1}k$.   □

---

[1] In [1], the transformed instance is in fact a MAX-c-SAT instance.

5   MAX-c-Permutation CSP above Average

In Section 3, we proved that every reduced instance of MAX-c-LIN-2 either has an assignment satisfying $W/2+k$ weight of equations, or has at most $(c(c+1)/2)k$ variables. This result can be applied to the problems MAX-c-PERMUTATION CSP for $c = 2, 3$ to obtain a bikernel with $O(k)$ variables. By Proposition 1, it suffices for us to focus on the problem 3-LINEAR ORDERING instead of considering general MAX-c-PERMUTATION CSP. Due to lack of space, we only have room to state the main theorems, and must defer the rest to the full paper. For 3-LINEAR ORDERING we give a series of reduction rules to run on a given instance, and we prove that "irreducible" instances (those instances which are unaffected by the rules) have special properties:

Theorem 6. *Let $I = (V, \mathcal{C}, k)$ be an irreducible instance of 3-LINEAR ORDERING. If I is a no-instance (i.e., less than $\rho W + k$ constraints in I can be simultaneously satisfied), then the number of variables in I is less than $15k$ variables.*

Corollary 3. *Let $I = (V, \mathcal{C}, k)$ be an irreducible instance of 2-LINEAR ORDERING. If I is a no-instance, we have $|V| < 10k$ variables.*

Theorem 7. *The problem MAX-c-PERMUTATION CSP ABOVE AVERAGE can be solved in time $2^{O(k)}$ for $c = 2, 3$.*

Proof. Using Proposition 1, MAX-c-PERMUTATION CSP ABOVE AVERAGE instances can be transformed into an equivalent instance of c-LINEAR ORDERING. Due to the result stated as Theorem 6 and Corollary 3, we either know the instance is a yes-instance or attain an equivalent instance with less than $15k$ variables for $c = 3$ (or an instance with less than $10k$ variables for $c = 2$). Hence, a $O^\star(2^n)$-algorithm to exactly compute the maximum number of satisfiable constraints on a n-variable instance of EXACT 3-LINEAR ORDERING will yield a desired result. Here, an instance I of EXACT 3-LINEAR ORDERING is given as a pair $(V, \mathcal{C})$ and the task is to find a linear ordering on V so as to *maximize* the number of satisfied constraints in $\mathcal{C}$. We give an exposition of such an algorithm for $c = 3$. An analogous observation applies to $c = 2$.

Bodlaender et. al [2] presents algorithms for VERTEX ORDERING problems which runs in $O^\star(2^n)$ time and $O^\star(2^n)$ space, or $O^\star(4^n)$ time and polynomial space. Let V be a set of elements, which may be vertices in graph problems or variables in our permutation CSP context. For a linear ordering π on V, we denote the set $\{w \in V : \pi(w) < \pi(v)\}$ by $\pi_{<,v}$. Consider a function f from the domain of triples (G, S, v) to an integer, where G is a graph, $S \subseteq V(G)$ and $v \in V(G)$. It is shown in [2] that if f is polynomial time computable, the value

$$\min_\pi \sum_{v \in V} f(G, \pi_{<,v}, v),$$

where π is taken over all possible linear orderings, can be computed either in $O^\star(2^n)$ time and $O^\star(2^n)$ space, or in $O^\star(4^n)$ time and polynomial space. Alerted readers might notice that the former uses dynamic programming in Held-Karp style and the latter employs recursion instead. We point out that although they describe the algorithms in the context of graph problems, the validity does not depend on whether the relations on V are of arity two or not. It remains to formulate the EXACT 3-LINEAR ORDERING problem to fit in the setting.

We take $f(I, S, v) := |\{e = (a, v, c) \in \mathcal{C} : a \in S, c \in V \setminus (S \cup \{v\})\}|$ and note that f can be computed in polynomial time given a triple $I = (V, \mathcal{C}), S \subseteq V$ and $v \in V$. To see that $\min_\pi \sum_{v \in V} -f(I, \pi_{<,v}, v) = \max_\pi \sum_{v \in V} f(I, \pi_{<,v}, v)$ equals the optimal value of the EXACT 3-LINEAR ORDERING instance I, it suffices to observe the followings: given a linear ordering π, (a) the family $\{e = (a, v, c) \in \mathcal{C}\}, v \in V$ *partitions* the constraint set $\mathcal{C}$, (b) a constraint $e = (a, v, c) \in \mathcal{C}$ contributes one to $f(I, \pi_{<,v}, v)$ if and only if π satisfies e. Finally, we note that the extension of the formulation to weighted instances and instances with (some) constraints of arity two is straightforward.   □

Closing this section, we point out that the recent work in [10], independently of our paper, also explores the idea of ensuring a monomial which represents a variable in the multilinear polynomial. They consider the c-LINEAR ORDERING problem in which every variable occurs in a bounded number of constraints and show that approximation beyond the random assignment threshold $1/c!$ is achievable. It is interesting to note as well that their motivation is to gain over the random assignment threshold.

Acknowledgement. The authors would like to thank Daniel Gonçalves for valuable discussion which inspired the results of Section 5.

References

1. Alon, N., Gutin, G., Kim, E.J., Szeider, S., Yeo, A.: Solving MAX-r-SAT above a tight lower bound. Algorithmica (2010) (to appear)
2. Bodlaender, H., Fomin, F., Koster, A., Kratsch, D., Thilikos, D.: A note on exact algorithms for vertex ordering problems on graphs. Theory of Computing Systems, 1–13 (2010), doi:10.1007/s00224-011-9312-0
3. Bodlaender, H.L.: Kernelization: New Upper and Lower Bound Techniques. In: Chen, J., Fomin, F.V. (eds.) IWPEC 2009. LNCS, vol. 5917, pp. 17–37. Springer, Heidelberg (2009)
4. Calabro, C., Impagliazzo, R., Paturi, R.: A duality between clause width and clause density for sat. In: IEEE Conference on Computational Complexity, pp. 252–260 (2006)
5. Charikar, M., Guruswami, V., Manokaran, R.: Every permutation CSP of arity 3 is approximation resistant. In: 24th Annual IEEE Conference on Computational Complexity, CCC 2009, pp. 62–73 (July 2009)
6. Crowston, R., Gutin, G., Jones, M., Kim, E.J., Ruzsa, I.Z.: Systems of Linear Equations over $\mathbb{F}_2$ and Problems Parameterized above Average. In: Kaplan, H. (ed.) SWAT 2010. LNCS, vol. 6139, pp. 164–175. Springer, Heidelberg (2010)
7. Flum, J., Grohe, M.: Parameterized Complexity Theory. Springer, Heidelberg (2006)

8. Guruswami, V., Håstad, J., Manokaran, R., Raghavendra, P., Charikar, M.: Beating the random ordering is hard: Every ordering csp is approximation resistant. Electronic Colloquium on Computational Complexity (ECCC) 18, 27 (2011)
9. Guruswami, V., Manokaran, R., Raghavendra, P.: Beating the random ordering is hard: Inapproximability of maximum acyclic subgraph. In: FOCS, pp. 573–582 (2008)
10. Guruswami, V., Zhou, Y.: Approximating bounded occurrence ordering CSPs (2011) (manuscript)
11. Gutin, G., Kim, E.J., Mnich, M., Yeo, A.: Betweenness parameterized above tight lower bound. J. Comput. Syst. Sci. 76(8), 872–878 (2010)
12. Gutin, G., Kim, E.J., Szeider, S., Yeo, A.: A probabilistic approach to problems parameterized above or below tight bounds. J. Comput. Syst. Sci. (2010) (to appear)
13. Gutin, G., van Iersel, L., Mnich, M., Yeo, A.: All Ternary Permutation Constraint Satisfaction Problems Parameterized above Average have Kernels with Quadratic Numbers of Variables. In: de Berg, M., Meyer, U. (eds.) ESA 2010, Part I. LNCS, vol. 6346, pp. 326–337. Springer, Heidelberg (2010)
14. Håstad, J.: Some optimal inapproximability results. J. ACM 48(4), 798–859 (2001)
15. Khot, S.: On the power of unique 2-prover 1-round games. In: Proceedings of the ACM symposium on Theory of Computing, pp. 767–775 (2002)
16. Mahajan, M., Raman, V.: Parameterizing above guaranteed values: MaxSat and MaxCut. J. Algorithms 31(2), 335–354 (1999)
17. Mahajan, M., Raman, V., Sikdar, S.: Parameterizing above or below guaranteed values. J. Comput. System Sci. 75(2), 137–153 (2009)
18. Niedermeier, R.: Invitation to fixed-parameter algorithms. Oxford Lecture Series in Mathematics and its Applications, vol. 31. Oxford University Press, Oxford (2006)
19. O'Donnell, R.: Some topics in analysis of boolean functions. In: STOC, pp. 569–578 (2008)
20. Vassilevska, V., Williams, R., Woo, S.L.M.: Confronting hardness using a hybrid approach. In: SODA, pp. 1–10 (2006)

On Polynomial Kernels for Structural Parameterizations of Odd Cycle Transversal*

Bart M.P. Jansen and Stefan Kratsch

Utrecht University, The Netherlands
{bart,kratsch}@cs.uu.nl

Abstract. The ODD CYCLE TRANSVERSAL problem (OCT) asks whether a given graph can be made bipartite (i.e., 2-colorable) by deleting at most ℓ vertices. We study structural parameterizations of OCT with respect to their polynomial kernelizability, i.e., whether instances can be efficiently reduced to a size polynomial in the chosen parameter. It is a major open problem in parameterized complexity whether ODD CYCLE TRANSVERSAL admits a polynomial kernel when parameterized by ℓ.

On the positive side, we show a polynomial kernel for OCT when parameterized by the vertex deletion distance to the class of bipartite graphs of treewidth at most w (for any constant w); this generalizes the parameter feedback vertex set number (i.e., the distance to a forest).

Complementing this, we exclude polynomial kernels for OCT parameterized by the distance to outerplanar graphs, conditioned on the assumption that NP $\not\subseteq$ coNP/poly. Thus the bipartiteness requirement for the treewidth w graphs is necessary. Further lower bounds are given for parameterization by distance from cluster and co-cluster graphs respectively, as well as for WEIGHTED OCT parameterized by the vertex cover number (i.e., the distance from an independent set).

1 Introduction

ODD CYCLE TRANSVERSAL (OCT), also called GRAPH BIPARTIZATION, is the task of making an undirected graph bipartite by deleting as few vertices as possible; such a set is a transversal of the odd-length cycles in the graph. The OCT problem has applications in computational biology [27,29], amongst others. It is NP-complete and admits a polynomial-time $\mathcal{O}(\sqrt{\log n})$-factor approximation algorithm [1]; no constant-factor approximation is possible unless Khot's Unique Games Conjecture fails [20,29].

In this work we study the parameterized complexity [9] of OCT, focusing on data reduction and kernelization. Parameterized analysis measures the complexity of an algorithm in two dimensions, the input size $|x|$ and an additional *parameter* $k \in \mathbb{N}$ which expresses some property of the instance, such as the size of the desired solution. A parameterized problem is a language $Q \subseteq \Sigma^* \times \mathbb{N}$, and Q is (strictly uniformly) *fixed-parameter tractable* (FPT) if there is an algorithm

---

* This work was supported by the Netherlands Organization for Scientific Research (NWO), project "KERNELS: Combinatorial Analysis of Data Reduction".

D. Marx and P. Rossmanith (Eds.): IPEC 2011, LNCS 7112, pp. 132–144, 2012.

that decides whether $(x, k) \in Q$ with running time bounded by $f(k)|x|^{\mathcal{O}(1)}$ for some computable function f.

In the standard parameterization of ODD CYCLE TRANSVERSAL which we call ℓ-OCT, the parameter $k := \ell$ measures the number of allowed vertex deletions ℓ: an instance is a tuple $((G, \ell), k := \ell)$ where G is a graph and $\ell \in \mathbb{N}$, and the question is whether there is a set $S \subseteq V(G)$ of size at most ℓ such that $G - S$ is bipartite. The ℓ-OCT problem has been very important to the development of parameterized algorithmics, since the algorithm given by Reed, Smith and Vetta [26] to solve ℓ-OCT in $\mathcal{O}(4^\ell \ell mn)$ time[1] introduced the technique of *iterative compression* which has turned out to be a key ingredient in finding FPT algorithms for DIRECTED FEEDBACK VERTEX SET [7] and MULTICUT [6,24], amongst others. There has been a significant amount of work on improved exact and parameterized algorithms for OCT and related problems [25,14,11,16,19,23].

Kernelization is an important subfield of parameterized complexity which studies polynomial-time preprocessing [15]. A *kernelization algorithm* (or *kernel*) for a parameterized problem Q is a polynomial-time algorithm which transforms an input $(x, k) \in \Sigma^* \times \mathbb{N}$ into an *equivalent* reduced instance (x', k') such that $|x'|, k' \leq f(k)$ for some computable function f, which is called the *size* of the kernel. All problems in FPT admit kernels for some suitable function f, but *polynomial kernels* (where $f(k) \in k^{\mathcal{O}(1)}$) are of particular interest. It is a famous open problem whether or not ℓ-OCT admits a polynomial kernel [16,14]. At the 2010 workshop on kernelization WORKER, this was stated as one of the two main open problems in kernelization to date (recent work of Kratsch and Wahlström [21] gives a randomized polynomial kernelization). Even finding a polynomial kernel for ℓ-OCT restricted to planar graphs was listed as an open problem by Bodlaender et al. in the full version of their work [3], despite the fact that planarity makes it significantly easier to obtain polynomial kernels.

Our Contribution. We study the existence of polynomial kernels for various *structural* parameterizations of the OCT problem. While we have not been able to settle the question of whether ℓ-OCT admits a polynomial kernel, we do give several upper- and lower bound results for kernel sizes that we believe are important steps towards resolving the main problem. All parameterized problems we consider fit into the following scheme, where $\mathcal{F}$ is a class of graphs:

Odd Cycle Transversal parameterized by vertex-deletion distance to $\mathcal{F}$ [$(\mathcal{F})$-OCT]
Input: A graph G, an integer ℓ and a set X such that $G - X \in \mathcal{F}$.
Parameter: $k := |X|$.
Question: Is there a set $S \subseteq V(G)$ of size at most ℓ such that $G - S$ is bipartite?

We give kernelization upper- and lower bounds for such parameterized problems.

Upper bounds. Our initial goal was to study OCT parameterized by the size of a feedback vertex set (FVS) of the input graph. Recall that a FVS can be defined

---

[1] Hüffner [16] re-analyzed the algorithm and showed it has time complexity $\mathcal{O}(3^\ell \ell mn)$.

as a set of vertices whose deletion turns the graph into a forest, and hence this is the (FOREST)-OCT problem. After having obtained a polynomial kernel for this problem, we considered generalizations and were able to extend our result significantly. Let BIP denote the class of all bipartite graphs, and let $\mathcal{G}_{\mathrm{TW}(w)}$ denote the graphs of treewidth at most w. It is well-known that FOREST $=$ BIP $\cap \mathcal{G}_{\mathrm{TW}(1)}$. We extended our result for feedback vertex number by showing that for every constant w, the problem (BIP $\cap \mathcal{G}_{\mathrm{TW}(w)}$)-OCT has a polynomial kernel. Using an approximation algorithm to compute the set X we can even drop the requirement that the set X is given in the input; the size of the reduced instance will then be bounded polynomially in the minimum-size of such a set X. Our result can therefore be stated as follows: for every fixed $w \geq 1$ there is a polynomial-time algorithm that transforms an instance (G, ℓ) of OCT into an equivalent instance whose size is bounded by a polynomial in $|X|$, where $X \subseteq V(G)$ is a smallest vertex set such that $G - X \in$ BIP $\cap \mathcal{G}_{\mathrm{TW}(w)}$.

We believe that the ingredients of our kernelization will be useful for solving the main open problem of whether ℓ-OCT admits a polynomial kernel. Our kernel uses several powerful techniques from the area of parameterized algorithmics; here is a brief overview. We introduce an annotated version of the problem and show that using these annotations the problem essentially reduces to a connectivity problem with respect to the vertices of the deletion set X. We give a lemma which shows that the main structure of the problem instance lies within an $|X|^{\mathcal{O}(1)}$-sized set of connected components of the graph $G - X$. Using a technique originating in the study of protrusion-based kernelization [3] we show the fact that $G - X \in \mathcal{G}_{\mathrm{TW}(w)}$ implies that the number of vertices from $V(G) \setminus X$ on the boundary of such regions can be bounded by a constant. We analyze the structure of a solution inside such a region in terms of combinatorial properties of separators in labeled graphs. Using the concept of *important separators* as introduced by Marx [22] we prove that the number of separators which are relevant to the problem can be bounded polynomially in $|X|$. To obtain the polynomial kernel we then show how to get rid of vertices which do not belong to any relevant separator.

Lower bounds. As described in the previous paragraph, we show the existence of polynomial kernels for (BIP $\cap \mathcal{G}_{\mathrm{TW}(w)}$)-OCT. We can also prove that the bipartite-ness condition cannot be dropped (under a reasonable complexity-theoretic assumption). Observe that $\mathcal{G}_{\mathrm{TW}(1)}$ coincides with the class of forests, and hence only contains bipartite graphs. But $\mathcal{G}_{\mathrm{TW}(2)}$ is the first class of bounded-treewidth graphs which contains non-bipartite graphs, and we prove using cross-composition [4] that $(\mathcal{G}_{\mathrm{TW}(2)})$-OCT does not admit a polynomial kernel unless NP $\subseteq$ coNP/poly, which implies a collapse of the polynomial-time hierarchy to the third level (PH $= \Sigma_3^p$) and further. We actually prove that (OUTERPLANAR)-OCT does not admit a polynomial kernel under this assumption, which is a stronger statement since OUTERPLANAR $\subseteq \mathcal{G}_{\mathrm{TW}(2)}$. We also show that if we take $\mathcal{F}$ to be a class of non-bipartite but very simply structured graphs (such as cluster graphs, the union of cliques, or their edge-complements co-cluster graphs) then we cannot obtain polynomial kernels: (CLUSTER)-OCT and (CO-CLUSTER)-OCT do not admit polynomial kernels unless NP $\subseteq$ coNP/poly. Since (co)cluster graphs have a very limited structure, the vertex-deletion distance to these graph classes will often be very large; our results show that even for such a large parameter one should not expect to find a polynomial kernel. Finally we look at the vertex-weighted version of OCT and prove that in the presence of vertex weights we cannot even obtain a polynomial kernel measured by the vertex deletion distance to an edgeless graph: (EDGELESS)-WEIGHTED ODD CYCLE TRANSVERSAL (which is equivalent to WEIGHTED ODD CYCLE TRANSVERSAL parameterized by the cardinality of a vertex cover) does not admit a polynomial kernel unless NP $\subseteq$ coNP/poly. All parameterizations for which we prove kernel lower bounds can be seen to be fixed-parameter tractable because the classes $\mathcal{F}$ have bounded cliquewidth [8] and therefore the cliquewidth of the input graphs is bounded by a function of the parameter.

Related Work. Recent work of Kratsch and Wahlström [21] gives a randomized polynomial kernel for ℓ-OCT, using matroid theory. However, this result is not combinatorial in the sense of not providing actual reduction rules. Instead, instances are converted into a (small) matroid representation relating to the vertex cuts between certain sets of terminal vertices. To the best of our knowledge no deterministic (and combinatorial) polynomial kernel is known for ℓ-OCT or for any non-trivial parameterizations of the OCT problem. Wernicke [29] used several reduction rules for OCT as part of his branch-and-bound algorithm, but these rules were not analyzed within the framework of kernelization and do not give provable bounds on the size of reduced instances with respect to any graph parameter. Kernelization with respect to structural parameterizations has been studied by a handful of authors, e.g., [10,4,5,17,28].

Organization. We start by giving some preliminaries. In Section 3 we give combinatorial bounds for separators in labeled graphs, which will be used in the kernelization algorithm. Section 4 presents the polynomial kernel for (BIP $\cap$ $\mathcal{G}_{\mathrm{TW}(w)}$)-OCT. We briefly discuss the kernelization lower bounds in Section 5 and conclude in Section 6.

2   Preliminaries

All graphs considered in this work are simple, undirected, and finite. If G is a graph then $V(G)$ and $E(G)$ denote the vertex- and edge set, respectively. We let *length* and *parity of a path* refer to the number of its vertices. For a vertex $v \in V(G)$ the open neighborhood is denoted by $N_G(v)$ and the closed neighborhood is $N_G[v] := N_G(v) \cup \{v\}$. The open neighborhood of a set $S \subseteq V(G)$ is $N_G(S) := \bigcup_{v \in S} N_G[v] \setminus S$. The graph $G - S$ is the result of removing all vertices in S and their incident edges from G. We use $[n]$ as a shorthand for $\{1, \ldots, n\}$. The term $\binom{X}{n}$ denotes the collection of all size-n subsets of the finite set X, whereas $\binom{X}{\leq n}$ represents the collection of size *at most* n subsets of X. The sizes of these collections are denoted by $\binom{|X|}{n}$ and $\binom{|X|}{\leq n}$, respectively.

3 Combinatorial Properties of Separators in Labeled Graphs

An important part of our kernelization relies on a combinatorial bound on the number of essentially distinct ways to separate terminals from labeled vertices in a graph: we prove that if the number of terminals and the size of the separators is taken as a constant, then the number of distinct ways to separate the labels grows polynomially with the number of labels. We believe this to be of independent interest. Some definitions are needed to formalize these claims.

Definition 1. *A* labeled graph *is a tuple* (G, L, f) *where* G *is a graph,* L *is a finite set of labels, and* $f\colon V(G) \to 2^L$ *is a labeling function which assigns to each vertex a (possibly empty) subset of the labels. For a subset* $S \subseteq V(G)$ *and terminal* $t \in V(G)$ *we denote by* $R(t, S)$ *the vertices of* $V(G) \setminus S$ *reachable from* t *in* $G - S$. *The labels reachable from* t *in* $G - S$ *are* $\mathcal{L}(t, S) := \bigcup_{v \in R(t,S)} f(v)$.

Definition 2. *Let* (G, L, f) *be a labeled graph and let* $T = t_1, \ldots, t_n$ *be a sequence of distinct terminal vertices in* G. *The* cut characteristic $\mathcal{K}(S, T)$ *of a set* $S \subseteq V(G)$ *with respect to the terminals* T *is an* n-*dimensional vector* $\mathcal{K}(S, T) := (\mathcal{L}(t_1, S), \mathcal{L}(t_2, S), \ldots, \mathcal{L}(t_n, S))$ *whose elements are subsets of* L. *The set of* distinct cut characteristics $\mathcal{K}^m(T)$ *for separators of size at most* $m \geq 1$ *is* $\mathcal{K}^m(T) := \left\{ \mathcal{K}(S, T) \,\middle|\, S \in \binom{V(G)}{\leq m} \right\}$.

Marx [22] introduced the notion of *important separators*, and proved their number to be bounded, independently of the graph size. An involved argument which relates important separators to distinct cut characteristics yields the following theorem; the interested reader is referred to the full version [18].

Theorem 1. *Let* $\kappa(n, m, r)$ *denote the maximum of* $|\mathcal{K}^m(T)|$ *over all labeled graphs* (G, L, f) *with* $|L| \leq r$ *and over all sets of terminals* $T = \{v_1, v_2, \ldots, v_n\} \subseteq V(G)$, *i.e., the maximum number of distinct cut characteristics induced by* m-*vertex separators in an* n-*terminal graph labeled with* r *different labels. Then* $\kappa(n, m, r) \in \mathcal{O}(m^{2n} \cdot r^{nm(m+3)/2} \cdot 4^{nm})$, *which is polynomial in* r *for fixed* n, m.

4 Polynomial Kernelization for (BIP ∩ $\mathcal{G}_{\mathrm{TW}(w)}$)-OCT

In this section we describe our polynomial kernelization for (BIP ∩ $\mathcal{G}_{\mathrm{TW}(w)}$)-OCT. Note that the definition of (BIP ∩ $\mathcal{G}_{\mathrm{TW}(w)}$)-OCT assumes a deletion set to be given in the input, and our kernelization will relate to its size. We will discuss the approximability of the deletion set at the end of the section, which will extend our kernelization to the case that no deletion set is given.

To simplify the formulation of the reduction process, we will actually work with an annotated version of the problem. To obtain the final reduced instance we will later undo these annotations at a small cost.

Annotated (BIP $\cap$ $\mathcal{G}_{\text{TW}(w)}$)-OCT

Input: A graph G, a set $X \subseteq V(G)$ such that $G - X \in (\text{BIP} \cap \mathcal{G}_{\text{TW}(w)})$, a set $M \subseteq \binom{X}{2}$, and an integer ℓ.
Parameter: $k := |X|$.
Question: Is there a set $S \subseteq V(G)$ of size at most ℓ such that $G - S$ is bipartite, and there is a proper 2-coloring c of $G - S$ such that $c(p) = c(q)$ for all $\{p, q\} \in M$?

We call vertex pairs $\{p, q\} \in M$ *monochromatic*, and these annotations allow us to easily talk about vertices which are constrained to have the same color in $G - S$. Observe that the dual notion, vertices $p, q \in X$ which must receive different colors in the bipartite graph $G - S$, is expressed simply through the existence of an edge $\{p, q\}$. We will therefore refer to vertices $p, q \in X$ which are adjacent as vertices to be annotated as *bichromatic*. There is no reason a priori that a pair $\{p, q\}$ cannot be constrained to be simultaneously bichromatic and monochromatic; this condition implies that any valid solution has to delete at least one vertex of the pair before a proper coloring 2-coloring can be found. A coloring is said to *respect all annotations* if it respects all edges between vertices of X as well as the monochromatic pairs given by the set M.

The following straightforward lemma will be used in a number of proofs throughout this section. It shows that any partial 2-coloring of a graph whose uncolored parts are bipartite can either be extended to a 2-coloring of the whole graph, or one finds a path between two already colored vertices whose parity does not match their colors (e.g., the path has an odd number of internal vertices but the color of the endpoints is different). Due to space restrictions all proofs of this section are deferred to the full version [18].

Lemma 1. *Let G be a graph, let $S \subseteq V(G)$ be such that $G - S$ is bipartite, and let $c \colon S \to \{0, 1\}$ be a proper 2-coloring of $G[S]$. Then in polynomial time one finds either an extension of c to a proper 2-coloring of G, or a connected component C of $G - S$ and vertices $p, q \in N_G(C) \subseteq S$ as well as a $p - q$ path P such that either*

- *P has an odd number of internal vertices and $c(p) \neq c(q)$, or*
- *P has an even number of internal vertices and $c(p) = c(q)$.*

Furthermore, all internal vertices of P are from $V(G) \setminus S$ and P is simple except possibly for $p = q$ (in the latter case P is in fact an odd cycle through p).

Now, for instructive purposes, consider an instance (G, X, M, ℓ) of the annotated problem and assume that there is a connected component C of $G - X$ such that the parity of all paths between vertices of X which run through C matches annotations: e.g., if there is an odd $p - q$ path, $p, q \in X$, with internal vertices from C then p and q are annotated as monochromatic, $\{p, q\} \in M$ (resp. for an even path we already have $\{p, q\} \in E(G)$). Since C is bipartite, Lemma 1 now implies that any 2-coloring of $G[X]$ that respects all annotations can be extended to a proper 2-coloring of $G[X \cup V(C)]$, i.e., extended onto C.

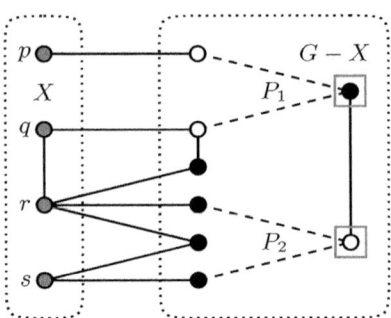

Fig. 1. A graph G and an odd cycle transversal X. Suppose $M = \{\{r, s\}\}$. The dashed path P_1 is an important $p - q$ X-path. The dashed path P_2 is a non-important $r - s$ X-path. Further, the two vertices marked by gray boxes intersect all important X-paths.

Thus, since the components of $G - X$ are already bipartite, we are only interested in paths between vertices of X that they provide, in particular in paths that do not match annotations. The following definition formalizes these as X-*paths* and *important* X-*paths*; see also Figure 1.

Definition 3. *An X-path of length r between (not necessarily distinct) vertices $p, q \in X$ in an instance of the annotated problem is a simple path $P = \{v_1, \ldots, v_r\}$ in $G - X$ such that there are distinct edges $\{p, v_1\}, \{v_r, q\} \in E(G)$. A $p - q$ X-path is important if (a) its length is odd, $p \neq q$, and $\{p, q\} \notin M$, or (b) its length is even and $\{p, q\} \notin E(G)$.*

Observe that the definition of a $p - q$ X-path excludes the possibility where $p = q$ and the odd $p - p$ X-path P consists of only one vertex $v_1 = v_r$, because in that case the edges $\{p, v_1\}$ and $\{v_r, q\}$ are not distinct.

With the following lemma we begin to explore the structure of the important X-paths. Given a graph G and a set X such that $G - X$ is bipartite we count vertex-disjoint odd and even length $p - q$ X-paths for all $p, q \in X$. For each pair and parity the lemma will provide in polynomial time either a small hitting set intersecting all important X-paths, or point out that the number of paths exceeds our budget of ℓ vertex deletions (this is indicated by the sets A, B, and C which will later be turned into annotations, edges, and vertex deletions); Algorithm 2 shows this in detail. We remark that both the lemma and the algorithm can also be applied to any other parameterization of OCT, given that X is a deletion set to any class of bipartite graphs (it is easy to see that $\ell < |X|$ in all interesting cases); in particular it can be applied to the standard parameterization whose kernelizability is still open.

Lemma 2. *Let G be a graph, ℓ be an integer, and $X \subseteq V(G)$ such that $G - X$ is bipartite. Then ComputeHittingSet(G, X, ℓ) computes sets $A, B \subseteq \binom{X}{2}$, a set $C \subseteq X$, and a set $H \subseteq V(G) \backslash X$ of size at most $4\ell \cdot |X|^2$ such that for all $\{u, v\} \in \binom{X}{2}$:*

1. *If $\{u, v\} \in A$ (resp. $\{u, v\} \in B$) then there are at least $\ell + 1$ vertex-disjoint X-paths of even (odd) length between u and v.*

Algorithm 1. VertexCut$(G, P, Q; u, v, S, T)$

Input: A graph G such that $G[P \cup Q]$ is bipartite with bipartition $P \cup Q$, vertices $u, v \in V(G) \setminus (P \cup Q)$, and sets $S, T \in \{P, Q\}$.

Output: A cut $Y \subseteq P \cup Q$ separating $N_G(u) \cap S$ from $N_G(v) \cap T$ in $G[P \cup Q] - Y$.

Let $G' := G[P \cup Q]$
Add a source s with $N_{G'}(s) := N_G(u) \cap S$ and a sink t with $N_{G'}(t) := N_G(v) \cap T$
Compute a minimum-size $s - t$ vertex-cut Y in G' using a flow algorithm
return Y

---

2. The set H intersects all even (odd) length $u - v$ X-paths with $\{u, v\} \notin A$ (resp. $\{u, v\} \notin B$).

If $v \in C$ then there are at least $\ell + 1$ vertex-disjoint even $v - v$ X-paths (i.e., odd cycles that intersect only in v), and H intersects all such paths for $v \in X \setminus C$.

Now, let us see how to turn the sets A, B, and C into annotations, edges, and vertex deletions such that H is a hitting set for all important X-paths in the resulting annotated instance, i.e., H will intersect each important X-path.

Lemma 3. Let (G, X, ℓ) be an instance of (BIP $\cap \mathcal{G}_{\text{TW}(w)}$)-OCT and let $A, B \subseteq \binom{X}{2}$, let $C \subseteq X$, and let $H \subseteq V(G) \setminus X$ as given by Lemma 2. Then in polynomial time one can find an equivalent instance (G', X', M, ℓ') with $X' \subseteq X$ and $\ell' \leq \ell$ of the annotated problem such that H intersects all important X'-paths in G'.

We will now turn our attention to the relation between the connected components of $G - X$ and the set H intersecting all important X-paths. It is obvious that no component of $(G - X) - H$ contains an important X-path. However, to use the fact that each such path needs to leave the component via a vertex of H and cross at least one other component before returning to X, we need to restrict the number of neighbors that any such component has in H. This is also the point from which on we need to use that $G - X$ has bounded treewidth. The following lemma, following along the lines of the protrusion partitioning lemma [3, Lemma 2] of Bodlaender et al., permits us to extend the set H slightly while decreasing the neighborhood size of the components obtained.

Lemma 4. Let G be a graph, let $\mathcal{T}$ be a tree decomposition of G of width w, and let $S \subseteq V(G)$. There is a polynomial-time algorithm that, given $(G, \mathcal{T}, S)$, computes a superset $S' \supseteq S$ of size at most $2(w + 1)|S|$ such that for each connected component C of $G - S'$ it holds that $|N_G(C) \cap S'| \leq 2w$.

The following lemma bounds the number of components of $(G - X) - H$, regardless of the structure of the set H; similar but simpler than Lemma 2.

Lemma 5. Let (G, X, M, ℓ) be an instance of the annotated problem and let H be a set of vertices of G. By deleting connected components of $(G - X) - H$ one can in polynomial time create an equivalent instance (G', X, M, ℓ) such that $(G' - X) - H$ has at most $2 \cdot (\ell + 1) \cdot (|X| + |H|)^2$ connected components.

Algorithm 2. ComputeHittingSet(G, X, ℓ)

Input: A graph G and vertex subset $X \subseteq V(G)$ such that $G - X$ is bipartite.
Output: Three sets of annotations A, B, and C as well as a hitting set H.

> Initialize $H, A, B, C := \emptyset$
> Let $P \cup Q$ be a bipartition of $G - X$ {Computable by BFS}
> **for each** $\{u, v\} \in \binom{X}{2}$ **do**
> $PP := \text{VertexCut}(G, P, Q; u, v, P, P)$
> $QQ := \text{VertexCut}(G, P, Q; u, v, Q, Q)$
> $PQ := \text{VertexCut}(G, P, Q; u, v, P, Q)$
> $QP := \text{VertexCut}(G, P, Q; u, v, Q, P)$
>   **if** $|PQ| > \ell$ or $|QP| > \ell$ **then** {$> \ell$ disjoint even-length $u - v$ X-paths}
> $A := A \cup \{\{u, v\}\}$
>   **else** {Set $PQ \cup QP$ intersects all even-length $u - v$ X-paths}
> $H := H \cup (PQ \cup QP)$
>   **if** $|PP| > \ell$ or $|QQ| > \ell$ **then**
> $B := B \cup \{\{u, v\}\}$
>   **else** {Set $PP \cup QQ$ intersects all odd-length $u - v$ X-paths}
> $H := H \cup (PP \cup QQ)$
> **for each** $v \in X$ **do**
> $PQ := \text{VertexCut}(G, P, Q; v, v, P, Q)$
>   **if** $|PQ| > \ell$ **then** {$> \ell$ disjoint even-length $v - v$ X-paths}
> $C := C \cup \{v\}$
>   **else** {Set PQ intersects all even-length $v - v$ X-paths}
> $H := H \cup PQ$
> **return** (A, B, C, H)

With the next lemma, we prepare the ground for applying the combinatorial bounds on the number of cut characteristics in labeled graphs. It formalizes and proves the fact that we may freely modify any given odd cycle transversal by replacing its intersection with a connected component with a separator of the same cut characteristic. It is crucial that all important paths must intersect the hitting set H and that each component is adjacent to only few vertices of H; the hitting set will correspond to terminals of certain labeled graphs, whose labels express adjacency to X.

Lemma 6 (Separator replacement lemma). *Let (G, X, M, ℓ) be an instance of the annotated problem. Let $H \subseteq V(G) \setminus X$ be a set of vertices that intersects all important X-paths of the instance. Let R be a solution to the problem, i.e., an odd cycle transversal such that $G - R$ has a proper 2-coloring respecting the annotations. Consider a connected component C of the graph $(G - X) - H$ and consider the terminal vertices $N_G(C) \setminus X$. Define D as the subgraph of G induced by the set $N_G[C] \setminus X$. Let $T = t_1, \ldots, t_n$ be a sequence containing the terminals $N_G(C) \setminus X$ in an arbitrary order. We define a labeling for the graph D as follows. The set of labels is the set of vertices in the modulator X augmented with one label per terminal in T, and the labeling function f is defined as follows for $v \in V(D)$:*

$$f(v) := \begin{cases} N_G(v) \cap X & \text{If } v \notin T. \\ (N_G(v) \cap X) \cup \{v\} & \text{If } v \in T. \end{cases}$$

Let $S := V(C) \cap R$ be the vertices from C chosen in the solution R. If $S' \subseteq V(C)$ is a subset such that S and S' have the same cut characteristic in the labeled graph $(D, X \cup T, f)$ with respect to the terminals T, then $R' := (R \setminus S) \cup S'$ is also a valid solution, or more formally: if $\mathcal{K}(S, T) = \mathcal{K}(S', T)$ with respect to the labeled graph $(D, X \cup T, f)$ then $G - R'$ has a proper 2-coloring respecting the annotations.

Now, we will use the Separator Replacement Lemma and the combinatorial bound on the number of cut characteristics (Theorem 1) to limit the choice of vertices that may be deleted from the connected components. The idea is that it suffices to have one separator for each cut characteristic; vertices outside these separators need not be considered for deletion. To this end we introduce a restricted version of the annotated odd cycle transversal problem. As an additional restriction a set Z of vertices is provided, and the task is to find a (small) odd cycle transversal that is a subset of Z.

Restricted Annotated (BIP $\cap \mathcal{G}_{\mathrm{TW}(w)}$)-OCT
Input: A graph G, a set $X \subseteq V(G)$ such that $G - X \in (\mathrm{BIP} \cap \mathcal{G}_{\mathrm{TW}(w)})$, a set $Z \subseteq V(G)$ of deletable vertices, a set $M \subseteq \binom{X}{2}$, and an integer ℓ.
Parameter: $k := |X|$.
Question: Is there a set $S \subseteq Z$ of size at most ℓ such that $G - S$ is bipartite, and there is a proper 2-coloring c of $G - S$ such that $c(p) = c(q)$ for all $\{p, q\} \in M$?

Lemma 7. *Let (G, X, M, ℓ) be an instance of* ANNOTATED *(*BIP $\cap \mathcal{G}_{\mathrm{TW}(w)}$*)-*OCT *and let $H \subseteq V(G) \setminus X$ be a set of vertices such that:*

1. *H intersects all important X-paths of G,*
2. *$(G - X) - H$ has at most α connected components,*
3. *and each connected component of $(G - X) - H$ has at most δ neighbors in H.*

For each fixed value of δ it is possible to compute in polynomial time an equivalent instance (G, X, M, ℓ, Z) of RESTRICTED ANNOTATED *(*BIP $\cap \mathcal{G}_{\mathrm{TW}(w)}$*)-*OCT *where $|Z| \leq |X| + |H| + \alpha \cdot \delta \cdot \kappa(\delta, \delta - 1, |X| + \delta)$, with κ as defined in Theorem 1.*

This final lemma provides the reduction from the restricted annotated problem back to (BIP$\cap \mathcal{G}_{\mathrm{TW}(w)}$)-OCT. The number of vertices in Z in the restricted instance determines the size of the vertex set in the new (equivalent) instance.

Lemma 8. *An instance (G, X, M, ℓ, Z) of* RESTRICTED ANNOTATED *(*BIP $\cap \mathcal{G}_{\mathrm{TW}(w)}$*)-*OCT *can be transformed in polynomial time into an equivalent instance (G', X', ℓ) of (*BIP $\cap \mathcal{G}_{\mathrm{TW}(w)}$*)-*OCT *with $|V(G')|$ bounded by $|Z| + (\ell + 1) \cdot |Z|^2$.*

Now we can wrap up our kernelization with the following theorem. The kernelization follows the lemmas and motivation given so far.

Theorem 2. *For each fixed integer $w \geq 1$ the problem (*BIP$\cap \mathcal{G}_{\mathrm{TW}(w)}$*)-*OCT *admits a polynomial kernel with $\mathcal{O}(k^{\mathcal{O}(w^3)})$ vertices.*

Approximating a Minimum-Size Deletion Set. For our kernelization we have assumed that a deletion set X to the class ($\text{BIP} \cap \mathcal{G}_{\text{TW}(w)}$) is given. If G is a graph for which the minimum size of such a deletion set is OPT, then we can compute in polynomial time a deletion set of size $\mathcal{O}(\text{OPT} \cdot \log^{3/2} \text{OPT})$ as follows. Observe that $\mathcal{G}_{\text{TW}(w)}$ is characterized by a finite set of forbidden minors, and excludes at least one planar graph as a minor. We can use the recent approximation algorithm by Fomin et al. [12] to approximate a deletion set $S_{\text{TW}(w)}$ to a graph of treewidth at most w. Then we may find a minimum-size odd cycle transversal S_{OCT} in the bounded-treewidth graph $G - S_{\text{TW}(w)}$ which can be computed in polynomial time using Courcelle's theorem, since w is a constant. The union $X := S_{\text{TW}(w)} \cup S_{\text{OCT}}$ is then a suitable deletion set, which we can use to run our kernelization. This procedure is formalized in the following lemma.

Lemma 9. *Let $w \geq 1$ be a fixed integer. There is a polynomial-time algorithm which gets as input a graph G, and computes a set $X \subseteq V(G)$ such that $G - X \in \text{BIP} \cap \mathcal{G}_{\text{TW}(w)}$ with $|X| \in \mathcal{O}(\text{OPT} \cdot \log^{3/2} \text{OPT})$, where OPT is the minimum size of such a deletion set.*

5   Lower Bounds for Kernelization

In this section we state the lower bound results for various structural kernelizations of OCT. All results use the recent notion of cross-composition introduced by Bodlaender et al. [4]. Cross-composition is a frontend to the lower bound framework via compositions based on work of Bodlaender et al. [2] as well as Fortnow and Santhanam [13]. It extends the notion of a composition, showing that a reduction of the OR of any NP-hard problem into an instance of the target parameterized problem with *small* parameter value excludes polynomial kernels, assuming that NP $\not\subseteq$ coNP/poly.

The following theorem states the obtained kernelization lower bounds (with proofs deferred to the full version [18]) ; all problems are FPT by having bounded cliquewidth, as briefly discussed in the introduction.

Theorem 3. *Assuming NP $\not\subseteq$ coNP/poly the following parameterized problems do not admit polynomial kernels:*

- (OUTERPLANAR)-OCT,
- (CLUSTER)-OCT,
- (CO-CLUSTER)-OCT,
- WEIGHTED ODD CYCLE TRANSVERSAL PARAMETERIZED BY THE SIZE OF A VERTEX COVER.

6   Conclusion

We have studied the existence of polynomial kernels for structural parameterizations of OCT. We have shown that in polynomial time the size of an instance (G, ℓ) of OCT can be reduced to a polynomial in the minimum number of vertex deletions needed to transform G into a bipartite graph of constant treewidth. We also gave several kernelization lower bounds when the parameter measures the vertex-deletion distance to a non-bipartite graph with a simple structure. These lower bounds show that even for very large parameters such as the deletion distance to a cluster graph, it is unlikely that OCT admits a polynomial kernel.

The important open problem remains to determine whether the natural parameterization ℓ-OCT admits a deterministic polynomial kernel. Encouraged by the recent randomized kernelization result [21], we believe this to be the case. We think that several components we introduced in this work, such as the notion of important X-paths and the algorithm to find a small hitting set for these paths, will be useful ingredients for a deterministic kernelization. These ingredients do not rely on our structural parameterization and are therefore directly applicable to the general ℓ-OCT problem.

References

1. Agarwal, A., Charikar, M., Makarychev, K., Makarychev, Y.: O(sqrt(log n)) approximation algorithms for min uncut, min 2cnf deletion, and directed cut problems. In: Proc. 37th STOC, pp. 573–581 (2005)
2. Bodlaender, H.L., Downey, R.G., Fellows, M.R., Hermelin, D.: On problems without polynomial kernels. J. Comput. Syst. Sci. 75(8), 423–434 (2009)
3. Bodlaender, H.L., Fomin, F.V., Lokshtanov, D., Penninkx, E., Saurabh, S., Thilikos, D.M.: (Meta) Kernelization. In: Proc. 50th FOCS, pp. 629–638 (2009)
4. Bodlaender, H.L., Jansen, B.M.P., Kratsch, S.: Cross-composition: A new technique for kernelization lower bounds. In: Proc. 28th STACS, pp. 165–176 (2011)
5. Bodlaender, H.L., Jansen, B.M.P., Kratsch, S.: Preprocessing for Treewidth: A Combinatorial Analysis through Kernelization. In: Aceto, L., Henzinger, M., Sgall, J. (eds.) ICALP 2011. LNCS, vol. 6755, pp. 437–448. Springer, Heidelberg (2011)
6. Bousquet, N., Daligault, J., Thomassé, S.: Multicut is FPT. In: Proc. 43rd STOC, pp. 459–468 (2011)
7. Chen, J., Liu, Y., Lu, S., O'Sullivan, B., Razgon, I.: A fixed-parameter algorithm for the directed feedback vertex set problem. J. ACM 55(5) (2008)
8. Courcelle, B., Makowsky, J.A., Rotics, U.: Linear time solvable optimization problems on graphs of bounded clique-width. Theory Comput. Syst. 33(2), 125–150 (2000)
9. Downey, R., Fellows, M.R.: Parameterized Complexity. Monographs in Computer Science. Springer, New York (1999)
10. Fellows, M.R., Lokshtanov, D., Misra, N., Mnich, M., Rosamond, F.A., Saurabh, S.: The complexity ecology of parameters: An illustration using bounded max leaf number. Theory Comput. Syst. 45(4), 822–848 (2009)
11. Fiorini, S., Hardy, N., Reed, B.A., Vetta, A.: Planar graph bipartization in linear time. Discrete Applied Mathematics 156(7), 1175–1180 (2008)
12. Fomin, F.V., Lokshtanov, D., Misra, N., Philip, G., Saurabh, S.: Hitting forbidden minors: Approximation and kernelization. In: Proc. 28th STACS, pp. 189–200 (2011)
13. Fortnow, L., Santhanam, R.: Infeasibility of instance compression and succinct PCPs for NP. J. Comput. Syst. Sci. 77(1), 91–106 (2011)

14. Guo, J., Gramm, J., Hüffner, F., Niedermeier, R., Wernicke, S.: Compression-based fixed-parameter algorithms for feedback vertex set and edge bipartization. J. Comput. Syst. Sci. 72(8), 1386–1396 (2006)
15. Guo, J., Niedermeier, R.: Invitation to data reduction and problem kernelization. SIGACT News 38(1), 31–45 (2007)
16. Hüffner, F.: Algorithm engineering for optimal graph bipartization. J. Graph Algorithms Appl. 13(2), 77–98 (2009)
17. Jansen, B.M.P., Bodlaender, H.L.: Vertex cover kernelization revisited: Upper and lower bounds for a refined parameter. In: Proc. 28th STACS, pp. 177–188 (2011)
18. Jansen, B.M.P., Kratsch, S.: On polynomial kernels for structural parameterizations of odd cycle transversal. CoRR, abs/1107.3658 (2011)
19. Kawarabayashi, K., Reed, B.A.: An (almost) linear time algorithm for odd cyles transversal. In: Proc. 21st SODA, pp. 365–378 (2010)
20. Khot, S.: On the power of unique 2-prover 1-round games. In: Proc. 34th STOC, pp. 767–775 (2002)
21. Kratsch, S., Wahlström, M.: Compression via matroids: A randomized polynomial kernel for odd cycle transversal. CoRR, abs/1107.3068 (2011)
22. Marx, D.: Parameterized graph separation problems. Theoretical Computer Science 351(3), 394–406 (2006); Parameterized and Exact Computation
23. Marx, D., O'Sullivan, B., Razgon, I.: Treewidth reduction for constrained separation and bipartization problems. In: Proc. 27th STACS, pp. 561–572 (2010)
24. Marx, D., Razgon, I.: Fixed-parameter tractability of multicut parameterized by the size of the cutset. In: Proc. 43rd STOC, pp. 469–478 (2011)
25. Raman, V., Saurabh, S., Sikdar, S.: Improved Exact Exponential Algorithms for Vertex Bipartization and other Problems. In: Coppo, M., Lodi, E., Pinna, G.M. (eds.) ICTCS 2005. LNCS, vol. 3701, pp. 375–389. Springer, Heidelberg (2005)
26. Reed, B.A., Smith, K., Vetta, A.: Finding odd cycle transversals. Oper. Res. Lett. 32(4), 299–301 (2004)
27. Rizzi, R., Bafna, V., Istrail, S., Lancia, G.: Practical Algorithms and Fixed-Parameter Tractability for the Single Individual SNP Haplotyping Problem. In: Guigó, R., Gusfield, D. (eds.) WABI 2002. LNCS, vol. 2452, pp. 29–43. Springer, Heidelberg (2002)
28. Uhlmann, J., Weller, M.: Two-Layer Planarization Parameterized by Feedback Edge Set. In: Kratochvíl, J., Li, A., Fiala, J., Kolman, P. (eds.) TAMC 2010. LNCS, vol. 6108, pp. 431–442. Springer, Heidelberg (2010)
29. Wernicke, S.: On the algorithmic tractability of single nucleotide polymorphism (SNP) analysis and related problems. Master's thesis, Wilhelm-Schickard-Institut für Informatik, Universität Tübingen (2003)

Kernel Bounds for Path and Cycle Problems*

Hans L. Bodlaender, Bart M.P. Jansen, and Stefan Kratsch

Utrecht University, P.O. Box 80.089, 3508 TB Utrecht, The Netherlands
{hansb,bart,kratsch}@cs.uu.nl

Abstract. Connectivity problems like k-PATH and k-DISJOINT PATHS relate to many important milestones in parameterized complexity, namely the Graph Minors Project, color coding, and the recent development of techniques for obtaining kernelization lower bounds. This work explores the existence of polynomial kernels for various path and cycle problems, by considering nonstandard parameterizations. We show polynomial kernels when the parameters are a given vertex cover, a modulator to a cluster graph, or a (promised) max leaf number. We obtain lower bounds via cross-composition, e.g., for HAMILTONIAN CYCLE and related problems when parameterized by a modulator to an outerplanar graph.

1 Introduction

Connectivity problems such as k-PATH and k-DISJOINT PATHS play important theoretical and practical roles in the field of parameterized complexity. On the practical side, k-PATH [16, ND29] has applications in computational biology [22] where the involved parameter is fairly small, thus giving an excellent opportunity to apply parameterized algorithms to find optimal solutions. On the theoretical side, these problems have triggered the development of very powerful algorithmic techniques. The k-DISJOINT PATHS problem [21] lies at the heart of the Graph Minors Algorithm, and is the source of the *irrelevant-vertex* technique. The *color coding* technique of Alon et al. [1] to solve k-PATH has found a wide range of applications and extensions [20,7], and new methods of solving k-PATH are still developing [2]. Despite the success stories of parameterized algorithms for these problems, the quest for polynomial kernels has resulted in mostly negative results. Indeed, the failure to find a polynomial kernel for k-PATH was one of the main motivations for the development of the kernelization lower-bound framework of Bodlaender et al. [3]. Using the framework it was shown that k-PATH does not admit a polynomial kernel unless $NP \subseteq coNP/poly$, even when restricted to very specific graph classes such as planar cubic graphs. It did not take long before related connectivity problems such as k-DISJOINT PATHS [6], k-DISJOINT CYCLES [6], k-CONNECTED VERTEX COVER [10], and restricted variants of k-CONNECTED DOMINATING SET [9] were also shown not to admit polynomial kernels unless $NP \subseteq coNP/poly$.

* This work was supported by the Netherlands Organization for Scientific Research (NWO), project "KERNELS: Combinatorial Analysis of Data Reduction".

D. Marx and P. Rossmanith (Eds.): IPEC 2011, LNCS 7112, pp. 145–158, 2012.
© Springer-Verlag Berlin Heidelberg 2012

Thus it seems that connectivity requirements in a problem form a barrier to polynomial kernelizability when it comes to the natural parameterization by solution size k. Driven by the desire to obtain useful preprocessing procedures for such problems, we may therefore investigate the kernelization complexity for nonstandard parameters. Early work by Fellows et al. [13] shows that such a different perspective can yield polynomial kernels: they proved that HAMILTO-NIAN CYCLE parameterized by the max leaf number of the input graph G, i.e., the maximum number of leaves in a spanning tree for G, admits a linear-vertex kernel. In this work we study the existence of polynomial kernels for various structural parameters such as the max leaf number, the size of a vertex cover, and the vertex-deletion distance to simple graph classes such as cluster graphs and outerplanar graphs. Our results:

1. We introduce a widely applicable technique based on matchings in bipartite graphs to show that the problems LONG CYCLE, its directed and path variants, DISJOINT PATHS, and DISJOINT CYCLES, admit kernels with $\mathcal{O}(|X|^2)$ vertices when parameterized by a vertex cover X.
2. For LONG CYCLE and LONG PATH we generalize to the stronger parameter "vertex-deletion distance to a cluster graph" (see Fig. 1) and obtain a polynomial kernel. An essential step in this kernelization is the use of a weighted version of the problem, using the observation that either the binary representation of the weights has size polynomial in the problem parameter, or we can solve the problem in polynomial time.
3. Using the same binary encoding trick we give a polynomial kernel for LONG CYCLE parameterized by the max leaf number, generalizing the result of Fellows et al. [13] for HAMILTONIAN CYCLE.
4. We give contrasting kernelization lower bounds using the recently introduced technique of cross-composition [4]: (a) DIRECTED HAMILTONIAN CYCLE PARAMETERIZED BY A MODULATOR TO BI-PATHS does not admit a polynomial kernel unless NP $\subseteq$ coNP/poly, where the parameter measures the vertex-deletion distance to a digraph whose underlying undirected graph is a path, and (b) we modify the construction to prove that HAMILTONIAN CYCLE PARAMETERIZED BY A MODULATOR TO OUTERPLANAR GRAPHS does not admit a polynomial kernel; both results assuming NP $\not\subseteq$ coNP/poly. These results carry over to LONG PATH, LONG CYCLE and related variants.
5. We initiate the parameterized complexity study of finding paths respecting forbidden pairs [16, GT54] under various parameterizations. We obtain W[1]-hardness proofs, FPT algorithms, kernel lower bounds and para-NP-completeness results.

Related Work. Chen, Flum and Müller studied various forms of kernelization lower bounds, and showed amongst others that k-POINTED PATH (with given startpoint) does not admit a parameter non-increasing polynomial kernelization unless P = NP, and that k-PATH does not have a polynomial kernel on connected planar graphs unless NP $\subseteq$ coNP/poly [8]. Very recently, Hermelin et al. [18] gave evidence that k-COLORED PATH does not have a polynomial Turing kernel by proving it complete for the class WK[1] of kernelization hardness.

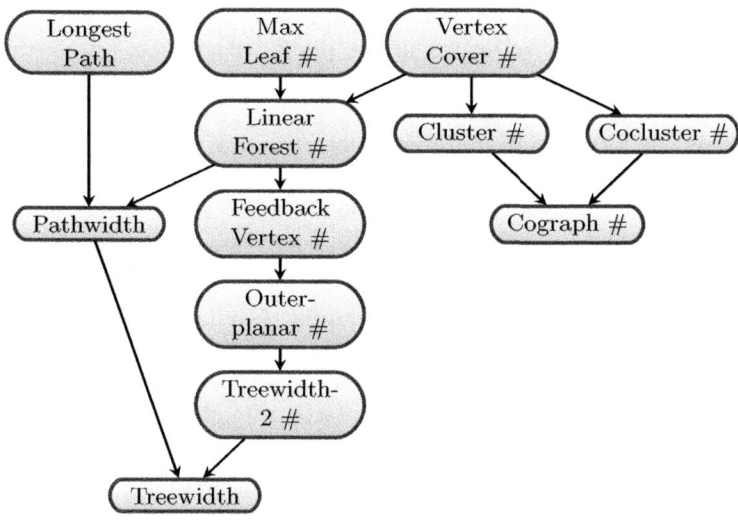

Fig. 1. The hierarchy of parameters used in this work. Arrows point from larger parameters to smaller parameters: an arc $P \rightarrow P'$ signifies that every graph G satisfies $P(G) + 2 \geq P'(G)$. For a graph class $\mathcal{G}$, the vertex-deletion distance from G to $\mathcal{G}$ (written as $\mathcal{G}$ #) is the minimum number of vertices whose removal from G results in a graph in $\mathcal{G}$. The vertex cover # is the size of a minimum vertex cover. Feedback vertex # and max leaf # are defined analogously.

Structural Parameterizations. Throughout this paper we use structural parameterizations of the path and cycle problems under consideration: we choose a graph parameter, and ask whether the size of an instance can efficiently be reduced to a polynomial in this parameter without changing the answer. To understand the relevance of our results it is important to consider the relationships between various graph parameters. Fig. 1 therefore organizes the relevant parameters in a hierarchy. Section 2 contains definitions of various graph classes.

Most proofs had to be omitted from this extended abstract due to space restrictions. They can be found in the full version of this work [5].

2 Preliminaries

Graphs. All graphs are finite and simple, unless indicated otherwise. An undirected graph G has a vertex set $V(G)$ and an edge set $V(G) \subseteq \binom{V(G)}{2}$. A directed graph D has a vertex set $V(D)$ and a set of directed arcs $A(D) \subseteq V(D)^2$. The minimum size of a vertex cover in a graph G is denoted by $\mathrm{VC}(G)$. A matching M in a graph covers a set of vertices U if each vertex in U is endpoint of an edge in M. For a digraph D and vertex v we write $N_D^+(v)$ and $N_D^-(v)$ for the in-neighbors and out-neighbors of v, respectively. The underlying undirected graph of a digraph D is the result of disregarding the orientation of the arcs and eliminating parallel edges. Let BI-PATHS (for bi-orientations of paths) be the class of digraphs whose underlying undirected graph is a path. Outerplanar graphs are those graphs which can be drawn in the plane without crossings such that all the vertices lie on the outer face; such graphs have treewidth at most two. Cluster graphs are disjoint unions of cliques; their edge-complements are cocluster graphs. A linear forest is an undirected forest of maximum degree at most two, i.e., a collection of paths. For a graph class $\mathcal{G}$ and a vertex set $X \subseteq V(G)$ of a graph G such that $G - X \in \mathcal{G}$, we say that X is a *modulator* to the class $\mathcal{G}$. We use $[n]$ as a shorthand for $\{1, 2, \ldots, n\}$. If X is a finite set then $\binom{X}{n}$ denotes the set of all size-n subsets of X.

Parameterized Complexity and Kernels. A parameterized problem Q is a subset of $\Sigma^* \times \mathbb{N}$, the second component being the *parameter* which expresses some structural measure of the input. A parameterized problem is (strongly uniform) *fixed-parameter tractable* if there exists an algorithm to decide whether $(x, k) \in Q$ in time $f(k)|x|^{\mathcal{O}(1)}$ where f is a computable function [11].

A *kernelization algorithm* (or *kernel*) for a parameterized problem Q is a polynomial-time algorithm which transforms an instance (x, k) into an equivalent instance (x', k') such that $|x'|, k' \leq f(k)$ for some computable function f, which is the *size* of the kernel. If f is a polynomial then this is a *polynomial kernel* [17].

Cross-Composition. To prove our lower bounds we use the framework of cross-composition [4], which builds on earlier work by Bodlaender et al. [3], and Fortnow and Santhanam [15].

3   A Property of Maximum Matchings in Bipartite Graphs

The following theorem simplifies the correctness proofs of our reduction rules.

Theorem 1. *Let $G = (X \cup Y, E)$ be a bipartite graph. Let $M \subseteq E(G)$ be a maximum matching in G. Let $X_M \subseteq X$ be the set of vertices in X that are endpoint of an edge in M. Then, for each $Y' \subseteq Y$, if there exists a matching M' in G that covers Y', then there exists a matching M'' in $G[X_M \cup Y]$ that covers Y'.*

Proof. Let G, M, and X_M be as stated in the theorem. Suppose the theorem does not hold for $Y' \subseteq Y$, and let M' be a matching in G that covers Y'. Over all such matchings M', take one that covers the largest number of vertices in X_M. By assumption M' is not a matching in $G[X_M \cup Y]$, so there is a vertex $y_0 \in Y'$ that is matched in M' to a vertex in $X \setminus X_M$, say x_0. We use an iterative process to derive a contradiction, maintaining the following invariants:

- $x_0 \notin X_M$.
- $\{x_j, y_j\} \in M'$ for $0 \leq j \leq i$.
- $\{y_j, x_{j+1}\} \in M$ for $0 \leq j < i$.
- The vertices x_j for $0 \leq j \leq i$ are distinct members of X, and vertices y_j for $0 \leq j \leq i$ are distinct members of Y.

It is easy to verify that given our choice of x_0, y_0 these invariants are initially satisfied for $i = 0$. We now continue the process based on a case distinction:

1. If y_i is not matched under M, then the sequence $(x_0, y_0, \ldots, x_i, y_i)$ is an M-augmenting path in G since x_0 and y_i are not matched under M, and all edges $\{y_j, x_{j+1}\}$ for $0 \leq j < i$ are contained in M. Hence $M'' := M \setminus \{\{y_j, x_{j+1}\} \mid 0 \leq j < i\} \cup \{\{x_j, y_j\} \mid 0 \leq j \leq i\}$ is a matching in G larger than M, contradicting that M is maximum.

2. In the remaining cases we may assume y_i is matched under M, say $\{y_i, x_{i+1}\} \in M$. If there is an index $0 \leq j \leq i$ such that $x_{i+1} = x_j$ then $j > 0$ (since $x_0 \notin X_M$) and the edges $\{y_i, x_{i+1}\}$ and $\{y_{j-1}, x_j\}$ are both contained in M and are distinct edges since $y_{j-1} \neq y_i$, contradicting the fact that M is a matching. Hence x_{i+1} is distinct from x_j for $0 \leq j \leq i$.

3. If x_{i+1} is not covered by M' then the matching $M'' := M' \setminus \{\{x_j, y_j\} \mid 0 \leq j \leq i\} \cup \{\{y_j, x_{j+1}\} \mid 0 \leq j \leq i\}$ contains as many edges as M' but covers more vertices of X_M, contradicting the choice of M'. Hence x_{i+1} is covered by M', say $\{x_{i+1}, y_{i+1}\} \in M'$. If there is an index $0 \leq j \leq i$ such that $y_{i+1} = y_j$ then $\{x_{i+1}, y_{i+1}\}$ and $\{x_j, y_j\}$ are two distinct edges in M' incident on y_{i+1}, contradicting that M' is a matching. Hence y_{i+1} is distinct from y_j for $0 \leq j \leq i$. Now observe that the invariant holds for $i + 1$, and we may proceed with the next step of the process.

By the last property of the invariant, the process must end. Hence the assumption that there is no matching in $G[X_M \cup Y]$ which covers Y' leads to a contradiction, which concludes the proof. □

4   Polynomial Kernels for Path and Cycle Problems

4.1   Long Cycle Parameterized by a Vertex Cover

In this section, we consider the LONG CYCLE problem parameterized by the size ℓ of a given vertex cover and present a kernel with $\mathcal{O}(\ell^2)$ vertices.

LONG CYCLE PARAMETERIZED BY A VERTEX COVER
Input: A graph G, an integer k, and a vertex cover $X \subseteq V(G)$ (which implies that $G - X$ is an independent set).
Parameter: $\ell := |X|$.
Question: Does G contain a cycle of length at least k?

We need only one reduction rule to get a kernelization, it uses a bipartite *connection graph* $H = H(G, k, X)$: One color class consists of the vertices in the independent set $I = V(G) \setminus X$, and the other consists of all (unordered) pairs of distinct vertices in X. We take an edge from a vertex $v \in I$ to a vertex representing the pair $\{p, q\} \subseteq X$, if and only if v is adjacent to p and to q.

Reduction Rule 1. *Given (G, k, X), if $k \leq 4$ then solve the problem (e.g. by the trivial $\mathcal{O}(n^4)$ algorithm) and return an equivalent dummy instance. Otherwise, construct the connection graph $H = H(G, k, X)$. Let M be a maximum matching in H. Let $J \subseteq I$ be the vertices touched by an edge in M. Remove all vertices in $I \setminus J$ and their incident edges from G. Let G' be the resulting graph, and return the instance (G', k, X).*

Observation 1. *In Rule 1, $|J|$ is at most the number of pairs of distinct vertices in X, and hence after applying the rule, G' has at most $\ell + \binom{\ell}{2} \in \mathcal{O}(\ell^2)$ vertices.*

Correctness of the rule follows from the following lemma.

Lemma 1. *Let (G, k, X) be an instance of* LONG CYCLE PARAMETERIZED BY A VERTEX COVER, *and let (G', k, X) be the instance returned by Rule 1. Then G has a cycle of length at least k if and only if G' has a cycle of length at least k.*

Proof. If $k \leq 4$ then the lemma holds trivially. Otherwise, we have that G' is an induced subgraph of G so cycles (in particular those of length at least k) in G' exist also in G. It remains to look at the converse.

Let C be a cycle of length at least $k \geq 5$ in G. Clearly, as $I = V \setminus X$ is an independent set, any vertices of I which are in C must be neighbored by vertices of X on C. Let $v_1, \ldots, v_r$ be all vertices of I contained in C and let p_i and q_i be the predecessor and successor of v_i on C, respectively (clearly $r \leq \ell$ but there might be far fewer vertices of I on C). Since C has length at least 5, it follows that $\{p_i, q_i\} \neq \{p_j, q_j\}$ for all $i, j \in [r]$ with $i \neq j$ (else it would have length 4). To show that G' contains a cycle of length at least k, it suffices to find replacements for all vertices v_i which are not in J (and hence not in G'); for this we will use the matching.

Clearly, in $H = H(G, k, X)$ there is a matching M covering $W := \{\{p_1, q_1\}, \ldots, \{p_r, q_r\}\}$, namely matching each pair to the corresponding vertex v_i. Further, by Rule 1, J is the set of endpoints in I of some maximum matching of H. Hence, by Theorem 1, there is a matching M' covering W in $H[J \cup W]$.

Let v_i' denote the vertex matched to $\{p_i, q_i\}$ by M', for $i \in \{1, \ldots, r\}$. It is easy to see that we may replace each v_i on C by v_i' since v_i' is adjacent to p_i and q_i in G, obtaining a cycle C' which intersects I only in vertices of J. Also, as all pairs $\{p_i, q_i\}$ are different, no vertex v_i' is required twice. Hence, C' is also a cycle of G', and of length at least k. □

The kernelization result now follows from Lemma 1 and Observation 1, and noting that Rule 1 can be easily performed in polynomial time.

Theorem 2. LONG CYCLE PARAMETERIZED BY A VERTEX COVER *has a kernel with $\mathcal{O}(\ell^2)$ vertices.*

4.2 Other Path and Cycle Problems Parameterized by Vertex Cover

The same technique can be used for a number of additional problems, all parameterized by the size of a vertex cover. For LONG PATH, DISJOINT PATHS, and DISJOINT CYCLES polynomial kernels with $\mathcal{O}(\ell^2)$ vertices with respect to the size ℓ of a given vertex cover can be obtained. The basic argument is that the matching strategy allows us to reroute any paths or cycles such that they use only matched vertices (which are kept).

For HAMILTONIAN PATH and HAMILTONIAN CYCLE it is easy to see that any vertex cover of a YES-instance must have size at least least $\lfloor \frac{|V|}{2} \rfloor$, since vertices of the remaining independent set cannot be adjacent on a Hamiltonian path or cycle. Thus all nontrivial instances (G, X) have $|V(G)| \leq 2|X| + 1$.

4.3   Parameterization by Max Leaf Number

In this section we consider path and cycle problems parameterized by the max leaf number, i.e., the maximum number of leaves in any spanning tree of the graph. Deviating slightly from the standard use, we will take the max leaf number of a disconnected graph to be the sum of max leaf numbers taken over all connected components. We will use LONG CYCLE as a running example, but as in Section 4.2 it is easy to generalize the arguments to further problems. As the max leaf number of a graph cannot be verified in polynomial time, we consider the parameterization in the sense of a promise problem, e.g.:

LONG CYCLE PARAMETERIZED BY MAX LEAF NUMBER (LCML)
Input: A graph G and two integers k and ℓ.
Parameter: ℓ.
Question: If G has max leaf number at most ℓ, then decide whether G contains a cycle of length at least k. Else the output may be arbitrary.

Although we need the concept of a promise-problem to cast LCML in a proper formal setting, knowing the exact value is not needed in practice. We can devise an algorithm that reduces the size of an instance with max leaf number ℓ to $\mathrm{poly}(\ell)$, by using our kernel with an approximation algorithm for MAX LEAF.

It is well known that a large graph having small max leaf number must contain long paths of degree two vertices and few vertices of degree at least three. The following bound was obtained by Fellows et al. [13] based on work by Kleitman and West [19]; it can be easily seen to hold for each connected component.

Lemma 2 ([13]). *If a graph G has max leaf number at most ℓ then it is a subdivision of some graph H of at most $4\ell - 2$ vertices. In particular, G has at most $4\ell - 2$ vertices of degree at least three.*

It is not hard to devise an FPT-algorithm for LCML.

Lemma 3. LONG CYCLE PARAMETERIZED BY MAX LEAF NUMBER *can be solved in time $2^{\mathcal{O}(\ell)} n^c$.*

The main idea for the kernelization is that one of two good cases must hold: Either all the path lengths are small enough such that a binary encoding of their length has size polynomial in ℓ, or the total number n of vertices is large enough such that the $2^{\mathcal{O}(\ell)} n^c$ is in fact polynomial in n.

Theorem 3. LONG CYCLE PARAMETERIZED BY MAX LEAF NUMBER *admits a polynomial kernel.*

Proof. Given an instance (G, k, ℓ) of LCML, we first check that k does not exceed the number of vertices and that there are at most 4ℓ vertices of degree at least three, or else return NO. If G has more than $2^{\mathcal{O}(\ell)}$ vertices (using the concrete bound resulting from an implementation of Lemma 3), then we solve the instance in time $2^{\mathcal{O}(\ell)} \cdot n^c = \mathcal{O}(n^{c+1})$, and answer YES or NO accordingly. Otherwise let B denote the set of vertices of degree at least three. If there are more than ℓ disjoint paths connecting any two vertices of B, then the max leaf number of G exceeds ℓ, and we return NO.

We replace each path connecting two vertices $b, b' \in B$, with internal vertices from $V(G) \setminus B$, by a single edge with an integer label denoting the number of internal vertices of the replaced path. We obtain a multigraph G' in which some edges have an integer label. It is easy to see that cycles in G correspond to cycles in G' of the same length, when taking the integer labels into account (i.e. labeled edges are simply worth as much as that many internal vertices). Clearly, each label can be encoded in binary by at most $\log 2^{\mathcal{O}(\ell)} = \mathcal{O}(\ell)$ bits. Furthermore, we delete all paths that start in a vertex of B, have internal vertices from $V(G) \setminus B$, and end in a vertex of degree one; clearly those cannot be used by cycles in G.

We obtain a multigraph G' with at most 4ℓ vertices in B and with at most ℓ edges between any two B-vertices. Thus we have $\mathcal{O}(\ell)$ vertices and $\mathcal{O}(\ell^3)$ integer labels of size $\mathcal{O}(\ell)$, for a total size of $\mathcal{O}(\ell^4)$ (this could be easily tightened, but it would not affect the result); clearly k can also be encoded in $\mathcal{O}(\ell)$ bits.

We obtain an equivalent instance of a slightly different LONG CYCLE problem on multigraphs in which some edges may be labeled, but which is in NP. By the implied Karp reduction to LCML we obtain the claimed polynomial kernel (cf. [6]). Deviating from Bodlaender et al. [6] we do not use the versions with parameter encoded in unary, but observe the following: All instances of LONG CYCLE with k exceeding the number of vertices are trivially NO and may be replaced by smaller dummy NO-instances, so the parameter value of the remaining instances is indeed polynomial in ℓ (as is the instance size, due to the Karp reduction).   □

Further Problems. A polynomial kernel for HAMILTONIAN CYCLE was already found by Fellows et al. [13]. Kernels for HAMILTONIAN PATH as well as DISJOINT CYCLES can be obtained in a similar way, by observing that the paths of degree-2 vertices can be reduced to having only one internal vertex. For LONG PATH it is again necessary to use the binary encoding trick for the path lengths.

Corollary 1. LONG PATH, HAMILTONIAN PATH, *and* DISJOINT CYCLES *parameterized by max leaf number admit polynomial kernels.*

For DISJOINT PATHS, i.e., finding k disjoint paths connecting k terminal pairs $(s_1, t_1), \ldots, (s_k, t_k)$, some more work is necessary on the paths between B vertices, and on paths between B vertices and the at most ℓ leaves.

Theorem 4. DISJOINT PATHS *parameterized by max leaf number admits a polynomial kernel.*

4.4   Parameterization by a Modulator to Cluster Graphs

In this section, we consider path and cycle problems parameterized by vertex-deletion distance from cluster graphs. To this end, alongside the input graph (and possibly further inputs) a modulator X is provided such that $G - X$ is a cluster graph. Technically this requires both a marking (or matching) strategy to identify a small set of important vertices and cliques together with the encoding trick, used in the previous section, to handle large cliques. For space reasons, we only state the main result of the section.

Theorem 5. LONG CYCLE PARAMETERIZED BY A MODULATOR TO CLUSTER GRAPHS *admits a polynomial kernel.*

5 Lower Bounds for Path and Cycle Problems

In this section we present kernelization lower bounds for the directed- and undirected variants of HAMILTONIAN CYCLE with structural parameters. The parameterizations we use are at least as large as the treewidth of the input graphs (or the underlying undirected graph, in the directed case) which shows that the parameterized problems for which we prove a kernel lower bound are indeed contained in FPT. Our proofs use the technique of cross-composition [4], in which a kernel lower bound is obtained by showing that the logical OR of a series of instances of an NP-hard problem, can be embedded in a single instance of the parameterized target problem at a small parameter cost.

5.1 Directed Hamiltonian Cycle with a Modulator to Bi-paths

We start by defining the NP-hard problem which we will use in the cross-composition.

> HAMILTONIAN $s - t$ PATH IN DIRECTED BIPARTITE GRAPHS
> **Input:** A bipartite digraph D with color classes $A = \{a_1, \ldots, a_{n_A}\}$ and $B = \{b_1, \ldots, b_{n_B}\}$ with $n_B = n_A + 1$ such that $N_D^-(b_1) = \emptyset$ and $N_D^+(b_{n_B}) = \emptyset$.
> **Question:** Does D contain a directed Hamiltonian path which starts in b_1 and ends in b_{n_B}?

It is not difficult to show that this problem is NP-complete. Now we formally define the parameterized problem for which we will prove a kernel lower bound.

> DIRECTED HAMILTONIAN CYCLE PARAMETERIZED BY A MODULATOR TO BI-PATHS
> **Input:** A digraph D and a modulator $X \subseteq V(D)$ such that $D - X \in$ BI-PATHS.
> **Parameter:** The size $|X|$ of the modulator.
> **Question:** Does D have a directed Hamiltonian cycle?

We can now give the cross-composition.

Theorem 6. DIRECTED HAMILTONIAN CYCLE PARAMETERIZED BY A MODULATOR TO BI-PATHS *does not admit a polynomial kernel unless NP $\subseteq$ coNP/poly.*

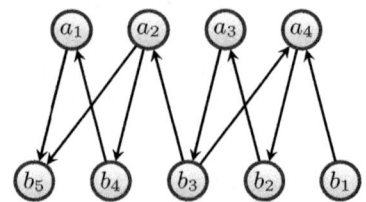

(a) Input instance (D_1, A_1, B_1) of HAMILTONIAN $s - t$ PATH IN DIRECTED BIPARTITE GRAPHS.

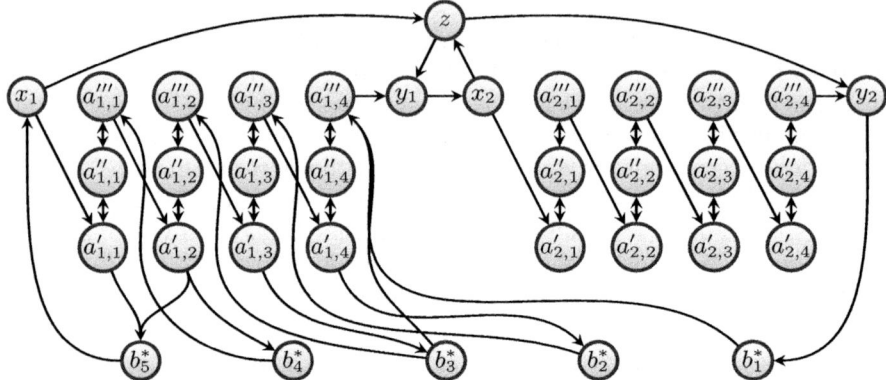

(b) Output instance of DIRECTED HAMILTONIAN CYCLE PARAMETERIZED BY A MODULATOR TO BI-PATHS.

Fig. 2. An example of the lower-bound construction of Theorem 6 when composing $r = 2$ inputs with $n_A = 4$ and $n_B = 5$. (a) The first input instance. (b) Resulting output instance. The arcs between $\{b_1^*, \ldots, b_5^*\}$ and $\{a_{2,j}', a_{2,j}'', a_{2,j}''' \mid j \in [4]\}$ which encode the second input (D_2, A_2, B_2) have been omitted for readability.

Proof. By an earlier result of the authors [4, Corollary 10] and the NP-completeness of the introduced classical problem, it is sufficient to show that HAMILTONIAN $s - t$ PATH IN DIRECTED BIPARTITE GRAPHS cross-composes into DIRECTED HAMILTONIAN CYCLE PARAMETERIZED BY A MODULATOR TO BIPATHS. We use a polynomial equivalence relationship (see [4, Definitions 3-4]) under which two well-formed instances (D_1, A_1, B_1) and (D_2, A_2, B_2) of HAMILTONIAN $s - t$ PATH IN DIRECTED BIPARTITE GRAPHS are equivalent if $|A_1| = |A_2|$ and $|B_1| = |B_2|$. It suffices to give an algorithm which composes a sequence of instances of HAMILTONIAN $s - t$ PATH IN DIRECTED BIPARTITE GRAPHS which are equivalent under $\mathcal{R}$ into one instance of DIRECTED HAMILTONIAN CYCLE PARAMETERIZED BY A MODULATOR TO BI-PATHS. In the remainder we may assume that the input contains r well-formed instances $(D_1, A_1, B_1), \ldots,$ (D_r, A_r, B_r), that $|A_i| = n_A$ and $|B_i| = n_B$ for $i \in [r]$ with $n_B = n_A + 1$. Label the vertices in each set A_i as $a_{i,1}, \ldots, a_{i,n_A}$ and the vertices of a set B_i as $b_{i,1}, \ldots, b_{i,n_B}$ for $i \in [r]$. Recall that instance i asks whether D_i has a Hamiltonian path from $b_{i,1}$ to b_{i,n_B}. We construct a digraph D^* as follows.

1. For $i \in [r]$, for $j \in [n_A]$ add vertices $a'_{i,j}, a''_{i,j}, a'''_{i,j}$ to D^*, and add arcs $(a'_{i,j}, a''_{i,j}), (a''_{i,j}, a'_{i,j}), (a''_{i,j}, a'''_{i,j}), (a'''_{i,j}, a''_{i,j})$.
2. As the next step we add one-directional arcs to connect adjacent triples. For $i \in [r]$, for $j \in [n_A - 1]$ add the arc $(a'''_{i,j}, a'_{i,j+1})$.
3. For each instance $i \in [r]$ add two special vertices x_i and y_i, together with arcs $(x_i, a'_{i,1})$ and (a'''_{i,n_A}, y_i). For $i \in [r-1]$ add the arcs (y_i, x_{i+1}).
4. Observe that at this stage, $D^* \in$ BI-PATHS. All vertices we add from this point on will go into the modulator X^* such that $D^* - X^*$ will be a member of BI-PATHS.
5. We add a special vertex z with arcs (x_i, z) and (z, y_i) for $i \in [r]$.
6. For $j \in [n_B]$ add a vertex b^*_j to the graph D^*, and let B^* be the set of these vertices. Add arcs (y_r, b^*_1) and $(b^*_{n_B}, x_1)$.
7. As the last step of the construction we re-encode the behavior of the input graphs D_i into the instance. For $i \in [r]$, for all arcs $(a_{i,j}, b_{i,h})$ in $A(D_i)$ add the arc (a'_j, b^*_h) to D^*. For all arcs $(b_{i,j}, a_{i,h}) \in A(D_i)$ add (b^*_j, a'''_h) to D^*. This concludes the description of D^*, which is illustrated in Fig. 2.

Now define $X^* := \{z\} \cup B^*$. The output of the cross-composition is the instance (D^*, X^*) of DIRECTED HAMILTONIAN CYCLE PARAMETERIZED BY A MODULATOR TO BI-PATHS. It is easy to verify that $D^* - X^* \in$ BI-PATHS, and that the construction can be carried out in polynomial time. The parameter $|X^*|$ is bounded by $1 + n_B$ which is sufficiently small. It remains to prove that D^* is YES if and only if one of the input instances is YES; the proof is deferred to the full version. □

The proof of Theorem 6 can be adapted to give a kernel lower bound for the variant where we are looking for a Hamiltonian path instead of a Hamiltonian cycle; these bounds in turn imply that the versions where we are looking for a long path or cycle (instead a Hamiltonian one) are at least as hard to kernelize, as is the case for finding a long $s - t$ path or a long cycle through a given vertex.

5.2 Hamiltonian Cycle with a Modulator to Outerplanar Graphs

Using a domino-type gadget to simulate the behavior of directed edges, we can modify the lower bound of the previous section to work for undirected graphs, at the expense of modulating to a slightly more complex graph class.

Theorem 7. HAMILTONIAN CYCLE PARAMETERIZED BY A MODULATOR TO OUTERPLANAR GRAPHS *admits no polynomial kernel unless* $NP \subseteq coNP/poly$.

6 Finding Paths with Respect to Forbidden Pairs

In this section we study multiple parameterizations of several path problems involving forbidden pairs. The first version we consider is defined as follows.

$s - t$ Path with Forbidden Pairs Parameterized by a Vertex Cover of G
Input: A graph G, distinct vertices $s, t \in V(G)$, a set $H \subseteq \binom{V(G)}{2}$ of forbidden pairs, and a vertex cover X of G.
Parameter: $\ell := |X|$.
Question: Is there an $s - t$ path in G which contains at most one vertex of each pair $\{u, v\} \in H$?

A straight-forward reduction from k-Multicolored Clique [12] shows that this problem is W[1]-hard. Let us now consider some related problems. In Shortest $s - t$ Path With Forbidden Pairs and Longest $s - t$ Path With Forbidden Pairs there is an extra integer k in the input and we are asking for an $s - t$ path containing at most or at least k vertices. In Longest Path With Forbidden Pairs we omit the inputs s and t, and are looking for *any* sufficiently long path, regardless of its endpoints. The related problem Shortest Path With Forbidden Pairs is not interesting, since its solution always consist of a path containing a single vertex.

The W[1]-hardness result easily carries over to all these variants. Clearly, the hardness of the path problems with forbidden pairs stems from the extra structure of the forbidden pairs H, which is not taken into account when considering structural parameters of G. In the following we consider the effect of parameterizing by the structure of the graph $G \cup H$ (i.e., G with an added edge for every forbidden pair).

Using the optimization version of Courcelle's Theorem applied to *structures* of bounded treewidth (cf. [14, Section 11.4]), it is not difficult to obtain an FPT result parameterized by the treewidth of $G \cup H$ by building a formula in Monadic Second Order Logic over an appropriate structure to test for the existence of an $s - t$ path respecting the forbidden pairs. Using standard extensions of MSOL we may also maximize or minimize the size of a set of edges which forms an $s - t$ path respecting forbidden pairs, extending the fixed-parameter tractability to the variants for short- and long paths.

For the case of Shortest $s - t$ Path With Forbidden Pairs the structure of G is actually not so important for the complexity of the problem: it is sufficient to parameterize by a vertex cover of the graph on the edge set H to obtain fixed-parameter tractability, by trying all ways in which the vertex set of the path could intersect the vertex cover. For Longest $s - t$ Path With Forbidden Pairs a parameterization by vc(H) is not fruitful, since the latter problem is already NP-complete when there are no forbidden pairs. We mention without proof that $s - t$ Path with Forbidden Pairs is NP-complete when the graph induced by H is a matching, showing that we cannot improve the parameterization by a vertex cover of H to the treewidth of H.

Finally let us consider the kernelization complexity of forbidden path problems. Using an intricate cross-composition we obtain a super-polynomial lower bound on the kernel size of $s - t$ Path with Forbidden Pairs Parameterized by a Vertex Cover of $G \cup H$. This hardness proof carries also carries over to the other problem variants. Table 1 contains a summary of the results.

Table 1. Complexity of path problems with forbidden pairs. Each column represents a different parameterization. F.P. abbreviates "WITH FORBIDDEN PAIRS". The classification "No poly" means "no polynomial kernel unless NP ⊆ coNP/poly", and "Para-NP-c" means "NP-complete for a constant value of the parameter". For a parameterization in FPT, we either list "FPT" or "No poly", depending on which of the two is more relevant: all parameterizations listed as "No poly" are in FPT, and none of the problems listed "FPT" admit polynomial kernels. SHORTEST PATH F.P. is trivially in P.

	$vc(G)$	$vc(H)$	$tw(H)$	$tw(G \cup H)$	$vc(G \cup H)$
$s - t$ PATH F.P.	W[1]-hard	FPT	Para-NP-c	FPT	No poly
SHORTEST $s - t$ PATH F.P.	W[1]-hard	FPT	Para-NP-c	FPT	No poly
LONGEST $s - t$ PATH F.P.	W[1]-hard	Para-NP-c	Para-NP-c	FPT	No poly
LONGEST PATH F.P.	W[1]-hard	Para-NP-c	Para-NP-c	FPT	No poly

7   Conclusion

In this work we have shown that for sufficiently strong structural parameterizations, many path and cycle problems admit polynomial kernels even though their natural parameterizations do not. The marking technique using bipartite matching yields quadratic-vertex kernels for many problems parameterized by the size of a vertex cover. We introduced a binary encoding trick which gives polynomial kernels for problems parameterized by the max leaf number. On the negative side, we also exhibited smaller structural parameters which provably do not lead to polynomial kernels for HAMILTONIAN CYCLE unless NP ⊆ coNP/poly. Let us reflect briefly on the parameters used for the upper- and lower bounds.

Recall that the vertex cover number of a graph can also be interpreted as the number of vertex-deletions needed to reduce the graph to an independent set, i.e., the vertex-deletion distance to a graph of treewidth 0. Hence Theorem 2 shows that LONG CYCLE admits a polynomial kernel parameterized by vertex-deletion distance to treewidth 0. On the other hand, Theorem 7 shows that if NP ⊄ coNP/poly then HAMILTONIAN CYCLE does *not* have a polynomial kernel parameterized by the deletion distance to treewidth two (since outerplanar graphs have treewidth at most two), and of course this carries over to the harder problem LONG CYCLE. It is interesting to settle what happens for treewidth one, i.e., forests: does HAMILTONIAN CYCLE parameterized by a feedback vertex set admit a polynomial kernel? To generalize the result of Theorem 5 by distance to a cluster graph, one could consider the distance to cographs.

The kernelization complexity of compound parameterizations remains largely unexplored: for example, how does the LONG CYCLE problem behave when parameterized by the solution size plus the vertex-deletion distance to an outerplanar graph? It follows from the work of Bodlaender, Thomassé and Yeo [6] that DISJOINT PATHS and DISJOINT CYCLES do not admit polynomial kernels parameterized by the target value k plus the deletion distance to a path. We hope that a search for polynomial kernels of structural parameterizations leads to reduction rules which are useful in practice.

References

1. Alon, N., Yuster, R., Zwick, U.: Color-coding. J. ACM 42(4), 844–856 (1995)
2. Björklund, A., Husfeldt, T., Kaski, P., Koivisto, M.: Narrow sieves for parameter-ized paths and packings. CoRR, abs/1007.1161 (2010)
3. Bodlaender, H.L., Downey, R.G., Fellows, M.R., Hermelin, D.: On problems without polynomial kernels. J. Comput. Syst. Sci. 75(8), 423–434 (2009)
4. Bodlaender, H.L., Jansen, B.M.P., Kratsch, S.: Cross-composition: A new technique for kernelization lower bounds. In: Proc. 28th STACS, pp. 165–176 (2011)
5. Bodlaender, H.L., Jansen, B.M.P., Kratsch, S.: Kernel bounds for path and cycle problems. CoRR, abs/1106.4141 (2011)
6. Bodlaender, H.L., Thomassé, S., Yeo, A.: Kernel Bounds for Disjoint Cycles and Disjoint Paths. In: Fiat, A., Sanders, P. (eds.) ESA 2009. LNCS, vol. 5757, pp. 635–646. Springer, Heidelberg (2009)
7. Chen, J., Lu, S., Sze, S.-H., Zhang, F.: Improved algorithms for path, matching, and packing problems. In: Proc. 18th SODA, pp. 298–307 (2007)
8. Chen, Y., Flum, J., Müller, M.: Lower bounds for kernelizations and other prepro-cessing procedures. Theory Comput. Syst. 48(4), 803–839 (2011)
9. Cygan, M., Pilipczuk, M., Pilipczuk, M., Wojtaszczyk, J.O.: Kernelization Hard-ness of Connectivity Problems in 2-Degenerate Graphs. In: Thilikos, D.M. (ed.) WG 2010. LNCS, vol. 6410, pp. 147–158. Springer, Heidelberg (2010)
10. Dom, M., Lokshtanov, D., Saurabh, S.: Incompressibility through Colors and IDs. In: Albers, S., Marchetti-Spaccamela, A., Matias, Y., Nikoletseas, S., Thomas, W. (eds.) ICALP 2009. LNCS, vol. 5555, pp. 378–389. Springer, Heidelberg (2009)
11. Downey, R., Fellows, M.R.: Parameterized Complexity. Monographs in Computer Science. Springer, New York (1999)
12. Fellows, M.R., Hermelin, D., Rosamond, F.A., Vialette, S.: On the parameterized complexity of multiple-interval graph problems. Theor. Comput. Sci. 410(1), 53–61 (2009)
13. Fellows, M.R., Lokshtanov, D., Misra, N., Mnich, M., Rosamond, F.A., Saurabh, S.: The complexity ecology of parameters: An illustration using bounded max leaf number. Theory Comput. Syst. 45(4), 822–848 (2009)
14. Flum, J., Grohe, M.: Parameterized Complexity Theory. Springer-Verlag New York, Inc. (2006)
15. Fortnow, L., Santhanam, R.: Infeasibility of instance compression and succinct PCPs for NP. J. Comput. Syst. Sci. 77(1), 91–106 (2011)
16. Garey, M.R., Johnson, D.S.: Computers and Intractability, A Guide to the Theory of NP-Completeness. W.H. Freeman and Company, New York (1979)
17. Guo, J., Niedermeier, R.: Invitation to data reduction and problem kernelization. SIGACT News 38(1), 31–45 (2007)
18. Hermelin, D., Kratsch, S., Soltys, K., Wahlström, M., Wu, X.: Hierarchies of inef-ficient kernelizability. CoRR, abs/1110.0976 (2011)
19. Kleitman, D.J., West, D.B.: Spanning trees with many leaves. SIAM J. Discret. Math. 4(1), 99–106 (1991)
20. Kneis, J., Mölle, D., Richter, S., Rossmanith, P.: Divide-and-Color. In: Fomin, F.V. (ed.) WG 2006. LNCS, vol. 4271, pp. 58–67. Springer, Heidelberg (2006)
21. Robertson, N., Seymour, P.D.: Graph minors. XIII. The disjoint paths problem. J. Comb. Theory, Ser. B 63(1), 65–110 (1995)
22. Scott, J., Ideker, T., Karp, R.M., Sharan, R.: Efficient algorithms for detecting signaling pathways in protein interaction networks. Journal of Computational Bi-ology 13(2), 133–144 (2006)

On the Hardness of Losing Width

Marek Cygan[1], Daniel Lokshtanov[2], Marcin Pilipczuk[1],
Michał Pilipczuk[1], and Saket Saurabh[3]

[1] Institute of Informatics, University of Warsaw, Poland
{cygan@,malcin@,mp248287@students.}mimuw.edu.pl
[2] University of California, San Diego, La Jolla, CA 92093-0404, USA
dlokshtanov@cs.ucsd.edu
[3] The Institute of Mathematical Sciences, Chennai - 600113, India
saket@imsc.res.in

Abstract. Let $\eta \geq 0$ be an integer and G be a graph. A set $X \subseteq V(G)$ is called a *η-transversal in G* if $G \setminus X$ has treewidth at most η. Note that a 0-transversal is a vertex cover, while a 1-transversal is a feedback vertex set of G. In the η/ρ-TRANSVERSAL problem we are given an undirected graph G, a ρ-transversal $X \subseteq V(G)$ in G, and an integer ℓ and the objective is to determine whether there exists an η-transversal $Z \subseteq V(G)$ in G of size at most ℓ. In this paper we study the kernelization complexity of η/ρ-TRANSVERSAL *parameterized* by the size of X. We show that for every fixed η and ρ that either satisfy $1 \leq \eta < \rho$, or $\eta = 0$ and $2 \leq \rho$, the η/ρ-TRANSVERSAL problem does not admit a polynomial kernel unless NP $\subseteq$ coNP/poly. This resolves an open problem raised by Bodlaender and Jansen in [STACS 2011]. Finally, we complement our kernelization lower bounds by showing that $\rho/0$-TRANSVERSAL admits a polynomial kernel for any fixed ρ.

Keywords: η-transversal, kernelization upper and lower bounds, polynomial parameter transformation.

1 Introduction

The last few years have seen a surge in the study of kernelization complexity of parameterized problems, resulting in a multitude of new results on upper and lower bounds for kernelization [1,2,6,7,9]. Bodlaender and Jansen [11] initiated the systematic study of the kernelization complexity of a problem parameterized by something else than the value of the objective function.

The problem (or parameter) that received the most attention in this regard is *vertex cover*. A vertex cover of a graph G is a vertex set S such that all edges of G have at least one endpoint in S, and the *vertex cover number* of G is the size of the smallest vertex cover in G. In the VERTEX COVER problem we are given a graph G and an integer k and asked whether the vertex cover number of G is at most k. Over the last year we have seen several studies of problems parameterized by the vertex cover number of the input graph [3,4,12], as well as a study of the VERTEX COVER problem parameterized by the size of the

D. Marx and P. Rossmanith (Eds.): IPEC 2011, LNCS 7112, pp. 159–168, 2012.

smallest feedback vertex set of the input graph G. A *feedback vertex set* of G is a set S such that $G \setminus S$ is acyclic and the feedback vertex number of G is the size of the smallest feedback vertex set in G.

The reason parameterizing VERTEX COVER by the feedback vertex number of the input graph is interesting is that while the feedback vertex number is always at most the vertex cover number, it can be arbitrarily smaller. In particular, in forests the feedback vertex number is zero, while the vertex cover number can be arbitrarily large. Hence a kernel of size polynomial in the feedback vertex number is always polynomial in the vertex cover number, yet it could also be much smaller. Bodlaender and Jansen [11] show that VERTEX COVER parameterized by the feedback vertex number admits a polynomial kernel. At this point a natural question is whether VERTEX COVER has a polynomial kernel when parameterized by even smaller parameters than the feedback vertex number of the input graph. Bodlaender and Jansen [11] ask a particular variant of this question; whether VERTEX COVER admits a polynomial kernel when parameterized by the size of the smallest ρ-transversal (see below) of the input graph, for any $\rho \geq 2$.

Definition 1. *Let $\eta \geq 0$ be an integer and G be a graph. A set $X \subseteq V(G)$ is called an η-transversal in G if $G \setminus X$ has treewidth at most η.*

Observe that a 0-transversals of G are vertex covers, while 1-transversals are feedback vertex sets. In the η-TRANSVERSAL problem we are given a graph G and integer ℓ and asked whether G has a η-transversal of size at most ℓ. In this paper we consider the kernelization complexity of η-TRANSVERSAL, when parameterized by the size of the smallest ρ-transversal of the input graph G, for fixed values of η and ρ. Specifically, we consider the following problem.

η/ρ-TRANSVERSAL	**Parameter:** $|X|$

Input: An undirected graph G, a ρ-transversal $X \subseteq V(G)$ in G, and an integer ℓ.
Question: Does there exist an η-transversal $Z \subseteq V(G)$ in G of size at most ℓ?

Fomin et al. [8] recently proved that ρ-transversal admits a $O((\log OPT)^{\frac{3}{2}})$ approximation. Therefore, we could relax the condition of giving the ρ-transversal X along with the graph, as the algorithm can always approximate this set. This shows equivalence of existence of polynomial kernels for η/η-TRANSVERSAL and the classical η-TRANSVERSAL parameterized by the solution size.

The result of Bodlaender and Jansen [11] can now be reformulated as follows; 0/1-TRANSVERSAL admits a polynomial kernel. We settle the kernelization complexity of η/ρ-TRANSVERSAL for a wide range of values of η and ρ. In particular we resolve the open problem of Bodlaender and Jansen [11] by showing that unless NP $\subseteq$ coNP/poly, 0/ρ-TRANSVERSAL does not admit a polynomial kernel for any $\rho \geq 2$. Finally, we complement our negative results by showing that $\rho/0$-TRANSVERSAL admits a polynomial kernel for every fixed ρ. A concise description of our results can be found in Table 1.

Table 1. Kernelization complexity of the η/ρ-TRANSVERSAL problem. YES means that the problem admits a polynomial kernel, NO means that the problem does not admit a polynomial kernel and ? means that the status of the kernelization complexity of the problem is unknown. Boldface indicates results proved in this paper.

$\eta \setminus \rho$	0	1	2	3	4	5	$\cdots$
0	YES	YES	NO	NO	NO	NO	NO $\cdots$
1	YES	YES	NO	NO	NO	NO	NO $\cdots$
2	**YES**	?	?	NO	NO	NO	NO $\cdots$
3	**YES**	?	?	?	NO	NO	NO $\cdots$
4	**YES**	?	?	?	?	NO	NO $\cdots$
5	**YES**	?	?	?	?	?	NO $\cdots$
$\vdots$	$\vdots$	$\vdots$	$\vdots$	$\vdots$	$\vdots$	$\vdots$	$\vdots \cdots$

The diagonal entries of the table - the η/η-TRANSVERSAL problems are particularly interesting. Note that 0/0-TRANSVERSAL and 1/1-TRANSVERSAL are equivalent to the classical VERTEX COVER and FEEDBACK VERTEX SET problems, respectively, parameterized by the solution size. Furthermore, let $\mathcal{F}$ be a finite set of graphs. In the $\mathcal{F}$-DELETION problem, we are given an n-vertex graph G and an integer k as input, and asked whether at most k vertices can be deleted from G such that the resulting graph does not contain any graph from $\mathcal{F}$ as a minor. It is well known that η/η-TRANSVERSAL can be thought of as a special case of the $\mathcal{F}$-DELETION problem, where $\mathcal{F}$ contains a planar graph. It is conjectured in [8] that $\mathcal{F}$-DELETION admits a polynomial kernel if and only if $\mathcal{F}$ contains a planar graph. Notice that a polynomial kernel for η/η-TRANSVERSAL automatically implies a polynomial kernel for η/ρ-TRANSVERSAL for $\eta \geq \rho$. The conjecture of [8] implies, if true, that η/η-TRANSVERSAL does admit polynomial kernel and that therefore, all the empty slots of Table 1 should be "YES".

Notation. All graphs in this paper are undirected and simple. For a graph G we denote its vertex set by $V(G)$ and edge set by $E(G)$. For a vertex $v \in V(G)$ we define its neighbourhood $N_G(v) = \{u : uv \in E(G)\}$ and closed neighbourhood $N_G[v] = N_G(v) \cup \{v\}$. If X is a set of vertices or edges of G, by $G \setminus X$ we denote the graph G with all vertices and edges in X deleted (when deleting a vertex, we delete its incident edges as well). We use a shortened notation $G \setminus v$ for $G \setminus \{v\}$. If $u, v \in V(G)$, $u \neq v$ and $uv \notin E(G)$, then $G \cup \{uv\}$ denotes the graph G with added edge uv. A set $S \subseteq V(G)$ is said to *separate* u from v, if $u, v \in V(G) \setminus S$ and u and v lie in different connected components of $G \setminus S$.

2 η-Transversal Parameterized by Vertex Cover

In this section we show that for any $\eta \geq 0$ the $\eta/0$-TRANSVERSAL problem has a kernel with $O(|X|^{\max(\eta+1,3)})$ vertices.

Let $\eta \geq 0$ be a fixed integer. We provide a set of reduction rules and assume that at each step we use an applicable rule with the smallest number. At each reduction rule we discuss its soundness, that is, we prove that the input and output instances are equivalent. All presented reductions can be applied in polynomial time in a trivial way. If no reduction rule can be used on an instance (G, X, ℓ), we claim that $|V(G)|$ is bounded polynomially in $|X|$.

Recall that in an $\eta/0$-TRANSVERSAL instance (G, X, ℓ) the set X is a vertex cover of G. As a vertex cover is an η-transversal for any $\eta \geq 0$, we obtain the following rule.

Reduction 1. If $|X| \leq \ell$, return a trivial YES-instance.

Thus, from this point we can assume that $|X| > \ell$.

Reduction 2. Let $x, y \in X$, $x \neq y$ and $xy \notin E(G)$. If $|N_G(x) \cap N_G(y)| \geq |X| + \eta$, then add an edge xy, that is, return the instance $(G \cup \{xy\}, X, \ell)$.

Lemma 2. *Reduction 2 is sound.*

Proof. Let $G' = G \cup \{xy\}$. First note that any η-transversal Z in G' is an η-transversal in G too, as $G \setminus Z$ is a subgraph of $G' \setminus Z$.

In the other direction, let Z be an η-transversal in G of size at most ℓ, and let $\mathcal{T}$ be a tree decomposition of $G \setminus Z$ of width at most η. If either $x \in Z$ or $y \in Z$ then clearly Z is also a transversal for G'. Hence we assume that $x, y \notin Z$. In this case we claim that there exists a bag that contains both x and y. If this is not the case, there exists a separator S of size at most η that separates x from y in $G \setminus Z$. Thus $S \cup Z$ separates x from y in G. Any such a separator needs to contain $N_G(x) \cap N_G(y)$. However,

$$|N_G(x) \cap N_G(y)| \geq |X| + \eta > \ell + \eta \geq |Z| + \eta \geq |S \cup Z|,$$

a contradiction. Thus there exists a bag with both x and y, and $\mathcal{T}$ is a tree decomposition of $G' \setminus Z$. □

Definition 3. *A vertex $v \in V(G) \setminus X$ is a simplicial vertex if $G[N_G(v)]$ is a clique.*

Observe that because of our definition a vertex $v \in X$ is *not* called simplicial even if $G[N_G(v)]$ is a clique.

Lemma 4. *Let (G, X, ℓ) be an $\eta/0$-TRANSVERSAL instance. There exists a minimum η-transversal in G that does not contain any simplicial vertex.*

Proof. Let Z be a minimum η-transversal in G with minimum possible number of simplicial vertices. Assume that there exists a simplicial vertex $v \in Z$. If $N_G(v) \subseteq Z$, then v is an isolated vertex in $G \setminus (Z \setminus \{v\})$ and $Z \setminus \{v\}$ is an η-transversal in G, a contradiction to the assumption that Z is minimum. Thus let $x \in N_G(v) \setminus Z$. Note that $x \in X$, as X is a vertex cover of G and $v \notin X$ by the definition of a simplicial vertex.

We claim that $Z' = Z \cup \{x\} \setminus \{v\}$ is an η-transversal in G. As v was simplicial, $N_G[v] \subseteq N_G[x]$. Let $\phi : V(G) \setminus Z' \to V(G) \setminus Z$, $\phi(v) = x$ and $\phi(u) = u$ if $u \neq v$.

Note that ϕ is an injective homomorphism of $G \setminus Z'$ into $G \setminus Z$, thus $G \setminus Z'$ is isomorphic to a subgraph of $G \setminus Z$. We infer that $G \setminus Z'$ has not greater treewidth than $G \setminus Z$, and Z' is a minimum η-transversal in G with smaller number of simplicial vertices than Z, a contradiction. $\qquad\square$

Reduction 3. For every set $A \subseteq X$ of size $\eta + 1$ such that $G[A]$ is a clique, let S_A be the set of simplicial vertices v satisfying $A \subseteq N_G(v)$. For every such A with nonempty S_A, mark one simplicial vertex from S_A (vertices can be marked multiple times). If there are any unmarked simplicial vertices, delete them, i.e., return the instance $(G \setminus U, X, \ell)$, where U is the set of unmarked simplicial vertices.

Lemma 5. *Reduction 3 is sound.*

Proof. We argue that deleting a single unmarked simplicial vertex v results in an equivalent instance. The claim follows by applying this argument consecutively for all the unmarked simplicial vertices.

Let $G' = G \setminus \{v\}$. First note that G' is a subgraph of G, so every η-transversal Z of G gives raise to an η-transversal $Z \setminus \{v\}$ of G' that is not larger.

In the other direction, let Z be an η-transversal of G' and let $\mathcal{T}$ be the tree decomposition of $G' \setminus Z$ of width at most η. By Lemma 4 we can assume that $Z \subseteq X$. Consider $R = N(v) \setminus Z$. Observe that R induces a clique in $G' \setminus Z$. Therefore, as $G' \setminus Z$ has treewidth at most η, it follows that $|R| \leq \eta+1$. Consider the case when $|R| = \eta + 1$. As R induces a clique of cardinality $\eta + 1$ in $G[X]$ and there is an unmarked simplicial vertex v such that $R \subseteq N(v)$, it follows that there exists another simplicial vertex v' with $R \subseteq N(v')$ that was actually marked for R. Recall that $Z \subseteq X$, so $v' \notin Z$. Thus, $R \cup \{v'\}$ induces a clique of size $\eta + 2$ in $G' \setminus Z$, a contradiction with $G' \setminus Z$ having treewidth at most η.

We conclude that $|R| \leq \eta$. As R induces a clique in $G' \setminus Z$, there exists a bag B in the decomposition $\mathcal{T}$ such that $R \subseteq B$. Consider tree decomposition $\mathcal{T}'$ obtained from $\mathcal{T}$ by introducing a bag $R \cup \{v\}$ as a leaf attached to the bag B. It is easy to check that $\mathcal{T}'$ is a tree decomposition of $G \setminus Z$, while its width is bounded by η due to $|R \cup \{v\}| \leq \eta + 1$. Therefore, Z is an η-transversal in G as well. $\qquad\square$

We now claim that if none of the above reduction rules are applicable, the remaining instance is small.

Lemma 6. *Let (G, X, ℓ) be an $\eta/0$-TRANSVERSAL instance. If Reductions 1–3 are not applicable, then*

$$|V(G)| \leq |X| + \binom{|X|}{2}(|X| + \eta - 1) + \binom{|X|}{\eta+1} = O(|X|^{\max(\eta+1,3)}).$$

Proof. Any vertex of G is of one of three types: either in X, or not in X and simplicial, or not in X and not simplicial. The number of vertices of the first type is trivially bounded by $|X|$.

Let $v \in V(G) \setminus X$ be a non-simplicial vertex. Then there exist $x, y \in N_G(v)$ such that $x \neq y$ and $xy \notin E(G)$. However, for fixed $x, y \in X$ with $x \neq y$ and $xy \notin E(G)$ we may have at most $|X| + \eta - 1$ vertices in $N_G(x) \cap N_G(y)$, since Reduction 2 is not applicable. We infer that there are at most $\binom{|X|}{2}(|X| + \eta - 1)$ vertices in $V(G) \setminus X$ that are not simplicial.

Since reduction 3 is not applicable, the number of simplicial vertices in the graph is bounded by the number of subsets of X of size $\eta + 1$. Therefore, there are at most $\binom{|X|}{\eta+1}$ simplicial vertices in the graph. $\qquad \square$

We conclude this section with the following theorem.

Theorem 7. *There exists a polynomial-time algorithm that takes as an input an $\eta/0$-TRANSVERSAL instance (G, X, ℓ) and outputs an equivalent instance (G', X, ℓ) with $|V(G')| \in O(|X|^{\max(\eta+1,3)})$.*

Proof. First note that our reductions do not change the set X nor the required size of the η-transversal, i.e., the integer ℓ. All our reductions work in polynomial time for fixed η and each of them either decreases the number of vertices of the graph or introduces new edges where there was no edge before. Therefore, the number of applications of the rules is bounded polynomially in the size of the graph. Lemma 6 provides the claimed bound on $|V(G)|$ when no reduction is applicable. $\qquad \square$

3   Lower Bounds

In this section we first prove that under reasonable complexity assumptions the $0/2$-TRANSVERSAL problem does not have a polynomial kernel, which resolves an open problem by Bodlaender et al. [11]. Next we generalize this result and prove that for any η, ρ such that $\rho \geq \eta + 1$ and $(\eta, \rho) \neq (0, 1)$ the η/ρ-TRANSVERSAL problem does not have a polynomial kernel. To prove the non-existence of a polynomial kernel we use the notion of polynomial parameter transformation.

Definition 8 ([5]). *Let P and Q be parameterized problems. We say that P is polynomial parameter reducible to Q, if there exists a polynomial time computable function $f : \Sigma^* \times \mathbb{N} \to \Sigma^* \times \mathbb{N}$ and a polynomial p, such that for all $(x, k) \in \Sigma^* \times \mathbb{N}$ the following holds: $(x, k) \in P$ iff $(x', k') = f(x, k) \in Q$ and $k' \leq p(k)$. The function f is called a polynomial parameter transformation.*

Theorem 9 ([5]). *Let P and Q be parameterized problems and $\tilde{P}$ and $\tilde{Q}$ be the unparameterized versions of P and Q respectively. Suppose that $\tilde{P}$ is NP-hard and $\tilde{Q}$ is in NP. Assume there is a polynomial parameter transformation from P to Q. Then if Q admits a polynomial kernel, so does P.*

To show that $0/2$-TRANSVERSAL does not have a polynomial kernel we show a polynomial parameter transformation from CNF-SAT parameterized by the number of variables.

$CNF - SAT_n$ **Parameter:** n
Input: A formula ϕ on n variables.
Question: Does there exist an assignment Φ satisfying the formula ϕ?

Theorem 10 ([10]). *The $CNF - SAT_n$ problem does not have a polynomial kernel unless $NP \subseteq coNP/poly$.*

Theorem 11. *The $0/2$-TRANSVERSAL problem does not have a polynomial kernel unless $NP \subseteq coNP/poly$.*

Proof. We show a polynomial parameter transformation from CNF-SAT parameterized by the number of variables. Let ϕ be a formula on n variables $x_1, \ldots, x_n$. Without loss of generality we may assume that each clause of ϕ consists of an even number of literals since we can repeat an arbitrary literal of each odd size clause. We create the following graph G. First, we add a set X of $2n$ vertices $x_i, \neg x_i$ for $1 \leq i \leq n$. Moreover, we add n edges connecting x_i with $\neg x_i$ for each $1 \leq i \leq n$. Furthermore, for each clause C of the formula ϕ we add a clause gadget $\widehat{C}$ to the graph G. Let $\{l_1, l_2, \ldots, l_c\}$ be the multiset of literals appearing in the clause C. For each literal l_i we make a vertex u_i. Next we add to the graph G two paths $P_1 = v_1, \ldots, v_c$ and $P_2 = v_1', \ldots, v_c'$ having c vertices each, and connect v_i with v_i' for every $1 \leq i \leq c$. We add a pendant vertex to both vertices v_1 and v_c. Finally, for each $1 \leq i \leq c$ we make the vertex u_i adjacent to v_i, v_i' and also to the vertex $x \in X$ corresponding to the negation of the literal l_i (see Fig. 1). We would also like to remark that the clause gadget used here is the same as the one used in [13], for showing algorithmic lower bounds on the running time of an algorithm for INDEPENDENT SET parameterized by the treewidth of the input graph.

Observe that $G \setminus X$ is of treewidth two and consequently (G, X, ℓ) is a proper instance of $0/2$-TRANSVERSAL, where we set $\ell = n + \sum_{C \in \phi} 2|C|$. We show that (G, X, ℓ) is a YES-instance of $0/2$-TRANSVERSAL iff ϕ is satisfiable. Let us assume that ϕ is satisfiable and let Φ be a satisfying assignment. Since $|V(G)| = \ell + n + \sum_{C \in \phi}(|C| + 2)$, instead of showing a vertex cover of size ℓ it is enough to show an independent set of size $n + \sum_{C \in \phi}(|C| + 2)$. For each variable we add to the set I one of the vertices $x_i, \neg x_i$ which is assigned a true value by Φ. For each clause $C = \{l_1, \ldots, l_c\}$ we add to the set I an independent set of vertices from $\widehat{C}$ containing one vertex u_{i_0} corresponding to the literal satisfying the clause C, two pendant vertices adjacent to v_1 and v_c, and exactly one vertex from $\{v_i, v_i'\}$ for each $1 \leq i \leq c$, $i \neq i_0$ (see Fig. 1). It is easy to check that I is an independent set in the graph G of size $n + \sum_{C \in \phi}(|C| + 2)$, which shows that (G, X, ℓ) is a YES-instance of the $0/2$-TRANSVERSAL problem.

In the other direction, assume that (G, X, ℓ) is a YES-instance of the $0/2$-TRANSVERSAL problem. Hence there exists an independent set I in G of size $n + \sum_{C \in \phi}(|C| + 2)$. Since for each clause C the independent set I contains at most $|C| + 2$ vertices from the clause gadget $\widehat{C}$, we infer that I contains exactly $|C| + 2$ vertices out of each gadget $\widehat{C}$ and exactly one vertex from each pair

$x_i, \neg x_i$. Let Φ be an assignment such that $\Phi(x_i)$ is true iff $x_i \in I$. Consider a clause $C = \{l_1, \ldots, l_c\}$ of the formula ϕ. Observe that since C has an even number of literals the set I, contains at least one vertex u_i from the clause gadget $\widehat{C}$. Since I is independent we infer that the vertex $\neg l_i \in X$ is not in I and hence $l_i \in I$, which shows that the clause C is satisfied by Φ.

Since CNF-SAT is NP-hard and 0/2-TRANSVERSAL is in NP, by Theorem 9 the claim follows.   □

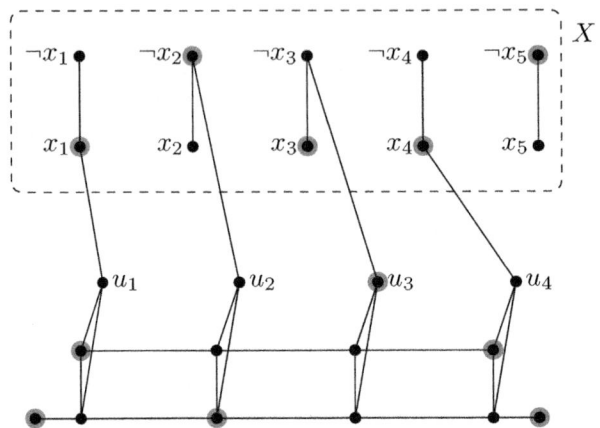

Fig. 1. A graph G for a formula consisting of a single clause $C = \{\neg x_1, x_2, x_3, \neg x_4\}$. The encircled vertices belong to an independent set I corresponding to an assignment setting to true literals $\{x_1, \neg x_2, x_3, x_4, \neg x_5\}$.

We generalize this result by showing a transformation from 0/2-TRANSVERSAL to η/ρ-TRANSVERSAL for $\eta \leq \rho + 1$ and $(\eta, \rho) \neq (0, 1)$.

Theorem 12. *For any non-negative integers η, ρ satisfying $\eta \leq \rho + 1$ and $(\eta, \rho) \neq (0, 1)$ the η/ρ-TRANSVERSAL problem does not admit a polynomial kernel unless $NP \subseteq coNP/poly$.*

Proof. Observe that by Theorem 11 and trivial transformations it is enough to prove the theorem for $\rho = \eta + 1$, where $\eta \geq 1$. We show a polynomial parameter transformation from 0/2-TRANSVERSAL to $\eta/(\eta+1)$-TRANSVERSAL. Let (G, X, ℓ) be a 0/2-TRANSVERSAL instance. Initially set $G' := G$. Now for each edge uv of the graph G we add to the graph G' a set of η vertices V_{uv} and make the set $V_{uv} \cup \{u, v\}$ a clique in G'.

First we show that (G', X, ℓ) is a proper instance of $\eta/\eta + 1$-TRANSVERSAL, that is we need to prove that $G' \setminus X$ has treewidth at most $\eta + 1$. Let $\mathcal{T}$ be a tree decomposition of width at most 2 of the graph $G \setminus X$. Consider each edge uv of the graph G. If $u, v \notin X$ then there exists a bag V_t of the tree decomposition $\mathcal{T}$ containing both u and v. We create a new bag $V_{t'} = \{u, v\} \cup V_{uv}$ and connect it, as a leaf, to the bag V_t. If $u, v \in X$, then we create a bag $V_{t'} = V_{uv}$ and connect it, as a leaf, to any bag of $\mathcal{T}$. In the last case w.l.o.g. we may assume that $u \in X$ and $v \notin X$. Then we create a new bag $V_{t'} = \{v\} \cup V_{uv}$ and connect it, as a leaf, to any bag of $\mathcal{T}$ containing the vertex v. After considering all edges of G the decomposition $\mathcal{T}$ is a proper tree decomposition of $G' \setminus X$ of width at most $\max(2, \eta + 1) = \eta + 1$.

Now we prove that (G, X, ℓ) is a YES-instance of 0/2-TRANSVERSAL iff (G', X, ℓ) is a YES-instance of $\eta/(\eta+1)$-TRANSVERSAL. Let Y be a vertex cover of G of size at most ℓ. Observe that each connected component of $G' \setminus Y$ contains exactly one vertex from the set $V(G)$ and after removing this vertex, this connected component decomposes into cliques of size η. For this reason $G' \setminus Y$ has treewidth at most η and consequently (G', X, ℓ) is a YES-instance of $\eta/(\eta+1)$-TRANSVERSAL.

Finally assume that there exists a set $Y \subseteq V(G')$ of size at most ℓ such that $G' \setminus Y$ has treewidth at most η. Let uv be an edge of the graph G. Recall that $V_{uv} \cup \{u, v\}$ is a clique in G' and hence $Y \cap (V_{uv} \cup \{u, v\})$ is nonempty. Observe that if $Y \cap V_{uv}$ is nonempty, then $Y \setminus V_{uv} \cup \{u\}$ is also a solution for (G', X, ℓ). Thus we may assume that for each edge uv we have $Y \cap \{u, v\} \neq \emptyset$, which means that Y is a vertex cover of G of size at most ℓ.

Since $\eta/(\eta+1)$-TRANSVERSAL is in NP and the unparameterized version of 0/2-TRANSVERSAL is NP-hard, the claim follows.   □

4   Conclusions and Perspectives

In this paper we showed that for every fixed η and ρ that either satisfy $1 \leq \eta < \rho$, or $\eta = 0$ and $2 \leq \rho$, the η/ρ-TRANSVERSAL problem does not admit a polynomial kernel unless NP $\subseteq$ coNP/poly. Finally, we complemented our negative result by showing that $\rho/0$-TRANSVERSAL admits a polynomial kernel for any fixed ρ. Several problems still remain open. The most notable ones are:

- Does η/η-TRANSVERSAL admit a polynomial kernel?
- Does $\mathcal{F}$-DELETION admit a polynomial kernel when $\mathcal{F}$ contains a planar graph?
- Does there exist a kernel for $\eta/0$-TRANSVERSAL of degree independent of η?

Another set of natural questions are obtained by restricting the input graphs. For example: does η/ρ-TRANSVERSAL admit a polynomial kernel on planar graphs, or on a graph class excluding a fixed graph H as a minor, or on graphs of bounded degree? Surprisingly, the answer to many of these questions is positive. One can easily show that the techniques from [9] imply that for every fixed η and ρ, η/ρ-TRANSVERSAL admits a linear kernel on H-minor free graphs. Moreover, going along the lines of [8] proves that η/ρ-TRANSVERSAL admits a linear vertex kernel on graphs of bounded degree or on graphs excluding $K_{1,t}$ as an induced subgraph. Here $K_{1,t}$ is a star with t leaves.

References

1. Bodlaender, H.L., Downey, R.G., Fellows, M.R., Hermelin, D.: On problems without polynomial kernels. J. Comput. Syst. Sci. 75(8), 423–434 (2009)
2. Bodlaender, H.L., Fomin, F.V., Lokshtanov, D., Penninkx, E., Saurabh, S., Thilikos, D.M.: (Meta) kernelization. In: FOCS, pp. 629–638 (2009)
3. Bodlaender, H.L., Jansen, B.M.P., Kratsch, S.: Cross-composition: A new technique for kernelization lower bounds. In: STACS, pp. 165–176 (2011)
4. Bodlaender, H.L., Jansen, B.M.P., Kratsch, S.: Preprocessing for Treewidth: A Combinatorial Analysis through Kernelization. In: Aceto, L., Henzinger, M., Sgall, J. (eds.) ICALP 2011. LNCS, vol. 6755, pp. 437–448. Springer, Heidelberg (2011)
5. Bodlaender, H.L., Thomasse, S., Yeo, A.: Analysis of data reduction: Transformations give evidence for non-existence of polynomial kernels, technical Report UU-CS-2008-030, Institute of Information and Computing Sciences, Utrecht University, Netherlands (2008)
6. Bodlaender, H.L., Thomassé, S., Yeo, A.: Kernel Bounds for Disjoint Cycles and Disjoint Paths. In: Fiat, A., Sanders, P. (eds.) ESA 2009. LNCS, vol. 5757, pp. 635–646. Springer, Heidelberg (2009)
7. Dom, M., Lokshtanov, D., Saurabh, S.: Incompressibility through Colors and IDs. In: Albers, S., Marchetti-Spaccamela, A., Matias, Y., Nikoletseas, S., Thomas, W. (eds.) ICALP 2009, Part I. LNCS, vol. 5555, pp. 378–389. Springer, Heidelberg (2009)
8. Fomin, F.V., Lokshtanov, D., Misra, N., Philip, G., Saurabh, S.: Hitting forbidden minors: Approximation and kernelization. In: STACS, pp. 189–200 (2011)
9. Fomin, F.V., Lokshtanov, D., Saurabh, S., Thilikos, D.M.: Bidimensionality and kernels. In: SODA, pp. 503–510 (2010)
10. Fortnow, L., Santhanam, R.: Infeasibility of instance compression and succinct PCPs for NP. In: STOC 2008: Proceedings of the 40th Annual ACM Symposium on Theory of Computing, pp. 133–142. ACM (2008)
11. Jansen, B.M.P., Bodlaender, H.L.: Vertex cover kernelization revisited: Upper and lower bounds for a refined parameter. In: STACS, pp. 177–188 (2011)
12. Jansen, B.M.P., Kratsch, S.: Data Reduction for Graph Coloring Problems. In: Owe, O., Steffen, M., Telle, J.A. (eds.) FCT 2011. LNCS, vol. 6914, pp. 90–101. Springer, Heidelberg (2011)
13. Lokshtanov, D., Marx, D., Saurabh, S.: Known algorithms on graphs on bounded treewidth are probably optimal. In: SODA, pp. 777–789 (2011)

Safe Approximation
and Its Relation to Kernelization

Jiong Guo[1,*], Iyad Kanj[2], and Stefan Kratsch[3,**]

[1] Universität des Saarlandes, Saarbrücken, Germany
jguo@mmci.uni-saarland.de
[2] DePaul University, Chicago, USA
ikanj@cs.depaul.edu
[3] Utrecht University, Utrecht, The Netherlands
kratsch@cs.uu.nl

Abstract. We introduce a notion of approximation, called *safe approximation*, for minimization problems that are subset problems. We first study the relation between the standard notion of approximation and safe approximation, and show that the two notions are different unless some unlikely collapses in complexity theory occur. We then study the relation between safe approximation and kernelization. We demonstrate how the notion of safe approximation can be useful in designing kernelization algorithms for certain fixed-parameter tractable problems. On the other hand, we show that there are problems that have constant-ratio safe approximation algorithms but no polynomial kernels, unless the polynomial hierarchy collapses to the third level.

1 Introduction

Studying the relation between parameterized complexity and approximation theory has attracted the attention of researchers from both areas. Cai and Chen initiated this study by showing that any optimization problem that has a fully polynomial time approximation scheme (FPTAS) is fixed-parameter tractable (FPT) [8]. This result immediately places a large number of optimization problems in the class FPT. Cesati and Trevisan [10] refined Cai and Chen's result by relaxing the condition that the problem has an FPTAS. A problem is said to have an *efficient polynomial time approximation scheme* (EPTAS), if the problem has a PTAS whose running time is of the form $f(1/\epsilon)n^{O(1)}$ (n is the input size and ϵ is the error bound). By definition, an FPTAS for a problem is also an EPTAS. Cesati and Trevisan [10] showed that having an EPTAS is a sufficient condition for a problem to be in FPT. Cai and Chen also showed in [8] that the class MaxSNP of maximization problems, defined by Papadimitriou and

---

* Supported by the DFG Excellence Cluster "Multimodal Computing and Interaction".
** Supported by the Netherlands Organization for Scientific Research (NWO), project "KERNELS: Combinatorial Analysis of Data Reduction".

D. Marx and P. Rossmanith (Eds.): IPEC 2011, LNCS 7112, pp. 169–180, 2012.
© Springer-Verlag Berlin Heidelberg 2012

Yannakakis [24], and the class MIN $F^+\Pi_1$ of minimization problems, defined by Kolaitis and Thakur [19], are subclasses of the class FPT.

In [13], Chen et al. introduced the notion of *efficient fixed-parameter tractability*, and gave a complete characterization of the relation between the class FPTAS and the class FPT. They showed that a parameterized problem has an FPTAS if and only if it is efficient fixed-parameter tractable [13], which complements the earlier result by Cai and Chen [8]. Moreover, to study the relation between EPTAS and FPT, Chen et al. [13] introduced the notion of the *planar W-hierarchy*, and showed that all problems in the *planar W-hierarchy*, which contains several known problems such as PLANAR VERTEX COVER and PLANAR INDEPENDENT SET, have EPTAS.

We note that the parameterized complexity framework has also been used to obtain negative approximation results (see [9,12,21], to name a few). For example, the above relations between approximation and parameterized complexity have been used to rule out the existence of EPTAS for certain problems that admit PTAS (see [9,21]). For an extensive overview on the relation of parameterized complexity and approximation, as well as on combinations of these two paradigms, we refer the interested reader to the recent survey of Marx [22].

More recently, Kratsch [20] studied the relation between kernelization and approximation. He showed that two large classes of problems having constant-ratio approximation algorithms, namely MIN $F^+\Pi_1$ and MAXNP, the latter including MAXSNP, admit polynomial kernelization for their parameterized versions. His result extends Cai and Chen's results [8] mentioned above.

In this paper we investigate further the relation between approximation and kernelization. We focus our attention on minimization problems that are subset problems (i.e., the solution is a subset of the search space), and define the notion of *safe approximation* for subset minimization problems. Informally speaking, an approximation algorithm for a subset minimization problem is *safe* if for every instance of the problem the algorithm returns a solution that is guaranteed to contain (subset containment) an optimal solution. We note that many natural subset minimization problems admit safe approximation algorithms. We start by showing that the notion of safe approximation is different from the standard notion of approximation, in the sense that there are problems that admit approximation algorithms with certain ratios but do not admit safe approximation algorithms even with much worse ratios, under standard complexity assumptions. For example, we show that there are natural problems that have PTASs but do not even have constant-ratio safe approximation algorithms unless $W[1] = \text{FPT}$. We then proceed to study the relation between safe approximation and kernelization. We demonstrate, through some nontrivial examples, that the notion of safe approximation can be very useful algorithmically: we show how safe approximation algorithms for certain problems can be used to design kernelization algorithms for their associated parameterized problems. On the other hand, we show that safe approximation does not imply polynomial kernelization by proving that there are problems that have constant-ratio safe approximation algorithms but whose associated parameterized problems do not have polynomial kernels, unless the polynomial hierarchy collapses to the third level.

Due to the lack of space, most proofs are deferred to the full version of the paper.

2   Preliminaries

Parameterized complexity and kernelization. A *parameterized problem* Q is a subset of $\Sigma^* \times \mathbb{N}$, where Σ is a finite fixed alphabet and $\mathbb{N}$ is the set of non-negative integers. Therefore, each instance of the parameterized problem Q is a pair (x, k), where the second component, i.e., the non-negative integer k, is called the *parameter*. We say that the parameterized problem Q is *fixed-parameter tractable* [16], shortly FPT, if there is an algorithm that decides whether an input (x, k) is a member of Q in time $f(k)|x|^{O(1)}$, where $f(k)$ is a recursive function of k. Let FPT denote the class of all fixed-parameter tractable problems. A parameterized problem Q is *kernelizable* if there exists a polynomial-time reduction, the *kernelization*, that maps instances (x, k) of Q to other instances (x', k') of Q such that: (1) $|x'| \leq g(k)$, (2) $k' \leq g(k)$, for some recursive function g, and (3) (x, k) is a yes-instance of Q if and only if (x', k') is a yes-instance of Q. The instance (x', k') is called the *kernel* of (x, k). A kernelization is *polynomial* if $g(k)$ is bounded by a polynomial in k.

A hierarchy of fixed-parameter intractability, *the W-hierarchy* $\bigcup_{t \geq 0} W[t]$, has been introduced. Here, $W[t] \subseteq W[t + 1]$ for all $t \geq 0$ and the 0-th level $W[0]$ is the class FPT. The hardness and completeness notions have been defined for each level $W[i]$ of the W-hierarchy, for $i \geq 1$ [16]. It is commonly believed that collapses in the W-hierarchy are unlikely (i.e., $W[i] \neq W[i - 1]$, for any integer $i \geq 1$), and in particular, $W[1] \neq$ FPT (see [16]).

NP-optimization problems and approximability. An *NP optimization problem* Q is a 4-tuple (I_Q, S_Q, f_Q, g_Q), where: I_Q is the set of input instances, which is recognizable in polynomial time. For each instance $x \in I_Q$, $S_Q(x)$ is the set of feasible solutions for x, which is defined by a polynomial p and a polynomial-time computable predicate π (p and π depend only on Q) as $S_Q(x) = \{y : |y| \leq p(|x|) \wedge \pi(x, y)\}$. The function $f_Q(x, y)$ is the objective function mapping a pair $x \in I_Q$ and $y \in S_Q(x)$ to a non-negative integer. The function f_Q is computable in polynomial time. The function g_Q is the *goal* function, which is one of the two functions $\{\max, \min\}$, and Q is called a *maximization problem* if $g_Q = \max$, or a *minimization problem* if $g_Q = \min$. We will denote by $opt_Q(x)$ the value $g_Q\{f_Q(x, z) \mid z \in S_Q(x)\}$, and if there is no confusion about the underlying problem Q, we will write $opt(x)$ to denote $opt_Q(x)$.

In this paper we restrict our attention to optimization problems in NP that are minimization problems. An algorithm A is an *approximation algorithm* for a minimization problem Q if for each input instance $x \in I_Q$ the algorithm A returns a feasible solution $y_A(x) \in S_Q(x)$. The solution $y_A(x)$ has an *approximation ratio* $r(|x|)$ if it satisfies the following condition:

$$f_Q(x, y_A(x))/opt_Q(x) \leq r(|x|).$$

The approximation algorithm A has an *approximation ratio* $r(|x|)$ if for every instance x in I_Q the solution $y_A(x)$ constructed by the algorithm A has an approximation ratio bounded by $r(|x|)$.

An optimization problem Q has a *constant-ratio approximation algorithm* if it has an approximation algorithm whose ratio is a constant (i.e., independent from the input size). An optimization problem Q has a *polynomial time approximation scheme* (PTAS) if there is an algorithm A_Q that takes a pair (x, ϵ) as input, where x is an instance of Q and $\epsilon > 0$ is a real number, and returns a feasible solution y for x such that the approximation ratio of the solution y is bounded by $1 + \epsilon$, and for each fixed $\epsilon > 0$, the running time of the algorithm A_Q is bounded by a polynomial of $|x|$. Finally, an optimization problem Q has a *fully polynomial time approximation scheme* (FPTAS) if it has a PTAS A_Q such that the running time of A_Q is bounded by a polynomial of $|x|$ and $1/\epsilon$.

Definition 1. Let $Q = (I_Q, S_Q, f_Q, g_Q)$ be a minimization problem. The *parameterized version of Q* is $Q_\leq = \{(x, k) \mid x \in I_Q \land opt_Q(x) \leq k\}$. A parameterized algorithm A_Q solves the parameterized version of Q if on any input $(x, k) \in Q_\leq$, A_Q returns "yes" with a solution y in $S_Q(x)$ such that $f_Q(x, y) \leq k$, and on any input not in $Q_\leq$, A_Q simply returns "no".

The above definition allows us to consider the parameterized complexity of a minimization problem Q, which is the parameterized complexity of $Q_\leq$.

The problems discussed in the current paper all share the property that they seek a subset, of a given set (a "search space"), that satisfies certain properties. We call such problems *subset problems*. Most of the problems studied in parameterized complexity and combinatorial optimization are subset problems.[1]

3   Safe Approximation

In this section we define a notion of approximation for subset minimization problems that we call *safe approximation*, and we study its relation to the standard notion of approximation.

Definition 2. Let Q be a subset minimization problem. An approximation algorithm $\mathcal{A}$ for Q is said to be *safe* if for every instance x of Q, $\mathcal{A}$ returns a solution $y_A(x)$ such that there exists an optimal solution $S_{opt}(x)$ of x satisfying $S_{opt}(x) \subseteq y_A(x)$. The notions of *constant-ratio safe approximation algorithm*, *safe PTAS*, and *safe FPTAS* are defined in a natural way.

Informally speaking, an approximation algorithm for a minimization subset problem is safe if the solution that it returns is guaranteed to contain an optimal solution.

---

[1] In the case of optimization problems the subset sought is one that minimizes/maximizes the objective function, among all subsets satisfying the required properties. For most problems considered in this paper, the objective function is the cardinality of the subset sought.

Some natural questions to ask are the following: (1) Are there (NP-hard) subset minimization problems that admit safe approximation algorithms with "small" ratios? (2) Does every problem that has a constant-ratio approximation algorithm (resp. PTAS/FPTAS) have a constant-ratio safe approximation algorithm (resp. safe PTAS/FPTAS)?

The answer to question (1) is positive: many minimization problems admit safe approximation algorithms with "small" ratios (e.g., constant ratios). Those problems include VERTEX COVER (follows from a well-known theorem of Nemhauser and Trotter [4,23]), many subset minimization problems on bounded-degree graphs (for many such problems we can simply return the whole set of vertices as the approximate solution), and many subset minimization problems on planar graphs (e.g., PLANAR DOMINATING SET).

We show next that, unless some unlikely collapses in complexity theory or parameterized complexity occur, the answer to question (2) is negative. First, we define the following problems.

A *vertex cover* in an undirected graph is a subset of vertices C such that every edge in the graph is incident to at least one vertex in C. The CONNECTED VERTEX COVER problem is: Given an undirected graph G, compute a subset of vertices C of minimum cardinality such that C is a vertex cover of G and the subgraph of G induced by C is connected.

A *dominating set* in an undirected graph is a subset of vertices D such that every vertex in the graph is either in D or has a neighbor in D. The DOMINATING SET problem is: Given an undirected graph G, compute a subset of vertices D of minimum cardinality such that D is a dominating set of G. A *unit disk graph* (UDG) is a graph on n points/vertices in the Euclidean plane such that there is an edge between two points in the graph if and only if their Euclidean distance is at most 1 (unit). The DOMINATING SET problem on UDGs, denoted UDG-DOMINATING SET, is the DOMINATING SET problem restricted to UDG's.

We answer question (2) negatively by showing that the DOMINATING SET problem, which has an approximation ratio $\lg n + 1$ [17] (n is the number of the vertices in the graph), is unlikely to have a safe approximation algorithm of ratio $c \lg n$, for any constant $c > 0$:

Theorem 1. *Unless $FPT = W[2]$,* DOMINATING SET *does not have a safe approximation algorithm of ratio $\rho \leq c \lg n$, for any constant $c > 0$.*[2]

Proof. Let (G, k) be an instance of DOMINATING SET$_{\leq}$. Suppose that DOMINATING SET has a safe approximation algorithm $\mathcal{A}$ of ratio $c \lg n$. We run $\mathcal{A}$ on G to obtain a solution D of G such that $|D|/|opt(G)| \leq c \lg n$. If $|D| > ck \lg n$, it follows that $opt(G) > k$, and we can reject the instance (G, k); so assume $|D| \leq ck \lg n$. Since $\mathcal{A}$ is a safe approximation algorithm, D contains a minimum dominating set. Therefore, in time $\sum_{i=1}^{k} \binom{ck \lg n}{i} n^2$ we can enumerate all subsets of D of size at most k, and check whether any of them is a dominating set.

---

[2] As a matter of fact, under the same complexity hypothesis, we can rule out (using a similar proof) the existence of a safe approximation algorithm of ratio $n^{o(1)}$ for DOMINATING SET.

If we find any, then we accept the problem instance (G, k); otherwise, we reject it. This shows that DOMINATING SET$_\leq$ is solvable in $f(k)n^c$ time for some constant c and completes the proof.   □

By a similar argument, it follows that there are problems that have a PTAS but that are unlikely to have even a safe constant-ratio approximation algorithm:

Theorem 2. *The* UDG-DOMINATING SET *problem admits a PTAS but does not admit a constant-ratio safe approximation algorithm unless* $W[1] = FPT$.

Finally, the CONNECTED VERTEX COVER problem, which has an approximation algorithm of ratio 2 [25], does not admit a constant-ratio safe approximation algorithm unless the polynomial time hierarchy collapses to the third level:

Theorem 3. *Unless the polynomial time hierarchy collapses to the third level, the* CONNECTED VERTEX COVER *problem does not have a constant-ratio safe approximation algorithm.*[3]

4   Kernelization and Safe Approximation

At the surface, the notion of safe approximation seems to be closely related to the notion of kernelization in parameterized complexity. We clarify some of the differences between the two notions in the following remark.

Remark 1. It seems intuitive that problems with a safe approximation algorithm should have kernels of matching size. Of course, the two notions are not equivalent: The safe approximation solution is not necessarily a kernel, as simply "forgetting" everything outside the solution cannot be guaranteed to give an equivalent instance. Furthermore, kernelizations are not restricted to subproblems of the original instance and, hence, do not have to return a safe approximation. Still, even if one aims to compute a safe approximation and cleverly reduce the part outside the solution to small size, it can be showed (Theorem 4) that there are problems with constant-factor safe approximation but without polynomial kernels (assuming that the polynomial hierarchy does not collapse).

Remark 2. If a subset minimization problem has a constant-ratio safe approximation algorithm (in fact, any ratio of the form $f(opt)$ suffices, where f is a nondecreasing efficiently computable function) then its parameterized version must be FPT (enumerate all subsets of the solution returned by the safe approximation algorithm in FPT time).

The HITTING SET problem is defined as follows. Given a pair $(S, \mathcal{F})$ where S is a set of elements and $\mathcal{F}$ is a family of subsets of S, compute a smallest subset H of S that intersects every set in $\mathcal{F}$.

---

[3] Under the same complexity hypothesis, we can strengthen this result to rule out the existence of a safe approximation algorithm of ratio $opt(G)^{O(1)}$ for CONNECTED VERTEX COVER.

Theorem 4. *Unless the polynomial-time hierarchy collapses to the third level, there are problems that have constant-ratio safe approximation algorithms but no polynomial kernels.*

Proof. Consider the following restriction of HITTING SET, denoted PAIRED-HS, consisting of the set of all instances of HITTING SET of the form $(S, \mathcal{F})$, where $|S| = 2N$ for some natural number N, and $\mathcal{F}$ contains, in addition to other sets, N pairwise disjoint sets, each of cardinality 2, whose union is S. It is not difficult to see that the instances of PAIRED-HS are recognizable in polynomial time (e.g., by computing maximum matching). Moreover, it follows easily from the definition of PAIRED-HS that it has a safe approximation algorithm of ratio 2 (the algorithm returns the set S as the solution to the instance $(S, \mathcal{F})$). Note also that PAIRED-HS$_\leq$ is FPT, since any instance in which the parameter is smaller than $|S|/2$ can be rejected immediately, otherwise, a brute force algorithm enumerating all subsets of S and checking whether each subset is a solution, is an FPT algorithm that solves the problem.

We claim that PAIRED-HS$_\leq$ does not have a polynomial kernel[4], unless the polynomial hierarchy collapses to the third level. To prove this claim, consider the d-SAT problem that consists of the set of instances of CNF-SAT in which each clause has at most d literals, where $d \geq 3$ is an integer constant. It was shown in [15] that, unless the polynomial time hierarchy collapses to the third level, the d-SAT problem parameterized by the number of variables n, has no oracle communication protocol of cost at most $O(n^{d-\epsilon})$, for any $\epsilon > 0$; this can be easily seen to exclude also kernels as well as compressions into instances of other problems of size $O(n^{d-\epsilon})$ (cf. [15]).

Now proceed by contradiction. Assume that PAIRED-HS$_\leq$ has a polynomial kernel of size $O(k^c)$ for some integer constant $c > 1$, and consider the d-SAT problem where $d = c + 1$. We can reduce d-SAT to PAIRED-HS as follows. For each instance F on n variables, construct the instance $(S, \mathcal{F}, n)$ (with parameter n) of PAIRED-HS$_\leq$ where S consists of the set of n variables in F and their negations; thus, S has $2n$ elements. For each variable in F we associate a set of two elements in $\mathcal{F}$ containing the variable and its negation. Finally, for each clause in F we associate a set in $\mathcal{F}$ containing the literals in the clause. Clearly, the resulting instance is an instance of PAIRED-HS$_\leq$. Moreover, F is a yes-instance of d-SAT if and only if $(S, \mathcal{F}, n)$ is a yes-instance of PAIRED-HS$_\leq$; the key observation is that the paired elements and the maximum size of n encode the selection of a truth assignment, the other sets check that it is satisfying. It follows that this reduction compresses instances of d-SAT into instances of size $O(n^c) = O(n^{d-1})$, which implies a collapse of the polynomial hierarchy to the third level. This completes the proof. □

In the remainder of this section we study further the relation between safe approximation and kernelization. We show that the notion of safe approximation can be useful for obtaining kernelization algorithms for FPT problems. The VERTEX COVER problem is a trivial example showing how a safe approximation

---

[4] The kernel size for HITTING SET$_\leq$ is the sum of the cardinalities of all sets in $\mathcal{F}$.

algorithm can be used to obtain a kernelization algorithm: no edge has both endpoints outside the safe approximation solution, and if an edge has one, we may safely take the other.[5] The NT-theorem [4,23], which is a local-ratio approximation algorithm of ratio 2 for VERTEX COVER, is at the same time a safe approximation algorithm. This algorithm has been used in [14] to obtain a kernel for VERTEX COVER of size at most $2k$, which currently stands as the best upper bound on the kernel size for VERTEX COVER.

It is not always as simple to get a kernelization from a safe approximation algorithm as in the case of VERTEX COVER. Therefore, it is interesting to investigate which safe approximation algorithms (for subset minimization problems) can be turned into kernelization algorithms. In addition to its theoretical importance, this question has an interesting algorithmic facet: given a solution to the instance that contains an optimal solution (the "important" part), can we "deal with" the remaining part of the instance (the "left overs")?

We illustrate next, through a few examples, how the existence of safe approximation algorithms implies the existence of kernelization algorithms for certain problems. These results should mainly be seen as illustrative examples of using safe approximation as a technique for obtaining kernelization algorithms; in most cases matching or better kernels are known. The problems under consideration are: EDGE MULTICUT, VERTEX MULTICUT, PLANAR DOMINATING SET, PLANAR FEEDBACK VERTEX SET, FEEDBACK VERTEX SET, and a generalization of FEEDBACK VERTEX SET, called FEEDBACK VERTEX SET WITH BLACKOUT VERTICES. Both PLANAR DOMINATING SET and PLANAR FEEDBACK VERTEX SET admit PTAS [3], and FEEDBACK VERTEX SET and its generalization with blackout vertices admit approximation algorithms of ratio 2 [2]. Both PLANAR DOMINATING SET$_\le$ [1] and PLANAR FEEDBACK VERTEX SET$_\le$ [6] have linear kernels, and FEEDBACK VERTEX SET has a quadratic kernel [26].

4.1   Planar Dominating Set

The PLANAR DOMINATING SET problem is the DOMINATING SET problem restricted to planar graphs. We show next that any safe approximation algorithm of ratio ρ for PLANAR DOMINATING SET can be used to design a kernelization algorithm for PLANAR DOMINATING SET$_\le$ that computes a kernel with at most $10\rho k$ vertices. For a vertex v in a graph, we denote by $N(v)$ the set of neighbors of v. Two vertices u and v in a graph are said to be *twins* if $N(u) = N(v)$.

Theorem 5. *If* PLANAR DOMINATING SET *has a safe approximation algorithm* $\mathcal{A}$ *of constant ratio* ρ *then* PLANAR DOMINATING SET$_\le$ *has a kernelization algorithm* $\mathcal{A}'$ *that computes a kernel with at most* $10\rho k$ *vertices.*

Proof. Given an instance (G, k) of PLANAR DOMINATING SET$_\le$, the kernelization algorithm $\mathcal{A}'$ starts by invoking the algorithm $\mathcal{A}$ to compute a set of vertices S of G whose cardinality is at most $\rho|opt(G)|$, and that contains a minimum

---

[5] This is also true for the d-HITTING SET; we may forget all elements that are outside the safe approximation solution, and shrink the sets accordingly.

dominating set of G. If $|S| > \rho k$ then clearly $opt(G) > k$ and the algorithm $\mathcal{A}'$ rejects the instance (G, k); so assume $|S| \leq \rho k$. Let $\overline{S} = V(G) \setminus S$. The algorithm $\mathcal{A}'$ applies the following reduction rules to G in the respective order.

Reduction Rule 1. Remove all the edges in $G[\overline{S}]$.

Reduction Rule 2. For any set of degree-1 vertices (degree taken in the current graph) in $\overline{S}$ that are twins, remove all of them except one vertex.

Reduction Rule 3. For any set of degree-2 vertices in $\overline{S}$ that are twins (i.e., all of them are twins), remove all of them except two vertices.

Let G' be the resulting graph from G after the application of the above rules. Note that $S \subseteq V(G')$. The algorithm $\mathcal{A}'$ returns the instance (G', k). Since S contains an optimal solution, it is not difficult to verify that (G', k) is an equivalent instance of (G, k). Next, we upper bound the number of vertices in G'.

Let $I = V(G') \setminus S$, and note that I is an independent set by Reduction Rule 1. We partition I into three sets: I_1 is the set of degree-1 vertices (degree taken in G'), I_2 is the set of degree-2 vertices, and $I_{\geq 3}$ is the set of vertices in I of degree at least 3. Next, we upper bound the cardinality of each of these three sets.

To upper bound the cardinality of $I_{\geq 3}$, we define the multihypergraph $\mathcal{H}$ as follows. The vertex-set of $\mathcal{H}$ is S. A subset of vertices e is an edge in $\mathcal{H}$ if and only if there exists a vertex $u \in I_{\geq 3}$ such that $N(u) = e$. Since the incidence graph of $\mathcal{H}$ is a subgraph of G', and hence is planar, the multihypergraph $\mathcal{H}$ is planar. It follows from Lemma 4.4 in [18] that $\mathcal{H}$ has at most $2|V(\mathcal{H})| - 4 = 2|S| - 4$ edges. Since the number of edges in $\mathcal{H}$ is exactly the number of vertices in $I_{\geq 3}$, it follows that $|I_{\geq 3}| \leq 2|S| - 4$. By Reduction Rule 2, we have $|I_1| \leq |S|$. To upper bound $|I_2|$, we construct a planar multigraph $\mathcal{G}$ whose vertex set is S, and such that there is an edge between two vertices u and v in $\mathcal{G}$ if and only if there exists a vertex $w \in I_2$ whose neighbors are u and v. Since G' is planar, $\mathcal{G}$ is planar, and by Reduction Rule 3, there are at most 2 edges between any two vertices in $\mathcal{G}$. It follows from Euler's formula that the number of edges in $\mathcal{G}$, and hence the number of vertices in I_2, is at most $2(3|V(\mathcal{G})| - 6) = 6|S| - 12$.

Thus $|V(G')| = |I| + |S| \leq 10|S| - 16 < 10\rho k$, completing the proof. $\square$

4.2 Feedback Vertex Set

Let G be an undirected graph. A set of vertices F in G is a *feedback vertex set* of G if the removal of F breaks all cycles in G, that is, if $G - F$ is acyclic. The FEEDBACK VERTEX SET problem is to compute a feedback vertex set of minimum cardinality in a given graph. The PLANAR FEEDBACK VERTEX SET problem is the restriction of the FEEDBACK VERTEX SET problem to planar graphs. We show first that a constant-ratio safe approximation for FEEDBACK VERTEX SET gives a kernel with a cubic number of vertices for FEEDBACK VERTEX SET$_\leq$, using only one reduction rule plus a simple marking procedure. We then consider a generalization of FEEDBACK VERTEX SET$_\leq$, which allows for the presence of blackout vertices, and asks for a feedback vertex set excluding all blackout vertices. We call this problem FEEDBACK VERTEX SET WITH BLACKOUT VERTICES, FVSBV for short. This problem was introduced by Bar-Yehuda [5], and has applications in Bayesian inference. We show that the cubic kernel can be improved for this generalization to match the known quadratic kernel by Thomassé [26], as the blackout annotation allows a more efficient processing of the trees that are outside the safe approximation, using simpler and different arguments. (Note that the quadratic upper bound does not carry to the standard FEEDBACK VERTEX SET$_\leq$ problem due to the presence of blackout vertices.) Finally, we show that a ratio ρ safe approximation for PLANAR FEEDBACK VERTEX SET gives a kernel with at most $3\rho k$ vertices for PLANAR FEEDBACK VERTEX SET$_\leq$.

Theorem 6. *If* FEEDBACK VERTEX SET *has a constant-ratio safe approximation, then* FEEDBACK VERTEX SET$_\leq$ *has a cubic kernel.*

Corollary 1. *If* FVSBV *has a constant-ratio safe approximation, then* FVSBV$_\leq$ *admits a quadratic kernel.*

Theorem 7. *If* PLANAR FEEDBACK VERTEX SET *has a safe approximation algorithm with constant ratio ρ then* PLANAR FEEDBACK VERTEX SET$_\leq$ *has a kernel with at most $3\rho k$ vertices.*

4.3   Multicut Problems

The EDGE MULTICUT problem is defined as follows: Given a graph $G = (V, E)$ and a set of pairs $T = \{(s_1, t_1), \ldots, (s_\ell, t_\ell)\}$ of vertices in G, compute a set of edges E' in G of minimum cardinality whose removal disconnects all pairs in T (i.e., there is no path from s_i to t_i, for $i = 1, \ldots, \ell$, in $(V, E \setminus E')$).

Theorem 8. *If* EDGE MULTICUT *has an $f(opt)$ safe approximation algorithm, where f is a nondecreasing efficiently computable function, then* EDGE MULTICUT$_\leq$ *has a polynomial kernel with at most $3f(k)$ vertices.*

A similar result holds for VERTEX MULTICUT$_\leq$, where the task is to delete at most k non-terminal vertices to disconnect all given terminal pairs.

Theorem 9. *If* VERTEX MULTICUT *has an $f(opt)$ safe approximation algorithm, where f is a nondecreasing efficiently computable function, then* VERTEX MULTICUT$_\leq$ *has a polynomial kernel with at most $2f(k)$ vertices.*

5   Conclusion

We presented the notion of safe approximation and studied its relation to the notion of kernelization in parameterized complexity.

Even though we have shown that the notions of safe approximation and kernelization are different for subset minimization problems, we illustrated through some nontrivial examples how safe approximation can be useful for obtaining kernelization algorithms. Some of those results imply linear kernelization algorithms for the problems under consideration. For example, it can be shown that PLANAR DOMINATING SET has a constant-ratio safe approximation algorithm, which, when combined with Theorem 5, gives a linear kernelization algorithm for PLANAR DOMINATING SET$_\leq$. Unfortunately, the obtained upper bound on the kernel size does not come close to the currently-best upper bound on the kernel size for PLANAR DOMINATING SET$_\leq$ [11]. This, however, may not be discouraging due to the mere fact that kernelization algorithms for PLANAR DOMINATING SET$_\leq$ have been extensively studied, whereas the notion of safe approximation was not considered before. Maybe a celebrated example that can be used to illustrate how safe approximation can be useful for designing kernelization algorithms is the example of VERTEX COVER. An approximation algorithm of ratio 2, the NT-theorem, for VERTEX COVER existed since 1975 [23]. Buss and Goldsmith [7], in 1993, presented a kernelization algorithm that gives a quadratic $(2k^2)$ kernel for VERTEX COVER$_\leq$. This upper bound on the kernel size was subsequently used in several parameterized algorithms for VERTEX COVER$_\leq$, until Chen et al. [14] observed in 2001 that the approximation algorithm given by the NT-theorem is *safe* (this notion was not defined at that point), and implies a $2k$ kernel for VERTEX COVER$_\leq$. We believe that the existence of the notion of safe approximation may bridge the gap between approximation and kernelization.

Several interesting questions arise from the current research. Many parameterized problems admit polynomial kernels and their optimization versions have constant-ratio approximation algorithms. Do these optimization versions admit constant-ratio safe approximation algorithms? For example, FEEDBACK VERTEX SET has a ratio 2 approximation algorithm [2] and a quadratic kernel [26], does it have a constant-ratio safe approximation algorithm? One can ask whether a sufficient condition (based on parameterized complexity) exists, such that if a problem satisfying this condition has an approximation algorithm then it must have a safe approximation algorithm.

Acknowledgment. The authors would like to thank Bart Jansen for suggesting the nice example used in the proof of Theorem 4.

References

1. Alber, J., Fellows, M., Niedermeier, R.: Polynomial-time data reduction for dominating set. J. ACM 51(3), 363–384 (2004)
2. Bafna, V., Berman, P., Fujito, T.: A 2-approximation algorithm for the Undirected Feedback Vertex Set problem. SIAM J. Discrete Math. 12(3), 289–297 (1999)
3. Baker, B.: Approximation algorithms for NP-complete problems on planar graphs. J. ACM 41(1), 153–180 (1994)
4. Bar-Yehuda, R., Even, S.: A local-ratio theorem for approximating the Weighted Vertex Cover problem. Annals of Discrete Mathematics 25, 27–46 (1985)
5. Bar-Yehuda, R., Geiger, D., Naor, J., Roth, R.: Approximation algorithms for the Feedback Vertex Set problem with applications to constraint satisfaction and bayesian inference. SIAM J. Comput. 27(4), 942–959 (1998)

6. Bodlaender, H.L., Penninkx, E.: A Linear Kernel for Planar Feedback Vertex Set. In: Grohe, M., Niedermeier, R. (eds.) IWPEC 2008. LNCS, vol. 5018, pp. 160–171. Springer, Heidelberg (2008)

7. Buss, J., Goldsmith, J.: Nondeterminism within P. SIAM J. Comput. 22, 560–572 (1993)

8. Cai, L., Chen, J.: Fixed parameter tractability and approximability of NP-hard optimization problems. J. Comput. Syst. Sci. 54, 465–474 (1997)

9. Cai, L., Fellows, M., Juedes, D., Rosamond, F.: The complexity of polynomial-time approximation. Theory Comput. Syst. 41(3), 459–477 (2007)

10. Cesati, M., Trevisan, L.: On the efficiency of polynomial time approximation schemes. Inf. Process. Lett. 64, 165–171 (1997)

11. Chen, J., Fernau, H., Kanj, I., Xia, G.: Parametric duality and kernelization: Lower bounds and upper bounds on kernel size. SICOMP 37(4), 1077–1106 (2007)

12. Chen, J., Huang, X., Kanj, I., Xia, G.: Linear FPT reductions and computational lower bounds. J. Comput. Syst. Sci. 72(8), 1346–1367 (2006)

13. Chen, J., Huang, X., Kanj, I., Xia, G.: Polynomial time approximation schemes and parameterized complexity. Discrete Appl. Mathematics 155(2), 180–193 (2007)

14. Chen, J., Kanj, I., Jia, W.: Vertex cover: further observations and further improvements. J. Algorithms 41, 280–301 (2001)

15. Dell, H., van Melkebeek, D.: Satisfiability allows no nontrivial sparsification unless the polynomial-time hierarchy collapses. In: STOC, pp. 251–260 (2010)

16. Downey, R., Fellows, M.: Parameterized Complexity. Springer, New York (1999)

17. Johnson, D.: Approximation algorithms for combinatorial problems. J. Comput. Syst. Sci. 9, 256–278 (1974)

18. Kanj, I., Pelsmajer, M., Xia, G., Schaefer, M.: On the induced matching problem. In: STACS. LIPIcs, vol. 08001, pp. 397–408 (2008)

19. Kolaitis, P., Thakur, M.: Approximation properties of NP minimization classes. J. Comput. Syst. Sci. 50, 391–411 (1995)

20. Kratsch, S.: Polynomial kernelizations for MIN $F^+\Pi_1$ and MAX NP. In: STACS. LIPIcs, vol. 3, pp. 601–612 (2009)

21. Marx, D.: Efficient Approximation Schemes for Geometric Problems? In: Brodal, G.S., Leonardi, S. (eds.) ESA 2005. LNCS, vol. 3669, pp. 448–459. Springer, Heidelberg (2005)

22. Marx, D.: Parameterized complexity and approximation algorithms. Comput. J. 51(1), 60–78 (2008)

23. Nemhauser, G., Trotter, L.: Vertex packing: structural properties and algorithms. Mathematical Programming 8, 232–248 (1975)

24. Papadimitriou, C., Yannakakis, M.: Optimization, approximation, and complexity classes. J. Comput. Syst. Sci. 43, 425–440 (1991)

25. Savage, C.: Depth-first search and the vertex cover problem. Inf. Process. Lett. 14(5), 233–237 (1982)

26. Thomassé, S.: A $4k^2$ kernel for feedback vertex set. ACM Transactions on Algorithms 6(2) (2010)

Simpler Linear-Time Kernelization
for Planar Dominating Set

Torben Hagerup

Institut für Informatik, Universität Augsburg, 86135 Augsburg, Germany
hagerup@informatik.uni-augsburg.de

Abstract. We describe a linear-time algorithm that inputs a planar graph G and outputs a planar graph of size $O(k)$ and with domination number k, where k is the domination number of G, i.e., the size of a smallest dominating set in G. In the language of parameterized computation, the new algorithm is a linear-time kernelization for the NP-complete PLANAR DOMINATING SET problem that produces a kernel of linear size. Such an algorithm was previously known (van Bevern et al., these proceedings), but the new algorithm and its analysis are considerably simpler.

1 Introduction

A current trend in the area of parameterized computation is the quest for kernelization algorithms for hard computational problems. Briefly stated, a kernelization algorithm is a polynomial-time procedure that transforms an instance of the problem under consideration into an equivalent instance whose size depends only on the value of the chosen parameter. More formally, a *kernelization* for a parameterized problem $L \subseteq \Sigma^* \times \mathbb{N}$, where Σ is an alphabet and $\mathbb{N} = \{1, 2, \ldots\}$, is an algorithm $\mathcal{A}$ for which there exists a polynomial $p : \mathbb{N} \to \mathbb{N}$ and a function $f : \mathbb{N} \to \mathbb{N}$ such that, applied to an instance $I = (G, k) \in \Sigma^* \times \mathbb{N}$, $\mathcal{A}$ computes, within $p(|I|)$ steps, an instance $I' = (G', k') \in \Sigma^* \times \mathbb{N}$ with $|I'| \leq f(k)$ and $k' \leq k$ such that $I \in L \Leftrightarrow I' \in L$. Here $|I|$ and $|I'|$ denote the number of symbols in the representation of I and I', respectively, according to some suitable encoding scheme.

A kernelization can be valuable in the solution of a hard parameterized problem because, used as a preprocessing routine, it allows the instance size to be reduced, perhaps significantly, before an exponential-time algorithm is applied to the remaining, reduced instance. From the outset, much importance was attached to the *kernel size*, $f(k)$, and for some problems a series of results successively lowered the smallest upper bound on the size of a kernel known to be computable in polynomial time. This was the case for the problem of interest in this paper, the NP-complete PLANAR DOMINATING SET problem, formally defined as the language

$$\{(G, k) : G \text{ is an undirected planar graph, } k \in \mathbb{N} \text{ and } Dom(G) \leq k\} \ ,$$

D. Marx and P. Rossmanith (Eds.): IPEC 2011, LNCS 7112, pp. 181–193, 2012.
© Springer-Verlag Berlin Heidelberg 2012

where $Dom(G)$ is the *domination number* of G, i.e., the size of a smallest dominating set in G, a smallest subset D of the vertex set V of G such that every vertex in $V \setminus D$ is adjacent in G to a vertex in D. Alber et al. [1] described a first kernelization for PLANAR DOMINATING SET, whose kernel is of size at most $335k$, where k is the domination number of the input graph. The bound was subsequently lowered to $67k$ by Chen et al. [4], who also proved that no kernelization can yield a kernel of size bounded by $(2 - \epsilon)k$, for arbitrary fixed $\epsilon > 0$, unless P = NP. Other, more general, approaches [3,5] yield kernels of linear size for several problems defined on planar graphs, including PLANAR DOMINATING SET.

Only more recently has the *kernelization time*, $p(|I|)$, come into focus. The kernelizations for the PLANAR DOMINATING SET problem mentioned above both operate in cubic time, and so are rather slow. Traditionally, this has been largely ignored, the reasoning being that since the polynomial-time preprocessing is followed by an exponential-time computation anyway, there is little point in worrying about the degree of the polynomial. Whereas there is much truth in this argument, it is hardly advisable to be dogmatic about the issue. From the standpoint of theory, one might object that the polynomial is applied to the original instance size, n, whereas exponential means exponential in the kernel size, which may be substantially smaller than n. In practice, one can observe that the running time of a polynomial-time procedure is not always a negligible fraction of the time consumed by an exponential-time computation. In summary, it seems a worthwhile goal to try to reduce *both* the kernel size and the kernelization time.

At present, linear-time kernelizations are known for only few problems. VERTEX COVER is a case in point: The classic Buss' kernelization that yields a kernel of size $O(k^2)$ works in linear time (more powerful kernelizations for VERTEX COVER that produce kernels with $O(k)$ vertices are known [9, Section 7.4], but none of these runs in linear time). Other problems for which linear-time kernelizations were described in the literature include certain problems on planar graphs [8], CLUSTER EDITING [11] and ROOTED LEAF OUTBRANCHING [7]; as in the case of VERTEX COVER, kernels of linear size are not obtained in linear time. An exception concerns the WEIGHTED MAX LEAF problem investigated by Jansen [6], who states without giving all details that a linear-sized kernel can be obtained in linear time. For the PLANAR DOMINATING SET problem, a first kernelization algorithm that yields a kernel of linear size in linear time was described by van Bevern et al. [2]. It is based on the *region decomposition* of the earlier cubic-time kernelization algorithm of [1] and can be viewed to some extent as a more efficient implementation of that algorithm. Here we take a fresh look at the problem and develop a second linear-time kernelization for PLANAR DOMINATING SET that yields a kernel of linear size. Compared with the result of van Bevern et al., the advantage of the new result is that the algorithm and its analysis, while not trivial, are considerably simpler. On the other hand, the approach based on region decomposition may have applications to other problems defined on planar graphs, while the techniques described here are more problem-specific.

In terms of kernel size, the new algorithm, at least with its present analysis, is not competitive with the earlier algorithms mentioned above. Moreover, there is a certain trade-off between simplicity and small kernel size. For these reasons, we refrain from calculating the exact kernel size and demonstrate only as simply as we can that it is $O(k)$. Of course, if the algorithm of [4] is applied after the new algorithm, a kernel of size at most $67k$ is obtained in $O(n + k^3)$ time.

Our result is slightly more than a kernelization. As is a standard requirement for kernelizations although usually not considered part of the formal definition, our arguments imply an efficient and in fact linear-time procedure for obtaining from an arbitrary dominating set in the kernel a dominating set of the same size in the original input graph. Moreover, similarly as the algorithm of van Bevern et al. [2], our algorithm actually does not need to know the parameter k. Rather, it inputs a graph G and computes a graph G' with $Dom(G') = Dom(G)$, so that $(G, k) \in$ PLANAR DOMINATING SET $\Leftrightarrow (G', k) \in$ PLANAR DOMINATING SET for every $k \in \mathbb{N}$. Correspondingly, a step in the kernelization that inputs a graph G and outputs another graph G' will be called *correct* if $Dom(G) = Dom(G')$.

Subsequently to the work described here, it was demonstrated that the same techniques can be employed to obtain linear-time kernelizations with linear kernel size for more general domination problems in planar graphs. These include PLANAR ANNOTATED DOMINATING SET, i.e., given an undirected planar graph $G = (V, E)$, a set $B \subseteq V$ and an integer $k \in \mathbb{N}$, decide whether there is a set $D \subseteq V$ with $|D| \le k$ such that every vertex in $B \setminus D$ is adjacent in G to a vertex in D. A few of the ideas in this paper suffice to give a linear-time kernelization with linear kernel size for the PLANAR EDGE DOMINATING SET problem, i.e., given an undirected planar graph $G = (V, E)$ and an integer $k \in \mathbb{N}$, decide whether there is a subset $D \subseteq E$ with $|D| \le k$ such that every edge in E shares an endpoint with an edge in D. This ongoing work will be reported elsewhere.

1.1   Overview of the New Algorithm

The starting point for the kernelization described here was the observation that if two vertices x and y in an n-vertex planar graph G with small domination number k have many joint neighbors, then most of these neighbors have no neighbors other than x and y that are not also neighbors of x or y. Then an optimal dominating set contains x or y, and most joint neighbors of x and y are without influence on the domination number and can be removed. It therefore makes sense to search for a small vertex set A such that many vertices have two neighbors in A. A good candidate for A is a smallest set P of vertices in G with total degree at least $n - |P|$.

It may not be the case that there are many vertices with two neighbors in P. Then, however, most vertices have exactly one neighbor in P, and the union Q of P with the set of vertices without neighbors in P is still small. Think of the vertices in Q as the centers of stars that include all vertices in G. If it is not optimal to dominate a star from its center, every vertex in the star must be dominated by a vertex that also dominates at least one vertex in a different star. By planarity, only few vertices dominate parts of three or more stars. As for vertices that dominate parts of exactly two stars, at most two such dominating vertices are needed for each pair of stars, since otherwise the two stars in question could be dominated more cheaply from their centers. We cannot know which two vertices are present in an optimal solution, but for each of the two stars we can pick a single vertex that dominates a maximum number of vertices in that star and at least one vertex in the other star. By planarity, adding the vertices that were picked and those that dominate parts of more than two stars to Q still results in a small set R.

Let S be the set of centers x of stars for which the vertices in $R \setminus \{x\}$ do not have enough edges to the star of x to dominate it completely. A dominating set can be assumed to be a superset of S, and therefore every vertex in the set T of vertices at distance 1 from S can be removed unless it is needed to dominate vertices outside of $S \cup T$. This is the case only if it has two or more neighbors outside of $S \cup T$, a condition that, by planarity, holds for only few vertices in T.

If most vertices belong to stars whose centers are not in S, there may not be many vertices in T to remove. In that case, however, R is a small set such that many vertices have two neighbors in R, and we can pick $A = R$ and remove many vertices outside of R as described above. In every situation, if G contains more than Ck vertices for a suitable constant C, a constant fraction of the vertices in G can be removed in linear time, and all that remains is to repeat the process until the number of vertices no longer drops as fast as established for graphs with more than Ck vertices.

2   Preliminaries

2.1   Definitions and Notation

Throughout this subsection, let $G = (V, E)$ be an undirected graph.

When $U \subseteq V$, $G[U]$ denotes, as usual, the subgraph of G induced by U, i.e., $(U, \{\{u, v\} \in E : u, v \in U\})$.

For $u \in V$, we denote by $N_G(u)$ the set $\{v \in V : \{u, v\} \in E\}$ of neighbors of u in G and write $N_G[u]$ for $N_G(u) \cup \{u\}$. For $U \subseteq V$, $N_G(U) = \bigcup_{u \in U} N_G(u)$ and $N_G[U] = \bigcup_{u \in U} N_G[u]$. A vertex $u \in V$ *dominates* the vertices in $N_G[u]$ in G, and a vertex set $U \subseteq V$ dominates the vertices in $N_G[U]$ in G and G itself if $N_G[U] = V$.

For $A \subseteq V$ and $i \geq 0$, we denote by $N_G^i(A)$ the set $\{u \in V \setminus A : |N_G(u) \cap A| = i\}$ of vertices in $V \setminus A$ that have precisely i neighbors in A, and $N_G^{\geq i}(A) = \bigcup_{j=i}^{\infty} N_G^j(A)$.

2.2   Properties of Planar Graphs

The facts stated in Lemma 1 below are well-known. For proofs see, e.g., [10]. The remaining lemmas in this section are straightforward consequences of Lemma 1.

Lemma 1. *Let $G = (V, E)$ be an undirected planar graph and take $n = |V|$ and $m = |E|$. Then*

(a) $m \leq 3n$.
(b) If G is bipartite, then $m \leq 2n$.
(c) No three distinct vertices in G have three common neighbors.

Lemma 2. *Let $G = (V, E)$ be an undirected planar graph and let $A \subseteq V$. Then*

(a) For $i = 3, 4, \ldots$, let $n_i = |N_G^i(A)|$. Then $\sum_{i=3}^{\infty} (i - 2) n_i \leq 2|A|$.
(b) $|N_G^{\geq 3}(A)| \leq 2|A|$.
(c) Let $\mathcal{F}$ be a set of faces in a planar embedding of G, the boundary of each of which contains 3 or more vertices in $N_G^{\geq 2}(A)$. Then at most $6|A|$ vertices in $N_G^{\geq 2}(A)$ lie on the boundary of one or more faces in $\mathcal{F}$.

Proof. (a) Applying Lemma 1(b) to the subgraph of G induced by the edges with one endpoint in A and one endpoint in $N_G^{\geq 3}(A)$ yields $\sum_{i=3}^{\infty} i n_i \leq 2(|A| + \sum_{i=3}^{\infty} n_i)$.

(b) Since $|N_G^{\geq 3}(A)| = \sum_{i=3}^{\infty} n_i \leq \sum_{i=3}^{\infty} (i - 2) n_i$, the claim follows from part (a).

(c) Apply part (a) to the graph obtained from G by, for each $F \in \mathcal{F}$, adding a new vertex v_F and a new edge from v_F to each vertex in $N_G^{\geq 2}(A)$ on the boundary of F. If the number of new edges is m, this shows that $m \leq 2(|A| + |\mathcal{F}|) \leq 2(|A| + m/3)$ and therefore that $m \leq 6|A|$.   □

Lemma 3. *Let $G = (V, E)$ be an n-vertex undirected planar graph, let $A \subseteq V$ and assume that $\sum_{x \in A} |N_G[x]| \geq n$. For $i = 0, 1, \ldots$, let $n_i = |N_G^i(A)|$. Then $n_0 \leq n_2 + 10|A|$.*

Proof. The total degree in G of the vertices in A is at least $n - |A|$ and, by Lemma 1(a), at most $3|A|$ edges have both endpoints in A. Therefore

$$\sum_{i=0}^{\infty} i n_i \geq n - |A| - 2 \cdot 3|A| = \sum_{i=0}^{\infty} n_i - 6|A|$$

and

$$n_0 \leq n_2 + \sum_{i=3}^{\infty} (i - 1) n_i + 6|A| \leq n_2 + 6|A| + 2 \sum_{i=3}^{\infty} (i - 2) n_i \ .$$

By Lemma 2(a), $\sum_{i=3}^{\infty} (i - 2) n_i \leq 2|A|$. The claim follows.   □

3   Near-Twin Reduction

A central part of our kernelization is a procedure called *near-twin reduction* that is applied to an undirected planar graph $G = (V, E)$ and parameterized by a vertex set $A \subseteq V$.

3.1   Description

With G and A as above, define the *A-neighborhood* of a vertex $u \in V$ as the set $N_G(u) \cap A$. For all $x, y \in A$ with $x \neq y$, call $u \in V$ *introspective* with *support* $\{x, y\}$ if $u \notin A$, the A-neighborhood of u is $\{x, y\}$, and the A-neighborhood of every vertex at distance at most 2 from u in $G[V \setminus A]$ is a nonempty subset of $\{x, y\}$. A *near-twin reduction* in G with *support* A returns the graph $G' = (V', E')$ obtained from G by the following operation: For all $\{x, y\} \subseteq A$ with $x \neq y$ such that the set $I_{\{x,y\}}$ of introspective vertices with support $\{x, y\}$ is of size at least 4, replace the vertices in $I_{\{x,y\}}$ by two new degree-2 vertices with neighbors x and y (see Fig. 1). G' is clearly planar. In Section 4, we write an application of the near-twin reduction with support A to G as a function call *NearTwin_Reduce*(G, A) that returns the graph G' resulting from the near-twin reduction.

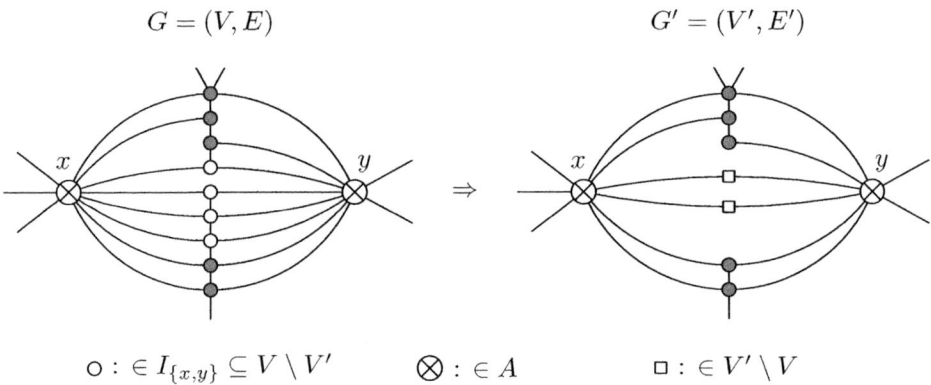

Fig. 1. A near-twin reduction with support A transforms G into G'

3.2   Correctness

Suppose that D is a dominating set in G and consider the set $D' \subseteq V'$ obtained from D as follows: For all $\{x, y\} \subseteq A$ with $x \neq y$ such that $|I_{\{x,y\}}| \geq 4$ and $D \cap N_G[I_{\{x,y\}}] \not\subseteq \{x, y\}$, replace the vertices in $D \cap N_G[I_{\{x,y\}}]$ by x and y. Note that $N_G[I_{\{x,y\}}] \cap N_G[I_{\{x',y'\}}] = \emptyset$ for all $\{x', y'\} \subseteq A$ with $x' \neq y'$ and $\{x, y\} \neq \{x', y'\}$, so that the replacement can actually be carried out as described. Since $N_G[N_G[I_{\{x,y\}}]] \subseteq N_G[\{x, y\}]$ for all $\{x, y\} \subseteq A$ with $x \neq y$, D' still dominates G, and since $D \cap N_G[I_{\{x,y\}}] \neq \emptyset$, it is easy to see that D' dominates G'. By Lemma 1(c), the vertices in a set $I_{\{x,y\}}$ with $|I_{\{x,y\}}| \geq 4$ cannot be dominated in G by any single vertex other than x or y. Therefore, if $D \cap N_G[I_{\{x,y\}}] \not\subseteq \{x, y\}$, then $|D \cap N_G[I_{\{x,y\}}]| \geq 2$. This shows that $|D'| \leq |D|$.

Suppose, conversely, that D' is a dominating set in G' and consider the set $D \subseteq V$ obtained from D' by replacing the vertices in $D' \cap (V' \setminus V)$ by their neighbors. D clearly dominates G, and $|D| \leq |D'|$ since for every vertex in $D' \cap (V' \setminus V)$ without neighbors in D' there is another vertex in $D' \cap (V' \setminus V)$ with the same two neighbors. Altogether, we have shown that $Dom(G) = Dom(G')$, i.e., the near-twin reduction is correct.

3.3   Execution Time

The near-twin reduction with support A can be applied to G in $O(|V|)$ time as follows: First each vertex $u \in V \setminus A$ computes the set U, where U is the A-neighborhood of u if the latter is of size 1 or 2 and $U = \{\bot\}$ for a special symbol $\bot$ otherwise. Then, in each of two successive rounds, each vertex in $V \setminus A$ broadcasts its set to all its neighbors in $V \setminus A$ and forms the new value of its own set as the union of the old value and the sets received, except that every union that contains $\bot$ or is of size 3 or more is replaced by $\{\bot\}$. A vertex is introspective exactly if its stored set is of size 2 throughout this process. The introspective vertices can be sorted by their supports in $O(|V|)$ time with radix sort, after which it is a simple matter to construct G'.

3.4   Reduction Progress

Lemma 4. *Let $G = (V, E)$ be an undirected planar graph, let $A \subseteq V$ and let $G' = (V', E')$ be the graph obtained by carrying out a near-twin reduction in G with support A. Then $|N_{G'}^2(A)| = O(|A| + Dom(G))$.*

Proof. We first show that in G', no four introspective vertices have a common support $\{x, y\}$. Otherwise some vertex $u \in (V \cap V') \setminus A$ with A-neighborhood $\{x, y\}$ would be introspective in G', but not in G. This can happen only if the near-twin reduction in G removes a vertex v at distance at most 2 from u in $G[V \setminus A]$ whose A-neighborhood is not $\{x, y\}$. But this is impossible since the A-neighborhood of u is $\{x, y\}$ and u is within distance 2 of v in $G[V \setminus A]$—the presence of u would prevent v from being introspective, and therefore from being removed.

Since clearly there is a planar graph on the vertex set A that contains an edge $\{x, y\}$ if there is an introspective vertex with support $\{x, y\}$, Lemma 1(a) shows that the number of introspective vertices in $U = N_{G'}^2(A)$ is bounded by $3 \cdot 3|A|$. To bound the number of the remaining vertices in U, fix a planar embedding of G' and its restriction ϕ to the subgraph of G' induced by the set of edges with one endpoint in A and one endpoint in U. Denote by $\mathcal{F}$ the set of faces of ϕ. We can clearly assume that $|U| \geq 3$. Then $\mathcal{F} = \mathcal{F}_2 \cup \mathcal{F}_{\geq 3}$, where $\mathcal{F}_2$ and $\mathcal{F}_{\geq 3}$ are the sets of faces in $\mathcal{F}$ whose boundary contains exactly 2 and ≥ 3 distinct vertices in U, respectively. Moreover, every face in $\mathcal{F}$ whose boundary contains three distinct vertices in A belongs to $\mathcal{F}_{\geq 3}$.

Let $x, y \in A$ be distinct and let u be a vertex in U whose A-neighborhood is $\{x, y\}$. If u is not introspective, there is a path π in $G[V \setminus A]$ of length at most 2 from u to a vertex v whose A-neighborhood is not a nonempty subset of $\{x, y\}$. Choose π as a shortest such path. Then one of the following holds (see Fig. 2):

- $v \in U$, and every face in $\mathcal{F}$ whose boundary contains both v and the last vertex in U that precedes v on π belongs to $\mathcal{F}_{\geq 3}$, since its boundary includes the A-neighborhoods of both u and v.

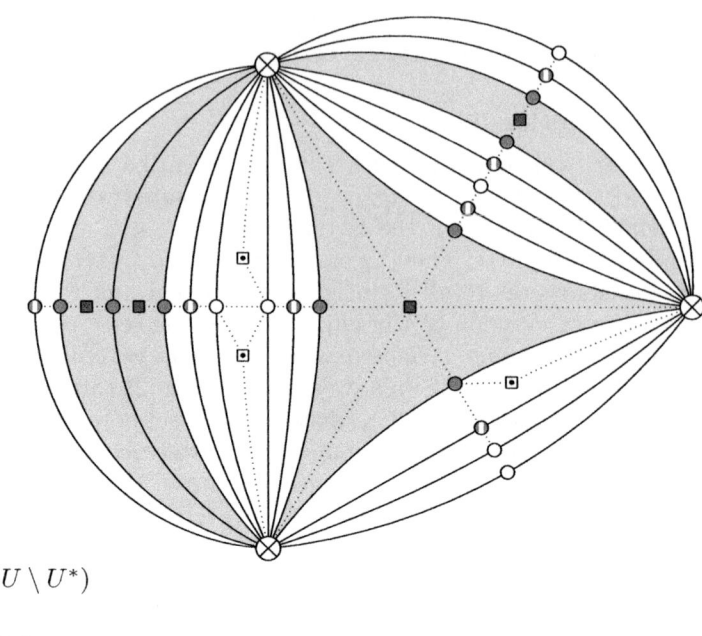

$\boxed{}$: $\in \mathcal{F}^*$

$\otimes$: $\in A$

$\bullet$: $\in U^*$

$\mathbf{o}$: $\in N_{G'}(U^*) \cap (U \setminus U^*)$

$\circ$: introspective

$\blacksquare$: $\in N_{G'}^0(A) \cup N_{\overline{G}'}^{\geq 3}(A)$ $\boxdot$: $\in N_{G'}^1(A)$

Fig. 2. Vertices in U sufficiently far from the faces in $\mathcal{F}^*$ are introspective

- $v \in N_{G'}^1(A)$, and the face in $\mathcal{F}$ whose interior contains v belongs to $\mathcal{F}_{\geq 3}$, since its boundary includes the A-neighborhoods of both u and v.
- $v \in N_{G'}^0(A) \cup N_{\overline{G}'}^{\geq 3}(A)$.

Let $\mathcal{F}^*$ be the union of $\mathcal{F}_{\geq 3}$ with the set of faces in $\mathcal{F}_2$ whose interior contains a vertex in $N_{G'}^0(A) \cup N_{\overline{G}'}^{\geq 3}(A)$ and let U^* be the set of vertices in U on the boundary of a face in $\mathcal{F}^*$. The considerations above show that every vertex in U that is not introspective is at distance at most 1 in G' from a vertex in U^*.

Let F be a face in $\mathcal{F}_2$ whose interior contains a vertex $v \in N_{G'}^0(A)$. A vertex that dominates v must lie in the interior of F or be one of its two boundary vertices in U, each of which can dominate vertices in at most one face other than F. This shows the number of vertices in U on the boundaries of faces in $\mathcal{F}_2$ whose interiors contain a vertex in $N_{G'}^0(A)$ to be at most $3Dom(G') = 3Dom(G)$. By Lemma 2(b), the number of vertices in U on the boundaries of faces in $\mathcal{F}_2$ whose interiors contain a vertex in $N_{\overline{G}'}^{\geq 3}(A)$ is bounded by $2 \cdot 2|A|$. And, finally, Lemma 2(c) shows the number of vertices in U on the boundaries of faces in $\mathcal{F}_{\geq 3}$ to be at most $6|A|$. In summary, $|U^*| \leq 10|A| + 3Dom(G)$. Each vertex in U^* lies on the boundary of at most one face in $\mathcal{F} \setminus \mathcal{F}^* \subseteq \mathcal{F}_2$ and therefore has at most one neighbor in $U \setminus U^*$. Altogether, U contains at most $9|A|$ introspective vertices and at most $2(10|A| + 3Dom(G))$ vertices that are not introspective. $\square$

4   Shrinking Iterations

4.1   Description

The main part of the kernelization is the *shrinking iteration* detailed below. It takes as input an undirected planar graph $G = (V, E)$, goes through three phases that successively transform G into G_1, G_2 and G_3, and returns G_3. Take $n = |V|$ and, for $i = 1, 2, 3$, write $G_i = (V_i, E_i)$.

The first two phases are simply near-twin reductions with different supports. The support used in Phase 1 is a smallest set $P \subseteq V$ with $\sum_{x \in P} |N_G[x]| \geq n$. To describe the support used in Phase 2, we need some additional notation.

First, let Q be the union of P with the set of vertices in V_1 without neighbors in P. Q dominates G_1, so we can form a partition $\{Star(x) : x \in Q\}$ of V_1 with $x \in Star(x) \subseteq N_{G_1}[x]$ for all $x \in Q$. For all $x \in Q$, we call $Star(x)$ a *star* and x its *center*. For all $u \in V_1$, let $L(u) = \{x \in Q : N_{G_1}[u] \cap Star(x) \neq \emptyset\}$ be the set of centers of stars represented in $N_{G_1}[u]$. Moreover, for all $x, y \in Q$ with $x \neq y$, let $w_y(x)$ be the supremum, over all $u \in V_1 \setminus \{x\}$ with $L(u) = \{x, y\}$, of $|N_{G_1}[u] \cap Star(x)|$ (equal to $-\infty$ if there are no such u) and let $g_y(x)$ be a vertex u that realizes the supremum (undefined if $w_y(x) = -\infty$). Phase 2 is a near-twin reduction with support

$$R = Q \cup \{g_y(x) : x, y \in Q, \, x \neq y \text{ and } w_y(x) > 0\} \cup \{u \in V_1 : |L(u)| \geq 3\} \ .$$

Let $w(x) = \sum_{y \in R \setminus \{x\}} |N_{G_1}[y] \cap Star(x)|$ for all $x \in Q$, $S = \{x \in Q : w(x) < |Star(x)|\}$ and $T = N_{G_2}(S) \setminus S$. Phase 3 consists in removing every vertex in T with at most one neighbor outside of $S \cup T$ and, for each $x \in S$, adding a new degree-1 vertex with neighbor x. A nonobvious point is that since the definition of S refers to G_1, S is best computed before the near-twin reduction in Phase 2.

The three phases are summarized below in pseudo-code.

```
// Phase 1
P := a smallest subset of V with ∑_{x∈P} |N_G[x]| ≥ n;
G_1 := (V_1, E_1) := NearTwin_Reduce(G, P);
```

```
// Phase 2
Q := P ∪ N⁰_{G_1}(P);
{Star(x) : x ∈ Q} := a partition of V_1 with x ∈ Star(x) ⊆ N_{G_1}[x] for all x ∈ Q;
R := Q ∪ {g_y(x) : x, y ∈ Q, x ≠ y and w_y(x) > 0} ∪ {u ∈ V_1 : |L(u)| ≥ 3},
    where L(u) = {x ∈ Q : N_{G_1}[u] ∩ Star(x) ≠ ∅} for all u ∈ V_1,
    w_y(x) = sup{|N_{G_1}[u] ∩ Star(x)| : u ∈ V_1 \ {x} and L(u) = {x, y}}
        for all x, y ∈ Q with x ≠ y and
    g_y(x) = some u ∈ V_1 \ {x} with L(u) = {x, y}
        and |N_{G_1}(u) ∩ Star(x)| = w_y(x)
        for all x, y ∈ Q with x ≠ y and w_y(x) > 0;
S := {x ∈ Q : w(x) < |Star(x)|},
    where w(x) = ∑_{y∈R\{x}} |N_{G_1}[y] ∩ Star(x)| for all x ∈ Q;
G_2 := (V_2, E_2) := NearTwin_Reduce(G_1, R);
```

// Phase 3
$T := N_{G_2}(S) \setminus S;$
$V_3' := V_2 \setminus \{u \in T : |N_{G_2}(u) \setminus (S \cup T)| \leq 1\};$
$G_3 := (V_3, E_3) :=$ the graph obtained from $G_2[V_3']$ by adding,
    for each $x \in S$, a new degree-1 vertex with neighbor x;
return G_3;

4.2   Correctness

It was proved already in Section 3.2 that near-twin reductions preserve the domination number. Demonstrating the correctness of a shrinking iteration therefore boils down to showing that $Dom(G_2) = Dom(G_3)$.

Let D_3 be a dominating set in G_3. Since $N_{G_3}(V_3 \setminus V_2) = S$ and $V_2 \setminus V_3 \subseteq N_{G_2}(S)$, $D_2 = (D_3 \cap V_2) \cup S$ is a dominating set in G_2. And since every vertex in S has a degree-1 neighbor in G_3 that does not belong to V_2, $|D_2| \leq |D_3|$.

Conversely, let D_1 be a minimum dominating set in G_1 (not in G_2), chosen to maximize $|D_1 \cap Q|$ among all such sets. D_1 does not contain any vertex $u \in V_1 \setminus Q$ with $|L(u)| = 1$, since every such vertex could be replaced in D_1 by the center of its star to obtain a minimum dominating set with one more element in Q. Likewise, for all $x, y \in Q$ with $x \neq y$, D_1 contains at most one vertex $u \in V_1 \setminus \{x\}$ with $L(u) = \{x, y\}$, since two such vertices could be replaced by x and y in D_1 to obtain a minimum dominating set with at least one more element in Q. R contains all $u \in V_1$ with $|L(u)| \geq 3$ and, for all $x, y \in Q$ with $x \neq y$ such that $L(u) = \{x, y\}$ for some $u \in V_1 \setminus \{x\}$, one such vertex u for which $|N_{G_1}[u] \cap Star(x)|$ is maximal. Because of this, it is not difficult to see that for all $x \in Q$, $w(x) = \sum_{y \in R \setminus \{x\}} |N_{G_1}[y] \cap Star(x)|$ is an upper bound on the number of vertices in $Star(x)$ that are dominated by $D_1 \setminus \{x\}$. If $w(x) < |Star(x)|$ for some $x \in Q$, therefore, $x \in D_1$. Hence $S \subseteq D_1$.

The argument in Section 3.2 that shows that $Dom(D_2) \leq Dom(D_1)$ also proves that there is a minimum dominating set D_2 in G_2 with $D_1 \cap R \subseteq D_2 \cap R$ and therefore $S \subseteq D_2$. Let D_2 be such a set and obtain $D_3 \subseteq V_3$ from D_2 by replacing each vertex $u \in D_2 \setminus V_3$ by the vertices in $N_{G_2}(u) \setminus (S \cup T)$. In G_3, D_3 dominates the vertices in $V_3 \cap (S \cup T)$ (because $S \subseteq D_3$), the vertices in $V_2 \setminus (S \cup T)$ (by the way D_3 is obtained from D_2), and the vertices in $V_3 \setminus V_2$ (again because $S \subseteq D_3$). Therefore D_3 is a dominating set in G_3. Since $|N_{G_2}(u) \setminus (S \cup T)| \leq 1$ for each $u \in V_2 \setminus V_3$, $|D_3| \leq |D_2|$. Thus $Dom(G_2) = Dom(G_3)$.

4.3   Execution Time

It was proved already in Section 3.3 that a near-twin reduction can be executed in linear time. To compute P, sort the vertices by their degrees and pick a suitable suffix of the sorted sequence. To compute Q, let each vertex determine membership in P for itself and its neighbors to know whether it should enter Q. The computation of T from S and of V_3' from S and T are similar, and the construction of G_3 from G_2, V_3' and S is easy.

In order to compute the partition $\{Star(x) : x \in Q\}$, let each vertex in $V_1 \setminus Q$ choose a neighbor $x \in Q$ and mark itself as belonging to $Star(x)$. Subsequently each vertex $u \in V_1$ can determine the number $|L(u)|$ of stars represented in $N_{G_1}[u]$ and enter R if $|L(u)| \geq 3$. If $|L(u)| = 2$, instead let u generate the tuples (x, y, r_x, u) (unless $u = x$) and (y, x, r_y, u) (unless $u = y$), where $L(u) = \{x, y\}$, $r_x = |N_{G_1}[u] \cap Star(x)|$ and $r_y = |N_{G_1}[u] \cap Star(y)|$. Sorting the tuples generated in this way lexicographically with radix sort allows us to compute the final constituent $\{g_y(x) : x, y \in Q, x \neq y \text{ and } w_y(x) > 0\}$ of R in $O(n)$ time: For all $x, y \in Q$ with $x \neq y$ and $w_y(x) > 0$, $g_y(x)$ is the fourth component of the last tuple in the sorted sequence with first component x and second component y. The computation of S, finally, is made easy by the alternative characterization $w(x) = \sum_{u \in Star(x)} |N_{G_1}[u] \cap (R \setminus \{x\})|$, for all $x \in Q$, which shows that $w(x)$ can be obtained for all $x \in Q$ by summing over the vertices in $Star(x)$ a quantity that is computable for all vertices in V_1 in $O(n)$ time. The application of a shrinking iteration to an n-vertex graph therefore takes $O(n)$ time.

4.4 Reduction Progress

Take $k = Dom(G)$ and note that $|P| \leq k$. Let $z = |V| - |V_1| = |N_G^2(P)| - |N_{G_1}^2(P)| \geq 0$ be the decrease in the number of vertices achieved by Phase 1. By Lemma 4, $|N_{G_1}^2(P)| = O(k)$, so $|N_G^2(P)| = z + O(k)$. Lemma 3 now shows that $|N_G^0(P)| \leq |N_G^2(P)| + O(k) = z + O(k)$. And then $|Q| = |P| + |N_G^0(P)| = z + O(k)$.

Let $R_{\geq 3} = \{u \in V_1 \setminus Q : |L(u)| \geq 3\}$. We want to show that $|R_{\geq 3}| = O(|Q|)$, which, by Lemma 1(a), will imply that $|R| \leq |Q| + 6|Q| + |R_{\geq 3}| = O(z + k)$. To this end, recall that for $t \in \mathbb{N}$, a t-coloring of an undirected graph $H = (V_H, E_H)$ is a mapping $h : V_H \to \{1, \ldots, t\}$ such that for every $\{u, v\} \in E_H$, $h(u) \neq h(v)$. Lemma 1(a) can easily be used to show that every undirected planar graph has a 6-coloring, and the famous four-color theorem (see, e.g., [10]) states that every undirected planar graph in fact has a 4-coloring. Therefore let h be a t-coloring of G_1 for some $t \leq 6$ and fix $j \in \{1, \ldots, t\}$. Consider the graph obtained from G_1 by merging each vertex $u \in V_1 \setminus Q$ with $h(u) \neq j$ into the center of its star. In the resulting graph every vertex $u \in R_{\geq 3}$ with $h(u) = j$ has at least 3 neighbors in Q. Therefore, by Lemma 2(b), $|\{u \in R_{\geq 3} : h(u) = j\}| \leq 2|Q|$, and altogether $|R_{\geq 3}| \leq 2t|Q| = O(|Q|)$.

Let $W = \sum_{x \in Q} w(x)$. Intuitively, W is (approximately) the number of vertices that, by virtue of having a neighbor in R other than the center of their star, can be "saved" from entering T. Under the assumption that z is small, so that Phase 1 removes only few vertices, the following holds: If W is small, T is large, and Phase 3 succeeds in removing a large number of vertices in T. On the other hand, if W is large, then many vertices have two neighbors in R, and the near-twin reduction with support R in Phase 2 removes many vertices. The rest of this section serves to formalize and prove these claims. First note that

$$|N_{G_1}[S]| \geq \sum_{x \in Q:\, w(x) < |Star(x)|} |Star(x)| \geq \sum_{x \in Q:\, w(x) < |Star(x)|} (|Star(x)| - w(x))$$

$$\geq \sum_{x \in Q} (|Star(x)| - w(x)) = |V_1| - W \ .$$

Thus at most W vertices in G_1 do not belong to $N_{G_1}[S]$. This implies that at most W vertices in G_2 do not belong to $N_{G_2}[S] = S \cup T$. And this in turn means that $|V_2 \setminus T| \leq W + |S|$. A vertex in T is included in V_3' only if it has at least two neighbors outside of T in addition to the center of its star. By Lemma 2(b), therefore, $|V_3'| \leq |V_2 \setminus T| + 2|V_2 \setminus T| \leq 3(W + |S|)$ and $|V_3| = |V_3'| + |S| \leq 3W + O(z + k)$. This concludes the analysis for small W. Take $U = V_1 \setminus R$. The analysis for large W depends on the following characterization of W.

$$W = \sum_{x \in Q} \sum_{y \in R \setminus \{x\}} |N_{G_1}[y] \cap Star(x)| = \sum_{y \in R} \sum_{x \in Q \setminus \{y\}} |N_{G_1}[y] \cap Star(x)|$$

$$= \sum_{y \in R} |N_{G_1}[y]| - \sum_{y \in Q} |Star(y)| = \sum_{y \in R} |N_{G_1}(y)| - |U| \ .$$

Let H be the bipartite subgraph of G_1 induced by the set of edges with one endpoint in R and one endpoint in U. The total degree $\sum_{y \in R} |N_{G_1}(y)|$ in G_1 of the vertices in R overcounts the number of edges in H by twice the number of edges between vertices in R, i.e., according to Lemma 1(a), by at most $6|R|$. Therefore H has at least $|U| + W - 6|R|$ edges. Every vertex in U has at least one incident edge in H, namely to the center of its star. With $n_i = |N_{G_1}^i(R)|$ for $i = 1, 2, \ldots$, this means that $\sum_{i=1}^{\infty} (i - 1)n_i \geq W - 6|R|$. But then, by Lemma 2(a),

$$\sum_{i=2}^{\infty} n_i \geq W - 6|R| - \sum_{i=3}^{\infty} (i - 2)n_i \geq W - 6|R| - 2|R| = W - O(z + k) \ .$$

After Phase 2, by Lemmas 4 and 2(b), the number of vertices with two or more neighbors in R is $O(z+k)$. Phase 2 therefore reduces the number of vertices with two or more neighbors in R from $W - O(z+k)$ to $O(z+k)$, i.e., by $W - O(z+k)$.

In summary, for a certain constant $c > 0$, Phase 1 and Phase 2 reduce the number of vertices by z and by at least $\max\{W - c(z+k), 0\}$, respectively, while the number of vertices left after Phase 3 is bounded by $3W + c(z + k)$. The complete shrinking iteration therefore reduces the number of vertices by at least

$$\max\{z, W - c(z + k), n - 3W - c(z + k)\} \ .$$

A simple case analysis shows that if $n \geq 16ck$, then one of the three arguments of the maximum above is $\Omega(n)$: Without loss of generality, $z \leq n/(16c)$ and therefore $c(z + k) \leq n/16 + n/16 = n/8$. If $W \geq n/4$, the second term is at least $n/4 - n/8 = n/8$, and if not, the third term is at least $n - 3n/4 - n/8 = n/8$.

A shrinking iteration that starts with at least $16ck$ vertices therefore reduces the number of vertices by at least a constant factor.

5   The Complete Algorithm

The complete kernelization inputs an undirected n-vertex planar graph G and consists in applying repeated shrinking iterations to G until the number of vertices no longer drops at least at the rate established in the analysis in the previous section. Since our upper bound on the running time of successive iterations forms a geometric series, the total running time is $O(n)$. Every iteration preserves the domination number, so the final graph has domination number $Dom(G)$, and it is planar and contains $O(Dom(G))$ vertices.

References

1. Alber, J., Fellows, M.R., Niedermeier, R.: Polynomial-time data reduction for DOMINATING SET. J. ACM 51(3), 363–384 (2004)
2. van Bevern, R., Hartung, S., Kammer, F., Niedermeier, R., Weller, M.: Linear-time computation of a linear problem kernel for Dominating Set on planar graphs. In: Marx, D., Rossmanith, P. (eds.) IPEC 2011. LNCS, vol. 7112, pp. 194–206. Springer, Heidelberg (2012)
3. Bodlaender, H.L., Fomin, F.V., Lokshtanov, D., Penninkx, E., Saurabh, S., Thilikos, D.M.: (Meta) kernelization. In: FOCS, pp. 629–638. IEEE Computer Society (2009)
4. Chen, J., Fernau, H., Kanj, I.A., Xia, G.: Parametric duality and kernelization: Lower bounds and upper bounds on kernel size. SIAM J. Comput. 37(4), 1077–1106 (2007)
5. Guo, J., Niedermeier, R.: Linear problem kernels for NP-hard problems on planar graphs. In: Arge, L., Cachin, C., Jurdziński, T., Tarlecki, A. (eds.) ICALP 2007. LNCS, vol. 4596, pp. 375–386. Springer, Heidelberg (2007)
6. Jansen, B.: Kernelization for Maximum Leaf Spanning Tree with positive vertex weights. In: Calamoneri, T., Díaz, J. (eds.) CIAC 2010. LNCS, vol. 6078, pp. 192–203. Springer, Heidelberg (2010)
7. Kammer, F.: A linear-time algorithm to find a kernel for the Rooted Leaf Out-branching problem (2011) (manuscript)
8. Kloks, T., Lee, C.M., Liu, J.: New algorithms for k-face cover, k-feedback vertex set, and k-disjoint cycles on plane and planar graphs. In: Kučera, L. (ed.) WG 2002. LNCS, vol. 2573, pp. 282–295. Springer, Heidelberg (2002)
9. Niedermeier, R.: Invitation to Fixed-Parameter Algorithms. Oxford University Press, Oxford (2006)
10. Nishizeki, T., Chiba, N.: Planar Graphs: Theory and Algorithms. Dover Publications, Inc., Mineola (1988)
11. Protti, F., da Silva, M.D., Szwarcfiter, J.L.: Applying modular decomposition to parameterized cluster editing problems. Theory Comput. Syst. 44(1), 91–104 (2009)

Linear-Time Computation of a Linear Problem Kernel for Dominating Set on Planar Graphs

René van Bevern[1,*], Sepp Hartung[1], Frank Kammer[2],
Rolf Niedermeier[1], and Mathias Weller[1,**]

[1] Institut für Softwaretechnik und Theoretische Informatik, TU Berlin,
Berlin, Germany
{rene.vanbevern,sepp.hartung,rolf.niedermeier,mathias.weller}@tu-berlin.de
[2] Institut für Informatik, Universität Augsburg, Augsburg, Germany
kammer@informatik.uni-augsburg.de

Abstract. We present a linear-time kernelization algorithm that transforms a given planar graph G with domination number $\gamma(G)$ into a planar graph G' of size $O(\gamma(G))$ with $\gamma(G) = \gamma(G')$. In addition, a minimum dominating set for G can be inferred from a minimum dominating set for G'. In terms of parameterized algorithmics, this implies a linear-size problem kernel for the NP-hard DOMINATING SET problem on planar graphs, where the kernelization takes linear time. This improves on previous kernelization algorithms that provide linear-size kernels in cubic time.

1 Introduction

This work lies in the intersection of two active lines of research:

1. NP-hard problems on planar graphs and the exploitation of their structural properties to obtain better algorithms (approximation or fixed-parameter) [4, 5, 14], and
2. polynomial-time data reduction and problem kernelization [6, 14], an important subfield of parameterized complexity analysis.

Indeed, planar graph problems played an important role in the development of several lines of research in parameterized complexity analysis. More specifically, the topic of subexponential time fixed-parameter algorithms was first studied for the DOMINATING SET problem on planar graphs [1, 10]. The linear-size problem kernel for DOMINATING SET on planar graphs [2] may be considered as a nucleus for the recent, rapid growth of results on problem kernels [6, 14], and planar graph problems led to the first tractability results for the local search paradigm in the context of parameterized complexity [11]. In our work, again DOMINATING SET serves as a starting example, now for studying the issue of *linear-time* kernelizability as a natural goal within polynomial-time data reduction.

In a nutshell, a kernelization algorithm transforms in polynomial time an instance of a (typically NP-hard) problem into an equivalent instance whose size

---

* Supported by the DFG, project AREG, NI 369/9.
** Supported by the DFG, project DARE, NI 369/11.

D. Marx and P. Rossmanith (Eds.): IPEC 2011, LNCS 7112, pp. 194–206, 2012.

is bounded from above by a function of a problem-specific parameter. Nowadays, it has become a standard challenge to minimize the size of problem kernels [6, 14]. As to DOMINATING SET on planar graphs (given an undirected planar graph G and a positive integer k, select at most k vertices such that each unselected vertex in G has at least one selected neighbor[1]), there first was a $335k$-vertex problem kernel [2], which was further refined into a $67k$-vertex problem kernel [8], both computable in cubic time.[2] In this previous work, the focus was on "engineering" data reduction rules in order to gain a small provable kernel size. Here, we aim at engineering the usage of known (and "established") ones in terms of improved time complexity instead of heading for new data reduction rules.

Following up the work on problem kernels for DOMINATING SET on planar graphs, we shift the focus from improving the kernel size to improving the running time (from cubic to linear) while still maintaining a linear-size problem kernel. Since this turns out to be a demanding task; for the sake of improving readability, we do not measure the constant factor for the problem kernel size.[3] In this sense, our work parallels other classification work related to DOMINATING SET, where the goal was to extend the considered graph class and/or class of problems [7, 12, 13]. Finally, we conjecture and already found some evidence that our data reduction approach also extends to other problems on planar graphs [13], again leading to linear-time linear-size problem kernels.

A similar result was achieved with a different approach by Hagerup [15]. He presents a linear problem kernel that can be computed in linear time by providing completely new reduction rules that are tailored for DOMINATING SET. On the one hand, this may lead to smaller constants in the running time or the kernel bound and less complex analysis thereof. On the other hand, our approach of recycling the old reduction rules of Alber et al. [2] has the advantage of greater versatility: It can likely be applied to a variety of NP-hard problems on planar graphs (see [13]) and bears the possibility of improving the kernel-size in analogy to the work of Chen et al. [8] and Wang et al. [16].

Our Contributions. Revisiting previous data reduction rules for DOMINATING SET on planar graphs [2], we shift focus to the execution time of data reduction, improving it from cubic to linear time. To this end, we "rework" the known rules and their mathematical analysis and carefully analyze their interaction.

---

[1] In what follows, knowledge of the parameter value k, that is, the maximum allowed size of a dominating set, will not be explicitly used in our algorithms. However, to formulate our results in a parameterized algorithmics setting, we need the parameter k.

[2] Experimental work showed that the corresponding data reduction rules are useful on several real-world data sets [3].

[3] A more or less standard analysis leads to large constants—however, on the one hand, with a more refined analysis they can be significantly improved and on the other hand, it is just a worst-case bound saying little about the effect on real-world instances. The goal of this paper is clearly of classification nature, affirmatively answering a question posed independently by Jiong Guo and Saket Saurabh at *WorKer'2010* held in Leiden, Netherlands; that is, a linear-time linear-size kernel for DOMINATING SET in planar graphs is possible.

Our central observation is that one can significantly gain efficiency by a non-exhaustive application of data reduction rules. More specifically, implementing the known data reduction rules [2] in the natural and straightforward way would "unavoidably" lead to cubic running time: The reason for this is that one has to inspect a quadratic number of so-called (potential) regions that a planar embedding of the graph may have. Thus, one of our major technical contributions is to restrict the region decomposition concept in such a way that the inspection (and, thus, the data reduction) can be done much faster. In this way, we achieve an $O(k)$-vertex problem kernel for DOMINATING SET on planar graphs in linear time. Notably, our kernel size analysis is not as fine-grained as the previous ones [2, 8], meaning that we did not analyze the constant factor for the upper bound on the number of problem kernel vertices. Note, however, since multiple kernelization algorithms can be run on top of each other (this makes them quite different from approximation algorithms), using our algorithm as spear-head in combination with Chen et al.'s algorithm [8], for an n-vertex planar graph we can trivially achieve a problem kernel with $67k$ vertices in $O(n + k^3)$ time, somewhat attenuating the quest for a sharper kernel size analysis. Due to the lack of space, most proofs are deferred to a full version of the paper.

Notation. We only consider undirected graphs $G = (V, E)$, where $V(G) := V$ is the set of vertices and $E(G) := E$ is the set of edges. Furthermore, let $n := |V|$ and $m := |E|$. The *open neighborhood* $N^G(v)$ of a vertex $v \in V$ in G is the set of vertices that are adjacent to v in G. The *closed neighborhood* $N^G[v]$ is $N^G(v) \cup \{v\}$. For a vertex set $S \subseteq V$ we set $N^G(S) := \bigcup_{v \in S} N^G(v)$. We use the *joint neighborhood* $N^G(v, w)$ of two vertices to denote $(N^G(v) \cup N^G(w)) \setminus \{v, w\}$ and the *closed joint neighborhood* $N^G[v, w] := N^G[v] \cup N^G[w]$. A v_1-v_ℓ-*path* in G is a sequence $\mathcal{P} := (v_1, v_2, \ldots, v_\ell) \in V^\ell$ of vertices with $\{v_i, v_{i+1}\} \in E$ for $i \in \{1, \ldots, \ell - 1\}$ and $v_i \neq v_j$ for $i \neq j$, where $\ell - 1$ is the *length* of the path. We use $V(\mathcal{P})$ to denote the set of vertices of the path $\mathcal{P}$. We call two vertices v and w *connected* in G if there is a v-w path in G. We use $\mathrm{dist}^G(v, w)$ to denote the length of a shortest path between v and w in G, also called *distance*. The superscript G is omitted if G is clear from the context. The *domination number* $\gamma(G)$ of a graph G is the size of a smallest dominating set of G.

For a language $L \subseteq \Sigma^* \times \mathbb{N}$, a *kernelization algorithm* takes as input an instance (x, k), where k is called *parameter* and, computes in time polynomial in $|x| + k$ an instance (x', k') such that $(x', k') \in L \Leftrightarrow (x, k) \in L$, $|x'| \leq f(k)$, and $k' \leq k$. Here, f is a computable function solely depending on k which measures the *size* of the *problem kernel* (x', k').

2   Comparison to Previous Kernelizations

To obtain a linear-size problem kernel for DOMINATING SET on planar graphs, we employ a framework developed by Alber et al. [2]. They showed that a planar graph G with domination number $\gamma(G)$ can be decomposed into $O(\gamma(G))$ so-called "regions". Data reduction ensures that each of these regions has constant size and that $O(\gamma(G))$ vertices are not contained in any region. We follow a similar approach, modifying data reduction rules to run in linear time.

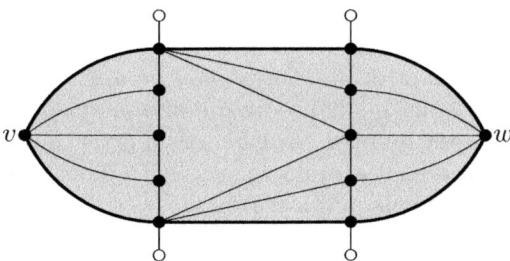

Fig. 1. Illustration of a region. The boundary path is shown as path of bold edges.

Basically, a region R of a planar graph G is a part of an embedding of G into the plane. Each region contains two vertices v and w such that $N[v, w]$ contains all vertices of R and a boundary that separates R from the rest of G, as illustrated in Figure 1.

Definition 1. *Let G be a plane graph.*[4] *A region $R(v, w)$ between two vertices v and w is a closed bounded subset S of the plane such that:*
  1. *the boundary of $R(v, w)$ is formed by two simple paths*[5] *between v and w, each of which has length at most three and*
  2. *all vertices inside S are from $N[v, w]$.*
We denote by $R(v, w)$ also the set of vertices in a region $R(v, w)$ and by $\partial R(v, w)$ the set of vertices on the boundary paths of a region $R(v, w)$. The vertices in $R(v, w) \setminus \partial R(v, w)$ are the inner vertices of $R(v, w)$.

Alber et al. [2] showed that each dominating set D of a planar graph G yields a so-called maximal D-region decomposition with $O(|D|)$ regions.

Definition 2. *For a plane graph G and $D \subseteq V(G)$, a D-region decomposition of G is a set $\mathcal{R}$ of regions between pairs of vertices in D such that*
  1. *$\forall v, w \subset D$ and $R(v, w) \in \mathcal{R}$, it holds that $D \cap R(v, w) = \{v, w\}$ and*
  2. *for two distinct regions $R_1, R_2 \in \mathcal{R}$, it holds that $(R_1 \cap R_2) \subseteq (\partial R_1 \cap \partial R_2)$.*
For a D-region decomposition $\mathcal{R}$, we define $V(\mathcal{R}) := \bigcup_{R \in \mathcal{R}} R$. A D-region decomposition $\mathcal{R}$ is maximal if there is no region $R \notin \mathcal{R}$ such that $\mathcal{R}' := \mathcal{R} \cup \{R\}$ is a D-region decomposition with $V(\mathcal{R}) \subsetneq V(\mathcal{R}')$.

Using data reduction, Alber et al. [2] shrink to constant size all regions that may potentially be part of a D-region decomposition for a minimum dominating set D of the input graph G. Since $|D| = \gamma(G)$, such a region decomposition comprises $O(\gamma(G))$ regions. Together with an $O(\gamma(G))$-bound on the number of vertices that are not in regions, this shows the linear size of the kernel.

Our goal is to modify the data reduction rules of Alber et al. [2] so that they can be applied in linear time instead of cubic time. Unfortunately, we have to make some sacrifices regarding the effectiveness of the data reduction rules, which we explain in Section 3. Alber et al. [2] employ two data reduction rules:

---

[4] A plane graph is a particular embedding of a planar graph.
[5] This also includes degenerated cases where the two paths have common vertices.

one that shrinks the neighborhood of vertices, and one that shrinks regions. To shrink the neighborhood of vertices, we use a slightly modified version of Reduction Rule 1 of Alber et al. [2] that can be applied exhaustively in linear time. To shrink regions, we show that we can find all regions whose inner vertices have at most a constant number of neighbors. Furthermore, we provide means to ensure that regions do not contain vertices with more neighbors. This enables us to find and shrink regions in linear time.

3   Data Reduction Rules

In this section, we first describe two data reduction rules and show that they are *correct*, that is, they maintain planarity and do not change the domination number of the input graph. Then, we show how to execute them in linear time. Whenever we introduce new vertices into a graph, we call them *dummy vertices*. Moreover, we assume that our data reduction rules can check in $O(1)$ time whether a vertex is a dummy vertex. This can be achieved by marking dummy vertices accordingly. Note that these marks will be removed from the final output graph in order to obtain a proper DOMINATING SET instance (where unmarked graphs are required as input).

3.1   Private Neighborhood Rule

As Alber et al. [2], we partition the neighborhood of a vertex v in a graph G into three subsets:

$$N_1^G(v) := \{u \in N^G(v) \mid N^G(u) \setminus N^G[v] \neq \emptyset\},$$
$$N_2^G(v) := \{u \in N^G(v) \setminus N_1^G(v) \mid N^G(u) \cap N_1^G(v) \neq \emptyset\},$$
$$N_3^G(v) := N^G(v) \setminus (N_1^G(v) \cup N_2^G(v)).$$

We now give our variant of Rule 1 of Alber et al. [2] for planar graphs.

Reduction Rule 1. Let $v \in V(G)$ be a vertex such that $|N_3^G(v)| > 1$ or $|N^G(N_3^G(v))| > 1$. Then, remove $N_3^G(v)$ from G and attach a new degree-one dummy vertex v' to v.

The correctness of Reduction Rule 1 follows from the correctness of Reduction Rule 1 of Alber et al. [2], as we only delete a subset of the vertices of which each was shown to be safely removable by Alber et al. [2]. By removing only a subset of the removable vertices, we can show that, for an exhaustive application of Reduction Rule 1, it is sufficient to apply Reduction Rule 1 once for every vertex. In this way, we can prove Lemma 1 below. In the following, we say that a graph G is *reduced* with respect to Reduction Rule 1 if Reduction Rule 1 is not applicable to G.

Lemma 1. *For planar graphs, Reduction Rule 1 can be applied exhaustively in $O(n)$ time.*

Proof. Let G be a planar graph, and let $v \in V(G)$ be a vertex that does not satisfy the conditions of Reduction Rule 1, that is, neither $|N_3^G(v)| > 1$ nor $|N^G(N_3^G(v))| > 1$. We show that, in the graph G' that results from applying Reduction Rule 1 to a vertex $u \in V(G) \setminus \{v\}$, the vertex v still does not satisfy the conditions of Reduction Rule 1. This implies that, in order to apply Reduction Rule 1 exhaustively to G, it is sufficient to apply Reduction Rule 1 at most once to each vertex. As shown by Alber et al. [2, Lemma 2], this can be done in $O(n)$ time for planar graphs.

Towards a contradiction, assume that Reduction Rule 1 is applicable to v in G'. Then, because Reduction Rule 1 does not add edges between vertices in $V(G)$, it must hold that $N_3^{G'}(v) \cap (N_1^G(v) \cup N_2^G(v)) \neq \emptyset$. However, we show that the contrary is true. To this end, recall that each vertex in $N_2^G(v)$ is adjacent to a vertex in $N_1^G(v)$. Thus, in order to show $N_3^{G'}(v) \cap (N_1^G(v) \cup N_2^G(v)) = \emptyset$, it is sufficient to show that for each vertex $x \in N_1^G(v)$ it holds that $N^G[x] \cap N_3^{G'}(v) = \emptyset$. We distinguish the following two cases:

First, assume that $x \in N_2^{G'}(v) \cup N_3^{G'}(v)$. This is only true if Reduction Rule 1, when applied to u, deletes all neighbors of x that are nonadjacent to v. Let $y \in N_3^G(u) \cap N^G(x) \setminus N^G[v]$ be one such neighbor. Since $y \in N_3^G(u)$, we know that $x \in N^G[u] \setminus N_1^G(u)$ and thus $N^{G'}[x] \subseteq N^{G'}[u]$. Hence, u is adjacent to v. Because u has a degree-one dummy neighbor in G', no vertex from $N^{G'}[u]$ is contained in $N_3^{G'}(v)$. Since $N^G(x) \cap V(G') \subseteq N^{G'}[u]$, this implies $N^G[x] \cap N_3^{G'}(v) = \emptyset$.

In the second case, assume that $x \notin N_2^{G'}(v) \cup N_3^{G'}(v)$. Thus, we only have to show that the vertices in $N^G(x)$ are not in $N_3^{G'}(v)$. If $x \in N_1^{G'}(v)$, then, obviously, none of the vertices in $N^G(x)$ is in $N_3^{G'}(v)$. Hence, consider the subcase where $x \notin N_1^{G'}(v)$. This implies $x \notin N^{G'}(v)$ and thus x would have been deleted by Reduction Rule 1. Therefore, we have $x \in N_3^G(u)$, that is, $N^G[x] \subseteq N^G[u]$, implying $v \in N^G(u)$. Again, because u has a degree-one dummy neighbor in G', no vertex from $N^{G'}[u]$ is contained in $N_3^{G'}(v)$, implying $N^G[x] \cap N_3^{G'}(v) = \emptyset$. □

3.2   Joint Neighborhood Rule

In this section, we present a data reduction rule that shrinks regions to constant size. The presented data reduction rule is based on Reduction Rule 2 by Alber et al. [2]. However, we modify it as follows: Reduction Rule 2 of Alber et al. [2] removes certain vertices from the sets $N(v, w)$ for vertices $v, w \in V$. Since we cannot compute $N(v, w)$ for all vertex pairs $v, w \in V$ in linear time, we will show that it is sufficient to only remove vertices from efficiently-computable subsets $N_0(v, w) \subseteq N(v, w)$ for a linear number of vertex pairs. More specifically, $N_0(v, w)$ contains vertices on short low-degree v-w-paths.

Definition 3. *A vertex v with $\deg(v) \leq 78$ is called* low-degree vertex.[6] *A v-w-path consisting only of v, w, and low-degree vertices is called* low-degree path.

Note that all low-degree paths of constant length c starting at some vertex v can be found in $O(\deg(v))$ time by starting a breadth-first search at v, only

---

[6] Herein, the constant 78 results from the mathematical analysis and can be improved using a more intricate analysis.

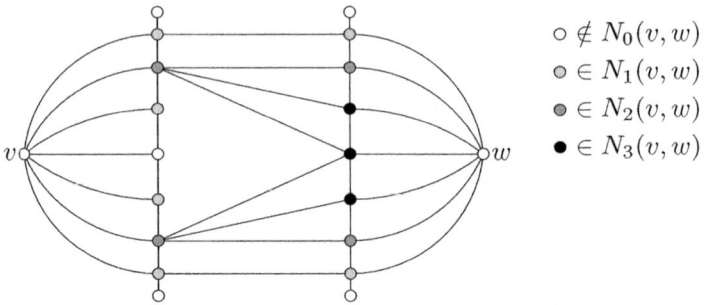

$$\circ \notin N_0(v, w)$$
$$\circ \in N_1(v, w)$$
$$\circ \in N_2(v, w)$$
$$\bullet \in N_3(v, w)$$

Fig. 2. An illustration of Definition 4

descending on low-degree vertices and stopping at depth c. It can be checked in $O(1)$ time whether a vertex is a low-degree vertex by iterating over its adjacency list, aborting when a 79th neighbor is found.

Observation 1. For a vertex v, all constant-length low-degree paths starting at v can be listed in $O(\deg(v))$ time.

To present our data reduction rule, we need the following definition of joint neighborhoods. It strongly resembles the definition used by Alber et al. [2, Section 2.2]. However, our sets $N_i^G(v, w)$ for $i \in \{1, 2, 3\}$ are defined with respect to $N_0^G(v, w)$ instead of $N^G(v, w)$. The following definition is illustrated in Figure 2.

Definition 4. *Let v, w be vertices in a planar graph G. We define*

$$N_0^G(v, w) := \{u \in N^G(v, w) \mid u \text{ is on a low-degree } v\text{-}w\text{-path of length at most}$$
$$\text{four that only consists of vertices in } N^G[v, w]\},$$
$$N_1^G(v, w) := \{u \in N_0^G(v, w) \mid N^G(u) \setminus N_0^G[v, w] \neq \emptyset\},$$
$$N_2^G(v, w) := \{u \in N_0^G(v, w) \setminus N_1^G(v, w) \mid N^G(u) \cap N_1^G(v, w) \neq \emptyset\},$$
$$N_3^G(v, w) := N_0^G(v, w) \setminus (N_1^G(v, w) \cup N_2^G(v, w)).$$

Using Definition 4, we present our second data reduction rule in form of Algorithm 1, which we now explain. Algorithm 1 basically corresponds to Reduction Rule 2 of Alber et al. [2]. For each pair $v, w \in V$ with $N_0^G(v, w) \neq \emptyset$, Algorithm 1 removes a subset of $N_0^G(v, w)$ from the graph G and, if applicable, attaches degree-one dummy vertices to v or w. We first explain the data reduction between lines 5 and 10 and then explain the purpose of the EnsurePaths procedure called in line 4. The set N_3 introduced in line 5 of Algorithm 1 is the set of vertices that may possibly be removed. We will see that N_3 can be efficiently computed and updated. Moreover, the choice of N_3 ensures the correctness of the data reduction executed by Algorithm 1, which can be seen by comparing it to Reduction Rule 2 of Alber et al. [2]: it is straightforward to observe that, if the condition in line 6 is satisfied, then the corresponding condition for Reduction Rule 2 of Alber et al. [2] is also satisfied. Moreover, in this case, we remove only a subset of vertices that Alber et al. [2] show to be safely removable. Note that the vertices z, z' chosen in line 7 exist:

Algorithm 1: Reduce Vertices in Regions

Input: A planar graph $G = (V, E)$.

Output: A planar graph $G' = (V', E')$.

1 compute $N_0^G(v, w)$ for all $v, w \in V$ with $N_0^G(v, w) \neq \emptyset$;

2 $G' \leftarrow G$;

3 **foreach** (v, w) such that v and w are non-dummy vertices and $N_0^G(v, w) \neq \emptyset$ **do**

4   EnsurePaths($G', v, w, N_0^G(v, w)$);

5      $N_3 \leftarrow N_3^{G'}(v, w) \cap N_0^G(v, w)$;

6   **if** $|N_3| \geq 4$ and N_3 cannot be dominated by a single vertex $u \notin \{v, w\}$ **then**

7   **if** $N_3 \subseteq N^{G'}(v) \cap N^{G'}(w)$ **then** remove the vertices in $N_3 \setminus \{z, z'\}$

        from G', for arbitrary $z, z' \in N_3$ with $(N^{G'}(z) \cap N^{G'}(z')) \setminus N_3 \subseteq \{v, w\}$

        and $\{z, z'\} \notin E(G')$;

8   **else if** $N_3 \subseteq N^{G'}(v)$ and $N_3 \nsubseteq N^{G'}(w)$ **then** remove N_3 from G' and

        (unless already done before) attach a degree-one dummy vertex v' to v;

9   **else if** $N_3 \nsubseteq N^{G'}(v)$ and $N_3 \subseteq N^{G'}(w)$ **then** remove N_3 from G' and

        (unless already done before) attach a degree-one dummy vertex w' to w;

10   **else if** $N_3 \nsubseteq N^{G'}(v)$ and $N_3 \nsubseteq N^{G'}(w)$ **then** remove N_3 from G' and

        (unless already done before) attach a degree-one dummy vertex v' to v

        and a new degree-one dummy vertex w' to w;

11 **return** G'

Lemma 2. *Let v, w be vertices of a planar graph G' and let $N_3 \subseteq N_3^{G'}(v, w) \cap N^{G'}(v) \cap N^{G'}(w)$ with $|N_3| \geq 4$. Then, N_3 contains vertices z, z' with $\{z, z'\} \notin E(G')$ and $(N^{G'}(z) \cap N^{G'}(z')) \setminus N_3 \subseteq \{v, w\}$.*

We now explain the EnsurePaths procedure called in line 4 for a vertex pair (v, w). Observe that the graph G' considered in the for-loop in line 3 of Algorithm 1 is not necessarily reduced with respect to Reduction Rule 1 since previous iterations of the loop might have deleted vertices. However, an application of Reduction Rule 1 might become necessary: it might happen that some vertex $u \in N_0^G(v, w)$ is not in $N_0^{G'}(v, w)$ when the pair (v, w) is considered in line 3. Such a situation is illustrated in Figure 3 and could arise if all vertices on u's low-degree v-w-paths are deleted by data reduction executed for some other vertex

Procedure. EnsurePaths($G', v, w, N_0^G(v, w)$)

// for $x, u \in V(G')$, let $B(x, u) := \{u' \in V(G') \setminus \{x\} \mid \text{dist}^{G' - \{x\}}(u, u') \leq 2\}$

1 $N_3^{\text{ldv}}(v) \leftarrow \{u \in N_3^{G'}(v) \cap N_0^G(v, w) \mid B(v, u)$ only has low-degree vertices$\}$;

2 $N_3^{\text{ldv}}(w) \leftarrow \{u \in N_3^{G'}(w) \cap N_0^G(v, w) \mid B(w, u)$ only has low-degree vertices$\}$;

3 **if** $|N^{G'}(v)| > 1 \wedge N_3^{\text{ldv}}(v) \neq \emptyset$ **then** remove $N_3^{\text{ldv}}(v)$ from G' and (unless v already has one) attach a new degree-one dummy vertex v' to v;

4 **if** $|N^{G'}(w)| > 1 \wedge N_3^{\text{ldv}}(w) \neq \emptyset$ **then** remove $N_3^{\text{ldv}}(w)$ from G' and (unless w already has one) attach a new degree-one dummy vertex w' to w;

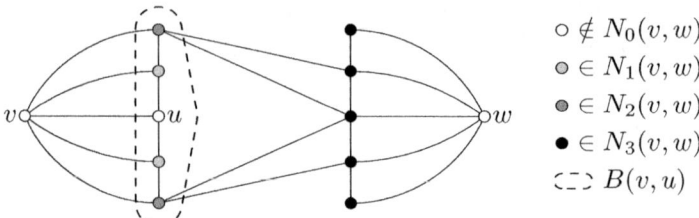

Fig. 3. The vertex u is not on a v-w-path of length at most four. Therefore, u and its neighbors are not deleted by a call of Algorithm 1. However, $u \in N_3(v)$. Hence, Reduction Rule 1 would delete u. Moreover, since $B(v, u)$ (surrounded by the dashed line) only contains low-degree vertices, EnsurePaths will delete u as well.

pair. As in Figure 3, this could prevent u or some of its neighbors from being removed from G'. In the situation shown, Reduction Rule 1 would delete u. Hence, in order to ensure that a vertex in G' that is in $N_0^G(v, w)$ has a low-degree v-w-path in G', it could help to apply Reduction Rule 1 to v and w. However, doing so for each considered pair (v, w) might be too time-consuming. This is the reason why EnsurePaths is employed, which in lines 3 and 4 deletes a subset of $N_3^{G'}(v)$ and $N_3^{G'}(w)$ in a case where Reduction Rule 1 would completely delete $N_3^{G'}(v)$ and $N_3^{G'}(w)$. Namely, those vertices $u \in N_3^{G'}(v)$ (or $N_3^{G'}(w)$) are deleted for which $B(v, u)$ (or $B(w, u)$, respectively) only contains low-degree vertices—a condition which merely ensures that we can efficiently check whether $u \in N_3^{G'}(v)$ or $u \in N_3^{G'}(w)$. Since EnsurePaths deletes only vertices which Reduction Rule 1 deletes, EnsurePaths is correct. Moreover, observe that, in Figure 3, the vertex u would be deleted from G' by EnsurePaths.

To execute Algorithm 1 one can compute all sets $N_0^G(v, w)$ in linear time using Algorithm 2. Also, one frequently has to check whether a vertex is in $N_3^{G'}(v, w)$. The following lemma shows that this can be done efficiently.

Lemma 3. *For a planar graph G and vertices u, v, w of G, it is $\mathrm{O}(1)$-time-decidable whether $u \in N_3^G(v, w)$. Moreover, all sets $N_0^G(v, w)$ can be enumerated in $\mathrm{O}(n)$ total time for all vertices $v, w \in V(G)$ with $N_0^G(v, w) \neq \emptyset$.*

Since we can efficiently check membership of a vertex in $N_3^G(v, w)$ and how to compute all sets $N_0^G(v, w)$, we have all ingredients to prove the running time of Algorithm 1.

Lemma 4. *On planar graphs, Algorithm 1 can be executed in $\mathrm{O}(n)$ time.*

4   Problem Kernel

This section presents our kernelization algorithm based on the data reduction rules shown in the previous section. We explain how, given a planar graph G, the algorithm computes a graph G' with $\gamma(G) = \gamma(G')$ whose size is linear

---

Algorithm 2: Compute $N_0^G(v,w)$ for all $v,w \in V(G)$ with $N_0^G(v,w) \neq \emptyset$

Input: A planar graph $G = (V,E)$ with vertices numbered $V := \{1, \ldots, n\}$.
Output: $N_0^G(v,w)$ for all $v,w \in V$ with $N_0^G(v,w) \neq \emptyset$.

1 $\mathcal{D} \leftarrow$ empty list;   /* $(v,w,u) \in \mathcal{D}$ will be equivalent to $u \in N_0^G(v,w)$ */
2 **for** $v \in V$ *and each low-degree path* p *of length at most four starting at* v **do**
3 $\quad$ $w \leftarrow$ ending vertex of p;
4 $\quad$ **if** *all vertices of* p *are in* $N^G[v,w]$ **then**
5 $\quad\quad$ **foreach** *vertex* $u \in V \setminus \{v,w\}$ *of* p **do** append (v,w,u) to $\mathcal{D}$;

6 sort $\mathcal{D}$ in lexicographical order using radix sort;
7 **foreach** $(v,w,u) \in \mathcal{D}$ *in lexicogr. order* **do**   /* collect $N_0^G(v,w)$ from $\mathcal{D}$ */
8 $\quad$ $(v',w',u') \leftarrow$ previous element in $\mathcal{D}$;
9 $\quad$ **if** $v \neq v' \vee w \neq w'$ **then**   /* we encounter the pair (v,w) the first time */
10 $\quad\quad$ new set $N_0^G(v,w) \leftarrow \{u\}$;

11 $\quad$ **else if** $u \neq u'$ **then** add u to $N_0^G(v,w)$;   /* avoids duplicates */
12 **return** $N_0^G(v,w)$ for all $v,w \in V$ with $N_0^G(v,w) \neq \emptyset$;

---

in $\gamma(G)$. The kernelization algorithm runs in three phases. Each phase applies Reduction Rule 1 or Algorithm 1 to finally output G'.

Phase 1. Exhaustively apply Reduction Rule 1 to G. Let G_1 denote the resulting graph. By Lemma 1, G_1 is computable in O(n) time and is reduced with respect to Reduction Rule 1.

Phase 2. Apply Algorithm 1 to G_1, then Reduction Rule 1 exhaustively, then again Algorithm 1 and, finally, Reduction Rule 1 exhaustively. Let the result be denoted by G_2. By Lemmas 1 and 4, G_2 is computable in O(n) time. We will see that most vertices in G_2 have degree at most 78.

Phase 3. Apply Algorithm 1 and exhaustively apply Reduction Rule 1 to G_2, resulting in a graph G_3. Using the fact that in G_2 most vertices have at most 78 neighbors and that G_2 is reduced with respect to Reduction Rule 1, we can show that, using Algorithm 1, Phase 3 removes enough parts from G_2 to obtain a linear-size problem kernel.

To show that G_3 has size O($\gamma(G)$), we transfer structural observations from G_1 to the graphs G_2 and G_3 using a maximal D-region decomposition of G_1 with respect to an arbitrary embedding. Herein, we choose D as a dominating set of size at most $2\gamma(G)$ for G_1 that contains the set D' of vertices in $V(G_1) \cap V(G_3)$ to which dummy vertices have been attached by data reduction rules, that is, $D' := \{v \in V(G_1) \cap V(G_3) \mid N^{G_3}(v) \setminus V(G_1) \neq \emptyset\}$. Note that we may not be able to efficiently compute D. However, D and the embedding of G_1 are only required for the analysis. We first show that D exists and, thereafter, we show how a maximal D-region decomposition helps us to transfer structure from G_1 to G_2 and G_3.

Lemma 5. *There is a dominating set* $D \supseteq D'$ *for* G_1 *with* $|D| \leq 2\gamma(G)$.

For the remainder of this section, we base our reasoning on a fixed maximal D-region decomposition $\mathcal{R}$ of G_1 with $D \supseteq D'$. Notice that if a degree-one dummy vertex is attached to a vertex v, then neither Reduction Rule 1 nor Algorithm 1 can delete v. Instead, v is in G_3, where it also has a (possibly different) dummy neighbor. The choice of $\mathcal{R}$ and the definition of a D-region-decomposition for G_1 then leads to the following:

Observation 2. Let $R(v, w) \in \mathcal{R}$ be a region. Reduction Rule 1 and Algorithm 1 do not attach dummy vertices to any vertex in $R(v, w) \setminus \{v, w\}$.

Observation 2 ensures that, in none of the graphs G_1, G_2, and G_3, the inner vertices of a region $R \in \mathcal{R}$ have neighbors that are not in R. In this way, Observation 2, to a certain extent, preserves the region structure of G_1 in G_2 and G_3.

As shown by Alber et al. [2, Proposition 1], $|\mathcal{R}| \in O(|D|) = O(\gamma(G))$. The proof of the problem kernel size now works as follows: Proposition 1 shows that Phase 2 shrinks the number of vertices that are not in any region of $\mathcal{R}$ to $O(\gamma(G))$.

Proposition 1. $|(V(G_1) \setminus V(\mathcal{R})) \cap V(G_2)| \in O(\gamma(G))$.

Lemma 6 shows that, as a result of Phase 2, inner vertices of regions in G_1 that are also present in G_2 have constant degree in G_2.

Lemma 6. *Let* $R(v, w) \in \mathcal{R}$, *and let* $u \in V(G_2) \setminus \{v, w\}$. *If* $\{v, w\} \subseteq V(G_2)$, *then* $|N^{G_2}(u) \cap R(v, w)| \leq 78$.

Exploiting Lemma 6, Proposition 2 shows that, using Algorithm 1, Phase 3 shrinks each region $R \in \mathcal{R}$ to $O(1)$ vertices.

Proposition 2. *Let* $R(v, w) \in \mathcal{R}$. *Then,* $|V(G_3) \cap R(v, w)| \in O(1)$.

Finally, we bound the number of vertices added by our kernelization algorithm (dummy vertices). Since $|D'| \in O(\gamma(G))$ is shown by Lemma 5 and to each such vertex at most one dummy vertex is added, it follows that there are at most $O(\gamma(G))$ vertices in $V(G_3) \setminus V(G_1)$. We conclude that G_3 consists of $O(\gamma(G))$ vertices, yielding our central theorem:

Theorem 1. *On planar graphs, a linear-size problem kernel for* DOMINATING SET *is computable in linear time.*

5   Conclusion

Our work is meant to provide a first case study, using the well-known kernelization of DOMINATING SET on planar graphs, on how known kernelization algorithms can be tuned for better time performance by carefully analyzing the interaction and costs of the underlying data reduction rules. Clearly, on the practical side it is important to further improve on the upper bound for the kernel size to be achieved in linear time.

Informally speaking, the analysis of our problem kernel shows that "knowing when to stop" may be a source of performance gain. This is particularly true for kernelization algorithms since faster algorithms with worse kernel bounds can be combined with slower algorithms yielding better kernel bounds in order to achieve an overall improvement.

As to future challenges, there are a lot of possibilities in revisiting known kernelization results and re-engineering them in terms of algorithmic efficiency. As to DOMINATING SET on planar graphs, we left open whether there is a linear-time problem kernel that can be achieved by an *exhaustive* application of data reduction rules (and not stopping their application for reasons of efficiency). In this work, we computed an $O(\gamma(G))$-size problem kernel in $O(n)$ time. For other problems, it might be helpful to alleviate the quest from $O(n)$-time computability to $O(p(k) + n)$-time or $O(p(k) \cdot n)$-time computability, where p is a polynomial solely depending on a parameter k. For instance, Chor et al. [9] presented an almost linear-time kernelization for a variant of the CLIQUE COVERING problem. Finally, we leave it as a challenge for future research to investigate the complexity classes of fixed-parameter tractable problems that allow for kernels of *any* size computable in (almost) linear running time.

References

[1] Alber, J., Bodlaender, H.L., Fernau, H., Kloks, T., Niedermeier, R.: Fixed parameter algorithms for Dominating Set and related problems on planar graphs. Algorithmica 33(4), 461–493 (2002)
[2] Alber, J., Fellows, M.R., Niedermeier, R.: Polynomial-time data reduction for dominating set. J. ACM 51(3), 363–384 (2004)
[3] Alber, J., Betzler, N., Niedermeier, R.: Experiments on data reduction for optimal domination in networks. Ann. Oper. Res. 146(1), 105–117 (2006)
[4] Baker, B.S.: Approximation algorithms for NP-complete problems on planar graphs. J. ACM 41(1), 153–180 (1994)
[5] Bateni, M., Hajiaghayi, M., Marx, D.: Approximation schemes for steiner forest on planar graphs and graphs of bounded treewidth. In. In: Proc. 42th STOC, pp. 211–220. ACM Press (2010)
[6] Bodlaender, H.L.: Kernelization: New Upper and Lower Bound Techniques. In: Chen, J., Fomin, F.V. (eds.) IWPEC 2009. LNCS, vol. 5917, pp. 17–37. Springer, Heidelberg (2009)
[7] Bodlaender, H.L., Fomin, F.V., Lokshtanov, D., Penninkx, E., Saurabh, S., Thilikos, D.M.: (Meta) kernelization. In: Proc. 50th FOCS, pp. 629–638. IEEE (2009)
[8] Chen, J., Fernau, H., Kanj, I.A., Xia, G.: Parametric duality and kernelization: Lower bounds and upper bounds on kernel size. SIAM J. Comput. 37(4), 1077–1106 (2007)
[9] Chor, B., Fellows, M., Juedes, D.W.: Linear Kernels in Linear Time, or How to Save k Colors in $O(n^2)$ Steps. In: Hromkovič, J., Nagl, M., Westfechtel, B. (eds.) WG 2004. LNCS, vol. 3353, pp. 257–269. Springer, Heidelberg (2004)
[10] Dorn, F., Fomin, F.V., Thilikos, D.M.: Subexponential parameterized algorithms. Computer Science Review 2(1), 29–39 (2008)
[11] Fellows, M.R., Rosamond, F.A., Fomin, F.V., Lokshtanov, D., Saurabh, S., Villanger, Y.: Local search: Is brute-force avoidable? In: Proc. 21st IJCAI, pp. 486–491 (2009)

[12] Fomin, F.V., Lokshtanov, D., Saurabh, S., Thilikos, D.M.: Bidimensionality and kernels. In: Proc. 21st SODA, pp. 503–510. ACM/SIAM (2010)

[13] Guo, J., Niedermeier, R.: Linear Problem Kernels for NP-Hard Problems on Planar Graphs. In: Arge, L., Cachin, C., Jurdziński, T., Tarlecki, A. (eds.) ICALP 2007. LNCS, vol. 4596, pp. 375–386. Springer, Heidelberg (2007)

[14] Guo, J., Niedermeier, R.: Invitation to data reduction and problem kernelization. SIGACT News 38(1), 31–45 (2007)

[15] Hagerup, T.: Linear-time kernelization for planar dominating set. In: Marx, D., Rossmanith, P. (eds.) IPEC 2011. LNCS, vol. 7112, pp. 181–193. Springer, Heidelberg (2012)

[16] Wang, J., Yang, Y., Guo, J., Chen, J.: Linear Problem Kernels for Planar Graph Problems with Small Distance Property. In: Murlak, F., Sankowski, P. (eds.) MFCS 2011. LNCS, vol. 6907, pp. 592–603. Springer, Heidelberg (2011)

Tight Complexity Bounds for FPT Subgraph Problems Parameterized by Clique-Width[*]

Hajo Broersma, Petr A. Golovach, and Viresh Patel

School of Engineering and Computing Sciences, Durham University,
Science Laboratories, South Road, Durham DH1 3LE, UK
{hajo.broersma,petr.golovach,viresh.patel}@durham.ac.uk

Abstract. We give tight algorithmic lower and upper bounds for some double-parameterized subgraph problems when the clique-width of the input graph is one of the parameters. Let G be an arbitrary input graph on n vertices with clique-width at most w. We prove the following results.

- The DENSE (SPARSE) k-SUBGRAPH problem, which asks whether there exists an induced subgraph of G with k vertices and at least q edges (at most q edges, respectively), can be solved in time $k^{O(w)} \cdot n$, but it cannot be solved in time $2^{o(w \log k)} \cdot n^{O(1)}$ unless the Exponential Time Hypothesis (ETH) fails.
- The d-REGULAR INDUCED SUBGRAPH problem, which asks whether there exists a d-regular induced subgraph of G, and the MINIMUM SUBGRAPH OF MINIMUM DEGREE AT LEAST d problem, which asks whether there exists a subgraph of G with k vertices and minimum degree at least d, can be solved in time $d^{O(w)} \cdot n$, but they cannot be solved in time $2^{o(w \log d)} \cdot n^{O(1)}$ unless ETH fails.

1 Introduction

The notion of clique-width introduced by Courcelle and Olariu [14] (we refer the reader to the survey [24] for further information on different width parameters) has now become one of the fundamental parameters in Graph Algorithms. Many problems which are hard on general graphs can be solved efficiently when the input is restricted to graphs of bounded clique-width. The meta-theorem of Courcelle, Makowsky, and Rotics [13] states that all problems expressible in MS_1-logic are *fixed parameter tractable* (FPT), when parameterized by the clique-width of the input graph (see the books of Downey and Fellows [18] and Flum and Grohe [21] for a detailed treatment of parameterized complexity). In other words, this theorem shows that any problem expressible in MS_1-logic can be solved for graphs of clique-width at most w in time $f(w) \cdot |I|^{O(1)}$, where $|I|$ is the size of the input and f is a computable function depending on the parameter w only. Here, the superexponential function f is defined by a logic formula, and it grows very fast.

The basic method for constructing algorithms for graphs of bounded clique-width is to use dynamic programming along an expression tree (the definition

---

[*] This work is supported by EPSRC (EP/G043434/1 and F064551/1).

D. Marx and P. Rossmanith (Eds.): IPEC 2011, LNCS 7112, pp. 207–218, 2012.

is given in Section 2). Computing clique-width is an NP-hard problem [20], but it can be approximated and a corresponding expression tree can be constructed in FPT-time [23,30]. In our paper it is always assumed that an expression tree is given. In this case dynamic programming algorithms can be relatively efficient: usually single-exponential in the clique-width. A natural question to ask is whether the running times of such algorithms are asymptotically optimal up to some reasonable complexity conjectures.

The Exponential Time Hypothesis has proved to be an effective tool for establishing tight complexity bounds for parameterized problems, but there are still not many results of this nature in the literature. The *Exponential Time Hypothesis (ETH)* [25] asserts that there does not exist an algorithm for solving 3-SAT running in time $2^{o(n)}$ on a formula with n variables; this is equivalent to the parameterized complexity conjecture that FPT $\neq$ M[1] [17,21]. Chen et al. [8,9,10] showed that there is no algorithm for k-CLIQUE running in time $f(k)n^{o(k)}$, for n-vertex graphs, unless ETH fails (on the other hand it is easily seen that k-CLIQUE can be solved in time $n^{O(k)}$). The lower bound on the k-CLIQUE problem can be extended to some other parameterized problems via linear FPT-reductions [9,10]. In particular, for problems parameterized by clique-width, Fomin et al. [22] proved that MAX-CUT and EDGE DOMINATING SET cannot be solved in time $f(w)n^{o(w)}$ on n-vertex graphs of clique-width at most w, unless ETH collapses. For FPT problems, Cai and Juedes [6] proved that the parameterized version of any MaxSNP-complete problem cannot be solved in time $2^{o(k)} \cdot |I|^{O(1)}$ if ETH holds. Here k is the natural parameter of an MaxSNP-complete problem with the instance I, i.e. the maximized function should have a value at least k.

Lokshtanov, Marx and Saurabh [28] considered several FPT problems solvable in time $2^{O(k \log k)} \cdot |I|^{O(1)}$ and showed that a $2^{o(k \log k)} \cdot |I|^{O(1)}$-time algorithm for these problems would violate ETH. To do this, they introduced special restricted versions of some basic problems like k-CLIQUE on graphs with k^2 vertices (and with some other restrictions) and proved that these problems cannot be solved in time $2^{o(k \log k)} \cdot k^{O(1)}$ unless ETH collapses. These results open the possibility of establishing algorithmic lower bounds for natural problems. We use this approach to prove asymptotically tight bounds for some double-parameterized subgraph problems when the clique-width of the input graph is one of the parameters. These results give the first known bounds for such types of problems parameterized by clique-width.

First, we consider the DENSE k-SUBGRAPH problem (also known as the k-CLUSTER problem). This problem asks whether, given a graph G and positive integers k and q, there exists an induced subgraph of G with k vertices and at least q edges. Clearly, DENSE k-SUBGRAPH is NP-hard since it is a generalization of the k-CLIQUE problem. It remains NP-hard, even when restricted to comparability graphs, bipartite graphs and chordal graphs [12], as well as on planar graphs [26]. Polynomial algorithms were given for cographs, split graphs [12], and for graphs of bounded tree-width [26]. Considerable work has been done on approximation algorithms for this problem [3,4,15,19,27].

Next, we consider some degree-constrained subgraph problems. The objective in such problems is to find a subgraph satisfying certain lower or upper bounds on the degree of each vertex. Typically it is necessary to either check the existence of a subgraph satisfying the degree constraints or to minimize (maximize) some parameter (usually the size of the subgraph).

The d-REGULAR SUBGRAPH problem, which asks whether a given graph contains a d-regular subgraph, has been intensively studied. We mention here only some complexity results. Chvátal et al. [11] proved that this problem is NP-complete for $d = 3$. It was shown that the problem with $d = 3$ remains NP-complete for planar bipartite graphs with maximum degree four, and that when $d \geq 3$, it is NP-complete even for bipartite graphs with maximum degree at most $d + 1$. Some further results were given in [7,32,33,34]. We consider a variant of this problem called d-REGULAR INDUCED SUBGRAPH, where we ask whether a given graph G contains a d-regular *induced* subgraph. This variant of the problem has also been studied. In particular, the parameterized complexity of different variants of the problem was considered by Moser and Thilikos [31] and by Mathieson and Szeider [29]. Observe that, trivially, d-REGULAR INDUCED SUBGRAPH can be solved in polynomial time for $d \leq 2$, and it easily follows from the known hardness results for d-REGULAR SUBGRAPH that d-REGULAR INDUCED SUBGRAPH is NP-complete for any fixed $d \geq 3$.

In [2] Amini et al. introduced the MINIMUM SUBGRAPH OF MINIMUM DEGREE AT LEAST d problem. This problem asks whether, given a graph G and positive integers d and k, there exists a subgraph of G with at most k vertices and minimum degree at least d. The parameterized complexity of the problem was considered in [2]. Some other hardness and approximation results can be found in [1].

Our Main Results and the Organization of the Paper. In Section 2 we give some basic definitions and some preliminary results. In Section 3 we consider the DENSE k-SUBGRAPH and SPARSE k-SUBGRAPH problems. The SPARSE k-SUBGRAPH problem is dual to DENSE k-SUBGRAPH and it asks whether, given a graph G and positive integers k and q, there exists an induced subgraph of G with k vertices and at most q edges. We prove that these problems can be solved in time $k^{O(w)} \cdot n$ for n-vertex graphs of clique-width at most w if an expression tree of width w is given, but they cannot be solved in time $2^{o(w \log k)} \cdot n^{O(1)}$ unless ETH fails even if an expression tree of width w is included in the input. In Section 4 we consider the d-REGULAR INDUCED SUBGRAPH and MINIMUM SUBGRAPH OF MINIMUM DEGREE AT LEAST d problems. We construct dynamic programming algorithms which solve these problems in time $d^{O(w)} \cdot n$ for n-vertex graphs of clique-width at most w if an expression tree of width w is given, and then prove that these problems cannot be solved in time $2^{o(w \log d)} \cdot n^{O(1)}$ unless ETH fails even if an expression tree of width w is provided. We conclude the paper with some open problems.

2   Definitions and Preliminary Results

Graphs. We consider finite undirected graphs without loops or multiple edges. The vertex set of a graph G is denoted by $V(G)$ and its edge set by $E(G)$. A set $S \subseteq V(G)$ of pairwise adjacent vertices is called a *clique*. For $v \in V(G)$, $E_G(v)$ denotes the set of edges incident with v. The *degree* of a vertex v is denoted by $d_G(v)$. For a non-negative integer d, a graph G is called *d-regular* if all vertices of G have degree d. For a graph G, the *incidence graph* of G is the bipartite graph $I(G)$ with vertex set $V(G) \cup E(G)$ such that $v \in V(G)$ and $e \in E(G)$ are adjacent if and only if v is incident with e in G. We denote by $\overline{G}$ the *complement* of a graph G, i.e. the graph with vertex set $V(G)$ such that any two distinct vertices are adjacent in $\overline{G}$ if and only if they are non-adjacent in G. For a set of vertices $S \subseteq V(G)$, $G[S]$ denotes the subgraph of G induced by S, and by $G - S$ we denote the graph obtained from G by the removal of all the vertices of S, i.e. the subgraph of G induced by $V(G) \setminus S$.

Clique-Width. Let G be a graph, and let w be a positive integer. A *w-graph* is a graph whose vertices are labeled by integers from $\{1, 2, \ldots, w\}$. We call the w-graph consisting of exactly one vertex v labeled by some integer i from $\{1, 2, \ldots, w\}$ an initial w-graph. The *clique-width* $\mathbf{cwd}(G)$ is the smallest integer w such that G can be constructed by means of repeated application of the following four operations: (1) *introduce*: construction of an initial w-graph with vertex v labeled by i (denoted by $i(v)$), (2) *disjoint union* (denoted by $\oplus$), (3) *relabel*: changing the labels of each vertex labeled i to j (denoted by $\rho_{i \to j}$) and (4) *join*: joining all vertices labeled by i to all vertices labeled by j by edges (denoted by $\eta_{i,j}$).

An *expression tree* of a graph G is a rooted tree T of the following form.

- The nodes of T are of four types: i, $\oplus$, η and ρ.
- Introduce nodes $i(v)$ are leaves of T, and they correspond to initial w-graphs with vertices v, which are labeled i.
- A union node $\oplus$ stands for a disjoint union of graphs associated with its children.
- A relabel node $\rho_{i \to j}$ has one child and is associated with the w-graph resulting from the relabeling operation $\rho_{i \to j}$ applied to the graph corresponding to the child.
- A join node $\eta_{i,j}$ has one child and is associated with the w-graph resulting from the join operation $\eta_{i,j}$ applied to the graph corresponding to the child.
- The graph G is isomorphic to the graph associated with the root of T (with all labels removed).

The *width* of the tree T is the number of different labels appearing in T. If a graph G has $\mathbf{cwd}(G) \leq w$ then it is possible to construct a rooted expression tree T of G with *width* w. Given a node X of an expression tree, the graph G_X is the graph formed by the subtree of the expression tree rooted at X.

Parameterized Reductions. We refer the reader to the books [18,21] for a detailed treatment of parameterized complexity. Here we only define the notion of parameterized (linear) reduction, which is the main tool for establishing our results. For parameterized problems A, B, we say that A is (uniformly many:1) FPT-*reducible* to B if there exist functions $f, g : \mathbb{N} \to \mathbb{N}$, a constant $\alpha \in \mathbb{N}$ and an algorithm Φ which transforms an instance (x, k) of A into an instance $(x', g(k))$ of B in time $f(k)|x|^\alpha$ so that $(x, k) \in A$ if and only if $(x', g(k)) \in B$. The reduction is called *linear* if $g(k) = O(k)$.

Capacitated Domination. For our reductions we use a variant of the CA-PACITATED DOMINATING SET problem. The parameterized complexity of this problem, with the tree-width of the input graph being the parameter, was considered in [5,16].

A *red-blue capacitated graph* is a pair (G, c), where G is a bipartite graph with a vertex bipartition into sets R and B, and $c \colon R \to \mathbb{N}$ is a *capacity* function such that $1 \le c(v) \le d_G(v)$ for every vertex $v \in R$. The vertices of the set R are called *red* and the vertices of B are called *blue*. A set $S \subseteq R$ is called a *capacitated dominating set* if there is a *domination mapping* $f \colon B \to S$ which maps every vertex in B to one of its neighbors such that the total number of vertices mapped by f to any vertex $v \in S$ does not exceed its capacity $c(v)$. We say that for a vertex $v \in S$, vertices in the set $f^{-1}(v)$ are *dominated by v*. The RED-BLUE CAPACITATED DOMINATING SET (or RED-BLUE CDS) problem asks whether, given a red-blue capacitated graph (G, c) and a positive integer k, there exists a capacitated dominating set S for G containing at most k vertices. A capacitated dominating set $S \subseteq R$ is called *saturated* if there is a domination mapping f which saturates all vertices of S, that is, $|f^{-1}(v)| = c(v)$ for each $v \in S$. The RED-BLUE EXACT SATURATED DOMINATING SET problem (RED-BLUE EXACT SATURATED CDS) takes a red-blue capacitated graph (G, c) and a positive integer k as an input and asks whether there exists a saturated capacitated dominating set with exactly k vertices.

The next proposition immediately follows from the results proved in [22].

Proposition 1. *The* RED-BLUE CDS *and* RED-BLUE EXACT SATURATED CDS *problems cannot be solved in time* $f(w) \cdot n^{o(w)}$, *where n is the number of vertices of the input graph G and w is the clique-width of the incidence graph $I(G)$, unless ETH fails, even if an expression tree of width w for $I(G)$ is given.*

The proof of Proposition 1 uses the result of Chen et al. [8,9,10] that there is no algorithm for k-CLIQUE (finding a clique of size k) running in time $f(k) \cdot n^{o(k)}$ unless there exists an algorithm for solving 3-SAT running in time $2^{o(n)}$ on a formula with n variables. Proposition 1 was proved via a linear reduction from the k-MULTI-COLORED CLIQUE problem (see [5,22]). The k-MULTI-COLORED CLIQUE problem asks for a given k-partite graph $G = (V_1 \cup \cdots \cup V_k, E)$, where $V_1, \ldots, V_k$ are sets of the k-partition, whether there is a k-clique in G. It should be noted that the construction of an expression tree of bounded width is part of the reduction and it is done in polynomial time. Lokshtanov, Marx and Saurabh [28] considered a special restricted variant of k-MULTI-COLORED CLIQUE called $k \times k$-CLIQUE. In this variant of the problem $|V_1| = \ldots = |V_k| = k$. They proved the following.

Proposition 2 ([28]). *The $k \times k$-CLIQUE problem cannot be solved in time $2^{o(k \log k)} \cdot n^{O(1)}$, where n is the number of vertices of the input graph G, unless ETH fails.*

By replacing k-MULTI-COLORED CLIQUE by the $k \times k$-CLIQUE problem in the reductions used for the proof of Proposition 1, we obtain the following corollary.

Corollary 1. *The RED-BLUE CDS and RED-BLUE EXACT SATURATED CDS problems cannot be solved in time $2^{o(w \log n)} \cdot n^{O(1)}$, where n is the number of vertices of the input graph G and w is the clique-width of the incidence graph $I(G)$, unless ETH fails, even if an expression tree of width w for $I(G)$ is given.*

Observe that Corollary 1 gives a slightly stronger claim than Proposition 1: while $o(w) \cdot \log n = o(w \log n)$, it is not so the other way around.

3   Sparse and Dense k-Subgraph Problems

In this section we consider the DENSE k-SUBGRAPH and SPARSE k-SUBGRAPH problems. The aim of this section is the proof of the following theorem.

Theorem 1. *The SPARSE k-SUBGRAPH problem can be solved in time $k^{O(w)} \cdot n$ on n-vertex graphs of clique-width at most w if an expression tree of width w is given, but it cannot be solved in time $2^{o(w \log k)} \cdot n^{O(1)}$ unless ETH fails, even if an expression tree of width w is given.*

Clearly, SPARSE k-SUBGRAPH and DENSE k-SUBGRAPH are dual, i.e. SPARSE k-SUBGRAPH is equivalent to DENSE k-SUBGRAPH for the complement of the input graph. Since for any graph G, $\mathbf{cwd}(\overline{G}) \le 2 \cdot \mathbf{cwd}(G)$ (see e.g. [14,35]), we can immediately get the following corollary.

Corollary 2. *The DENSE k-SUBGRAPH problem can be solved in time $k^{O(w)} \cdot n$ on n-vertex graphs of clique-width at most w if an expression tree of width w is given, but it cannot be solved in time $2^{o(w \log k)} \cdot n^{O(1)}$ unless ETH fails, even if an expression tree of width w is given.*

3.1   Algorithmic Upper Bounds for Sparse k-Subgraph

We sketch a dynamic programming algorithm for solving SPARSE k-SUBGRAPH in time $k^{O(w)} \cdot n$ on graphs of clique-width at most w. We describe what we store in the tables corresponding to the nodes in an expression tree.

Let G be a graph with n vertices and let T be an expression tree for G of width w. For a node X of T, let $U_1(X), \ldots, U_w(X)$ be the sets of vertices of G_X labeled $1, \ldots, w$, respectively. The table of data for the node X contains entries which store a positive integer $p \le q$ and a vector $(s_1, \ldots, s_w)$ of non-negative integers such that $s = s_1 + \ldots + s_w \le k$ for $i \in \{1, \ldots, w\}$, for which p is the minimum number of edges of an induced subgraph H with s vertices such that for $i \in \{1, \ldots, w\}$, $s_i = |U_i(X) \cap V(H)|$. If X is the root node of T

then G contains an induced subgraph with k vertices and at most q edges if and only if the table for X contains an entry with the parameter $p \leq q$ and vector $(s_1, \ldots, s_w)$ such that $s_1 + \ldots + s_w = k$.

The details how the tables are created and updated are omitted here because of the space restrictions. Correctness of the algorithm follows from the description of the procedure.

Since for each X, the table for X contains at most $(k+1)^w$ vectors and for each vector only one value of the parameter p is stored, the algorithm runs in time $k^{O(w)} \cdot n$. This proves that SPARSE k-SUBGRAPH can be solved in time $k^{O(w)} \cdot n$ on graphs of clique-width at most w.

3.2   Lower Bounds

To prove our lower bounds we give a reduction from the RED-BLUE CDS problem parameterized by the clique-width of the incidence graph of the input graph.

Construction. Let (G, c, k) be an instance of RED-BLUE CDS with $R = \{u_1, \ldots, u_n\}$ being the set of red vertices and $B = \{v_1, \ldots, v_r\}$ being the set of blue vertices. Let m be the number of edges of G. We assume without loss of generality that G has no isolated vertices. Hence, $m \geq n, r$.

First, we construct the auxiliary gadget $F(l)$ for a positive integer l.

Auxiliary gadget $F(l)$: Construct an $l+m+1$-partite graph $K_{2,\ldots,2}$ and denote by x_{i1}, x_{i2} the vertices of the i-th set of the partition (see Figure 1).

Reduction: Now we describe our reduction.

1. A copy of a gadget $F(k)$ is constructed. Denote this graph by F_R and let $V(F_R) = \{x_{i1}^R, x_{i2}^R | 1 \leq i \leq k+m+1\}$.
2. For each $i \in \{1, \ldots, n\}$, a copy of a gadget $F(c(u_i))$ is created. Denote this graph by F_{u_i} and let $V(F_{u_i}) = \{x_{j1}^{u_i}, x_{j2}^{u_i} | 1 \leq j \leq c(u_i) + m + 1\}$.
3. For each $i \in \{1, \ldots, r\}$, a copy of a gadget $F(1)$ is created. Denote this graph by F_{v_i} and let $V(F_{v_i}) = \{x_{j1}^{v_i}, x_{j2}^{v_i} | 1 \leq j \leq m + 2\}$.
4. For each $e \in E(G)$, the vertex w_e is constructed.
5. For each $i \in \{1, \ldots, n\}$, let $\{e_1, \ldots, e_{d_i}\} = E(u_i)$ for $d_i = d_G(u_i)$. We consider the vertices $w_{e_1}, \ldots, w_{e_{d_i}}$; these vertices are joined by edges to the vertices x_{i1}^R, x_{i2}^R of F_R, and for each $j \in \{1, \ldots, d_i\}$, w_{e_j} is joined by edges to the vertices $x_{j1}^{u_i}, x_{j2}^{u_i}$ of F_{u_i}.
6. For each $i \in \{1, \ldots, r\}$, let $\{e_1, \ldots, e_{d_i}\} = E(v_i)$ for $d_i = d_G(v_i)$. We consider the vertices $w_{e_1}, \ldots, w_{e_{d_i}}$ and for each $j \in \{1, \ldots, d_i\}$, w_{e_j} is joined by edges to the vertices $x_{j1}^{v_i}, x_{j2}^{v_i}$ of F_{v_i}.
7. Create $2m + 1$ vertices $z_1, \ldots, z_{2m+1}$ and join them to all vertices w_e for $e \in E(G)$.

Denote the obtained graph by H (see Figure 1).

Due the space restrictions the proof of the following lemmas are omitted.

Lemma 1. *The red-blue graph G has a capacitated dominating set of size at most k if and only if H contains an induced subgraph with $2(m+1)(n+r+1) + 2m + 1 + r$ vertices and at most $2m(m+1)(n+r+1) + r(2m+1)$ edges.*

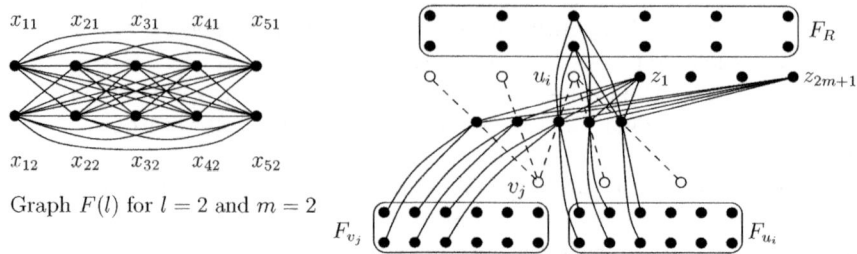

Graph $F(l)$ for $l = 2$ and $m = 2$

Fig. 1. Construction of H

We prove an upper bound for the clique-width of H as a linear function in the clique-width of the incidence graph $I(G)$ of G.

Lemma 2. *We have* $\mathbf{cwd}(H) \leq 9 \cdot \mathbf{cwd}(I(G)) + 1$ *and an expression tree of width at most* $9 \cdot \mathbf{cwd}(I(G)) + 1$ *for* H *can be constructed in polynomial time given an expression tree of width* $\mathbf{cwd}(I(G))$ *for* $I(G)$.

To complete the proof of Theorem 1, notice that the number of vertices of H and the parameter k are polynomial in $n + r$. Therefore, $\log k$ is linear in $\log(n + k)$, and if we could solve SPARSE k-SUBGRAPH in time $2^{o(\mathbf{cwd}(H)\log k)} \cdot |V(H)|^{O(1)}$ then RED-BLUE CDS could be solved in time $2^{o(\mathbf{cwd}(I(G))\log|V(G)|))} \cdot |V(G)|^{O(1)}$. By Corollary 1, it cannot be done unless ETH fails.

4 Degree-Constrained Subgraph Problems

The first aim of this section is the proof of the following theorem.

Theorem 2. *The* d-REGULAR INDUCED SUBGRAPH *problem can be solved on* n-vertex graphs of clique-width at most w in time $d^{O(w)} \cdot n$ *if an expression tree of width* w *is given for the input graph, but it cannot be solved in time* $2^{o(w \log d)} \cdot n^{O(1)}$ *unless ETH fails, even if an expression tree of width* w *is given.*

Proof. The algorithmic upper bounds are proved by constructing a dynamic programming algorithm for solving d-REGULAR INDUCED SUBGRAPH in time $d^{O(w)} \cdot n$ on graphs of clique-width at most w. To prove our complexity lower bound, we give a reduction from the RED-BLUE EXACT SATURATED CDS problem, parameterized by the clique-width of the incidence graph of the input graph, to the d-REGULAR INDUCED SUBGRAPH problem. The proof is organized as follows: we first give a construction, then prove its correctness and finally bound the clique-width of the transformed instance.

Construction. Let (G, c, k) be an instance of RED-BLUE EXACT SATURATED CDS with $R = \{u_1, \ldots, u_n\}$ being the set of red vertices and $B = \{v_1, \ldots, v_r\}$ being the set of blue vertices. Let $d = n + r + 1$ if $n + r$ is even and let $d = n + r + 2$ otherwise; notice that d is odd. We need an auxiliary gadget.

Auxiliary gadget $F(x)$***:*** Let x be a vertex. We construct $\frac{d-1}{2}$ copies of K_{d+1}, subdivide one edge of each copy, and glue (identify) all these vertices of degree two into one vertex y. Finally we join x and y by an edge. We are going to attach gadgets $F(x)$ to other parts of our construction through the vertex x. This vertex is called the *root* of $F(x)$. The gadget $F(x)$ for $d = 5$ is illustrated in Figure 2.

Reduction: Now we describe our reduction. Let $s = d - r - 1$ and $t = d - k - 1$.

1. Vertices $u_1, \ldots, u_n$ are created.
2. A clique of size r with vertices $v_1, \ldots, v_r$ is constructed.
3. For each edge $e = u_i v_j$ of G, a vertex w_e is added, joined by edges to u_i and v_j, and $d - 2$ copies of $F(w_e)$ are constructed.
4. A clique of size s with vertices $a_1, \ldots, a_s$ is created, all vertices a_i are joined to vertices $v_1, \ldots, v_r$, and for each $i \in \{1, \ldots, s\}$, a copy of $F(a_i)$ is added.
5. A vertex x is introduced and joined by edges to $v_1, \ldots, v_r$ and $a_1, \ldots, a_s$.
6. A vertex y is added and joined by an edge to x, and $k - 1$ copies of $F(y)$ are added.
7. A clique of size t with vertices $b_1, \ldots, b_t$ is constructed, the vertex y is joined by edges to all vertices of the clique, and for each $j \in \{1, \ldots, t\}$, k copies of $F(b_i)$ are added.
8. A vertex z is introduced and joined by edges to vertices y and $b_1, \ldots, b_t$.
9. For each $i \in \{1, \ldots, n\}$, we let $l_i = d - c(u_i) - 1$ and do the following:
   - Add a vertex p_i, join it to z by an edge, and construct $c(u_i) - 1$ copies of $F(p_i)$.
   - Construct a clique of size l_i with vertices $c_{i1}, \ldots, c_{il_i}$, join them to the vertex p_i by edges, and for each $j \in \{1, \ldots, l_i\}$, introduce $c(u_i)$ copies of $F(c_{ij})$.
   - Join the vertex u_i to the vertices p_i and $c_{i1}, \ldots, c_{il_i}$ by edges.

Denote the obtained graph by H. The construction of H is illustrated in Figure 2. The proof of the following lemmas are omitted.

Lemma 3. *The red-blue graph G has an exact saturated capacitated dominating set of size k if and only if H contains an induced d-regular subgraph.*

Now we show that the clique-width of H is bounded from above by a linear function in the clique-width of the incidence graph $I(G)$ of G.

Lemma 4. *We have that $\mathbf{cwd}(H) \leq 3 \cdot \mathbf{cwd}(I(G)) + 6$ and an expression tree of width at most $3 \cdot \mathbf{cwd}(I(G)) + 6$ for H can be constructed in polynomial time assuming we are given an expression tree of width $\mathbf{cwd}(I(G))$ for $I(G)$.*

To conclude this part of the proof of Theorem 2, we observe that the number of vertices of H and the parameter d are polynomial in $n + r$, and therefore if we could solve d-REGULAR INDUCED SUBGRAPH in time $2^{o(\mathbf{cwd}(H) \log d)} \cdot |V(H)|^{O(1)}$ then the RED-BLUE EXACT SATURATED CDS could be solved in time $2^{o(\mathbf{cwd}(I(G)) \log |V(G)|)} \cdot |V(G)|^{O(1)}$. By Corollary 1, this cannot be done unless ETH fails. □

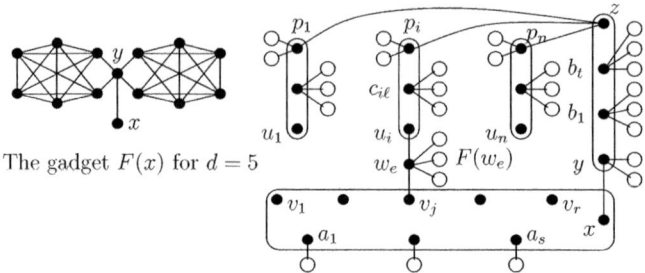

Fig. 2. Construction of H

In the d-REGULAR INDUCED SUBGRAPH problem we ask about the existence of a d-regular induced subgraph for a given graph. It is possible to get similar results for some variants of this problem. The MINIMUM d-REGULAR INDUCED SUBGRAPH problem and the MAXIMUM d-REGULAR INDUCED SUBGRAPH problem are respectively the problems of finding a d-regular induced subgraph of minimum and maximum size. For the COUNTING d-REGULAR INDUCED SUBGRAPH problem, we are interested in the number of induced d-regular subgraphs of the input graph. Using Theorem 2 we get the following corollary.

Corollary 3. *The* MINIMUM d-REGULAR INDUCED SUBGRAPH, MAXIMUM d-REGULAR INDUCED SUBGRAPH *and* COUNTING d-REGULAR INDUCED SUBGRAPH *problems can be solved on n-vertex graphs of clique-width at most w in time $d^{O(w)} \cdot n$ if an expression tree of width w is given, but they cannot be solved in time $2^{o(w \log n)} \cdot n^{O(1)}$ unless ETH fails, even if an expression tree of width w is given.*

We conclude this section by considering the MINIMUM SUBGRAPH OF MINIMUM DEGREE AT LEAST d problem.

Theorem 3. *The* MINIMUM SUBGRAPH OF MINIMUM DEGREE AT LEAST d *problem can be solved on n-vertex graphs of clique-width at most w in time $d^{O(w)} \cdot n$ if an expression tree of width w is given, but it cannot be solved in time $2^{o(w \log d)} \cdot n^{O(1)}$ unless ETH fails, even if an expression tree of width w is given.*

5   Conclusion

We established tight algorithmic lower and upper bounds for some double-parameterized subgraph problems when the clique-width of the input graph is one of the parameters. We believe that similar bounds could be given for other problems. Another interesting task is to consider problems parameterized by other width-parameters. Throughout the paper, in all our results we assumed that an expression tree of the given width is part of the input. This is crucial, since — unlike the case of tree-width — to date we are unaware of an efficient

(FPT or polynomial) algorithm for computing an expression tree with a constant factor approximation of the clique-width. The algorithm given by Oum and Seymour in [30] provides a constant factor approximation for another graph parameter — *rank-width* [24,30]. Hence, it is natural to ask whether it is possible to establish tight algorithmic bounds for DENSE k-SUBGRAPH, d-REGULAR INDUCED SUBGRAPH and MINIMUM SUBGRAPH OF MINIMUM DEGREE AT LEAST d parameterized by the rank-width of the input graph. Also it would be interesting to consider problems parameterized by the tree-width. For example, it can be shown that d-REGULAR INDUCED SUBGRAPH and MINIMUM SUBGRAPH OF MINIMUM DEGREE AT LEAST d can be solved in time $d^{O(t)} \cdot n$ for n-vertex graphs of tree-width at most t. Is this bound asymptotically tight?

References

1. Amini, O., Peleg, D., Pérennes, S., Sau, I., Saurabh, S.: Degree-Constrained Subgraph Problems: Hardness and Approximation Results. In: Bampis, E., Skutella, M. (eds.) WAOA 2008. LNCS, vol. 5426, pp. 29–42. Springer, Heidelberg (2009)
2. Amini, O., Sau, I., Saurabh, S.: Parameterized Complexity of the Smallest Degree-Constrained Subgraph Problem. In: Grohe, M., Niedermeier, R. (eds.) IWPEC 2008. LNCS, vol. 5018, pp. 13–29. Springer, Heidelberg (2008)
3. Arora, S., Karger, D.R., Karpinski, M.: Polynomial time approximation schemes for dense instances of p-hard problems. In: STOC, pp. 284–293. ACM (1995)
4. Asahiro, Y., Iwama, K., Tamaki, H., Tokuyama, T.: Greedily Finding a Dense Subgraph. In: Karlsson, R., Lingas, A. (eds.) SWAT 1996. LNCS, vol. 1097, pp. 136–148. Springer, Heidelberg (1996)
5. Bodlaender, H.L., Lokshtanov, D., Penninkx, E.: Planar Capacitated Dominating Set is $W[1]$-Hard. In: Chen, J., Fomin, F.V. (eds.) IWPEC 2009. LNCS, vol. 5917, pp. 50–60. Springer, Heidelberg (2009)
6. Cai, L., Juedes, D.W.: On the existence of subexponential parameterized algorithms. J. Comput. Syst. Sci. 67, 789–807 (2003)
7. Cheah, F., Corneil, D.G.: The complexity of regular subgraph recognition. Discrete Applied Mathematics 27, 59–68 (1990)
8. Chen, J., Chor, B., Fellows, M., Huang, X., Juedes, D., Kanj, I.A., Xia, G.: Tight lower bounds for certain parameterized NP-hard problems. Information and Computation 201, 216–231 (2005)
9. Chen, J., Huang, X., Kanj, I.A., Xia, G.: On the computational hardness based on linear FPT-reductions. J. Comb. Optim. 11, 231–247 (2006)
10. Chen, J., Huang, X., Kanj, I.A., Xia, G.: Strong computational lower bounds via parameterized complexity. J. Comput. Syst. Sci. 72, 1346–1367 (2006)
11. Chvátal, V., Fleischner, H., Sheehan, J., Thomassen, C.: Three-regular subgraphs of four-regular graphs. J. Graph Theory 3, 371–386 (1979)
12. Corneil, D.G., Perl, Y.: Clustering and domination in perfect graphs. Discrete Appl. Math. 9, 27–39 (1984)
13. Courcelle, B., Makowsky, J.A., Rotics, U.: Linear time solvable optimization problems on graphs of bounded clique-width. Theory Comput. Syst. 33, 125–150 (2000)
14. Courcelle, B., Olariu, S.: Upper bounds to the clique width of graphs. Discrete Appl. Math. 101, 77–114 (2000)

15. Demaine, E.D., Hajiaghayi, M.T., Kawarabayashi, K.: Algorithmic graph minor theory: Decomposition, approximation, and coloring. In: FOCS, pp. 637–646. IEEE Computer Society (2005)

16. Dom, M., Lokshtanov, D., Saurabh, S., Villanger, Y.: Capacitated Domination and Covering: A Parameterized Perspective. In: Grohe, M., Niedermeier, R. (eds.) IWPEC 2008. LNCS, vol. 5018, pp. 78–90. Springer, Heidelberg (2008)

17. Downey, R.G., Estivill-Castro, V., Fellows, M.R., Prieto, E., Rosamond, F.A.: Cutting up is hard to do: the parameterized complexity of k-cut and related problems. Electr. Notes Theor. Comput. Sci. 78 (2003)

18. Downey, R.G., Fellows, M.R.: Parameterized complexity. Monographs in Computer Science. Springer, New York (1999)

19. Feige, U., Peleg, D., Kortsarz, G.: The dense k-subgraph problem. Algorithmica 29, 410–421 (2001)

20. Fellows, M.R., Rosamond, F.A., Rotics, U., Szeider, S.: Clique-Width is NP-Complete. SIAM J. Discrete Math. 23, 909–939 (2009)

21. Flum, J., Grohe, M.: Parameterized complexity theory. Texts in Theoretical Computer Science. An EATCS Series. Springer, Berlin (2006)

22. Fomin, F.V., Golovach, P.A., Lokshtanov, D., Saurabh, S.: Algorithmic lower bounds for problems parameterized by clique-width. In: SODA, pp. 493–502 (2010)

23. Hlinený, P., Oum, S.-I.: Finding branch-decompositions and rank-decompositions. SIAM J. Comput. 38, 1012–1032 (2008)

24. Hlinený, P., Oum, S.-I., Seese, D., Gottlob, G.: Width parameters beyond treewidth and their applications. Comput. J. 51, 326–362 (2008)

25. Impagliazzo, R., Paturi, R., Zane, F.: Which problems have strongly exponential complexity? J. Comput. Syst. Sci. 63, 512–530 (2001)

26. Keil, J.M., Brecht, T.B.: The complexity of clustering in planar graphs. J. Combin. Math. Combin. Comput. 9, 155–159 (1991)

27. Kortsarz, G., Peleg, D.: On choosing a dense subgraph (extended abstract). In: FOCS, pp. 692–701. IEEE (1993)

28. Lokshtanov, D., Marx, D., Saurabh, S.: Slightly superexponential parameterized problems. In: SODA (2011)

29. Mathieson, L., Szeider, S.: The parameterized complexity of regular subgraph problems and generalizations. In: CATS. CRPIT, vol. 77, pp. 79–86. Australian Computer Society (2008)

30. Oum, S.-I., Seymour, P.: Approximating clique-width and branch-width. J. Combin. Theory Ser. B 96, 514–528 (2006)

31. Moser, H., Thilikos, D.M.: Parameterized complexity of finding regular induced subgraphs. J. Discrete Algorithms 7, 181–190 (2009)

32. Stewart, I.A.: Deciding whether a planar graph has a cubic subgraph is np-complete. Discrete Mathematics 126, 349–357 (1994)

33. Stewart, I.A.: Finding regular subgraphs in both arbitrary and planar graphs. Discrete Applied Mathematics 68, 223–235 (1996)

34. Stewart, I.A.: On locating cubic subgraphs in bounded-degree connected bipartite graphs. Discrete Mathematics 163, 319–324 (1997)

35. Wanke, E.: k-NLC graphs and polynomial algorithms. Discrete Appl. Math. 54, 251–266 (1994)

Finding Good Decompositions for Dynamic Programming on Dense Graphs*

Eivind Magnus Hvidevold, Sadia Sharmin, Jan Arne Telle, and Martin Vatshelle

Department of Informatics, University of Bergen, Norway

Abstract. It is well-known that for graphs with high edge density the tree-width is always high while the clique-width can be low. Boolean-width is a new parameter that is never higher than tree-width or clique-width and can in fact be as small as logarithmic in clique-width. Boolean-width is defined using a decomposition tree by evaluating the number of neighborhoods across the resulting cuts of the graph. Several NP-hard problems can be solved efficiently by dynamic programming when given a decomposition of boolean-width k, e.g. Max Weight Independent Set in time $O(n^2 k 2^{2^k})$ and Min Weight Dominating Set in time $O(n^2 + nk2^{3k})$. Finding decompositions of low boolean-width is therefore of practical interest. There is evidence that computing boolean-width is hard, while the existence of a useful approximation algorithm is still open. In this paper we introduce and study a heuristic algorithm that finds a reasonably good decomposition to be used for dynamic programming based on boolean-width. On a set of graphs of practical relevance, specifically graphs in TreewidthLIB, the best known upper bound on their tree-width is compared to the upper bound on their boolean-width given by our heuristic. For the large majority of the graphs on which we made the tests, the tree-width bound is at least twice as big as the boolean-width bound, and boolean-width compares better the higher the edge density. This means that, for problems like Dominating Set, using boolean-width should outperform dynamic programming by tree-width, at least for graphs of edge density above a certain bound. In view of the amount of previous work on heuristics for tree-width these results indicate that boolean-width could in the future outperform tree-width in practice for a large class of graphs and problems.

1 Introduction

Many NP-hard graph problems become polynomial-time solvable when restricted to graphs of bounded tree-width or bounded clique-width. These algorithms usually have two stages, a first stage finding a decomposition of width k of the input graph, and a second stage of dynamic programming along the decomposition. The dynamic programming is typically exponential in k, e.g. given a decomposition of tree-width k it solves Maximum Weight Independent set in time $O(n2^k)$ and Minimum Weight Dominating set in time $O(n3^k k^2)$ [20]. It is therefore important to have fast algorithms for the first stage, i.e. to find decompositions of small width. For clique-width such algorithms are not known, apart from the 2^{OPT} approximation achieved through

---

* Supported by the Norwegian Research Council, project PARALGO.

D. Marx and P. Rossmanith (Eds.): IPEC 2011, LNCS 7112, pp. 219–231, 2012.

rank-width [13]. For tree-width there is an $O(f(n)2^{O(k^3)})$ algorithm for finding a decomposition of tree-width k, if it exists [3]. This algorithm is not practical [17], but much work has been done on finding decompositions of low tree-width in practical settings, see the overviews [5,4]. The web site TreewidthLIB [19] has been established to provide a benchmark and to join the efforts of people working in experimental settings to solve graph problems using tree-width and branch-width [12,16]. This includes problems from computational biology [18,21,22], constraint satisfaction [9,11], and probabilistic networks [15]. However, tree-width and branch-width are unsuitable for non-sparse graphs, as a decomposition of tree-width or branch-width k means the graph has $O(k^2 n)$ edges. Clique-width, on the other hand, can be low for dense graphs, but so far no experimental study has been done for clique-width or similar notions. To our knowledge this paper is the first case of an experimental study on computing a notion of width that works also for non-sparse graphs.

Boolean-width is a recently introduced graph parameter motivated by algorithms [8]. It is defined by a decomposition tree that minimizes the number of different unions of neighbourhoods across resulting cuts of the graph. This decomposition is natural to solve problems where vertex sets having the same neighborhoods across the cuts can be treated as equivalent. This includes problems related to Independent Set, Dominating Set, Perfect Code, Induced k-Bounded Degree Subgraph, H-Homomorphism, H-Covering, H-Role Assignment etc [1]. Similarly to treewidth, dynamic programming algorithms to solve these problems using boolean-width employ a table at each node of the decomposition tree, to store solutions to partial problems. In contrast to treewidth, the dynamic programming for boolean-width involves a non-negligible pre-processing phase computing indices of the tables, the so-called 'representatives'. Regardless, the total runtimes are in many cases close to those for treewidth, e.g. given a decomposition of boolean-width k Max Weight Independent Set is solved in time $O(n^2 k 2^{2k})$ and Min Weight Dominating Set in time $O(n^2 + nk2^{3k})$ [8]. These boolean-width-based algorithms are straightforward and have been implemented in Java, without much effort, using only the description in [8]. Let us compare dynamic programming based on tree-width versus boolean-width, to solve Independent Set and Dominating Set, with focus on exponential factors. For Independent Set the exponential factor in the runtimes are 2^{tw} versus 2^{2boolw}, given decompositions of treewidth tw or boolean-width $boolw$, and boolean-width becomes preferable when $tw > 2boolw$. For Dominating Set the exponential factor in the runtime is 3^{tw} versus 2^{3boolw} and the cutoff is a bit lower, i.e. when $tw \geq 1.9boolw$.

It is known that boolean-width is never higher than tree-width or clique-width and it can be as low as logarithmic in clique-width [8]. For example, any interval graph or permutation graph has boolean-width $O(log n)$ [2] while there exist such graphs of clique-width $\Omega(\sqrt{n})$ and tree-width $\Omega(n)$. Also, a random graph with constant edge probability will almost surely have boolean-width $\Theta(\log^2 n)$ [1] but linear clique-width and tree-width. While these theoretical results favor boolean-width over tree-width, the cutoff $tw \geq 2boolw$ that we arrived at above applies when we are given a decomposition of treewidth tw or boolean-width $boolw$, as the output of a first stage algorithm. It is unknown if computing boolean-width is FPT or W-hard. In this paper we give a heuristic for the first stage, taking as input a graph G and finding a decomposition of G

having reasonably low boolean-width. We tried various heuristics and present the one with best performance, which is a local search algorithm where the search for new solutions is based on interweaving between greedy choices and random choices. Theoretical evidence that random choices are useful for boolean-width, at least for random graphs, comes from the analysis of [1] showing that any decomposition of a random graph is expected to be a decomposition of relatively low boolean-width. On a set of graphs of practical relevance, specifically graphs in TreewidthLIB, the best known upper bound on their tree-width is compared to the upper bound on their boolean-width given by our heuristic. For 78% of those graphs in TreewidthLIB where both tree-width and boolean-width upper bounds were encountered, the tree-width bound is at least twice the boolean-width bound, thus meeting the $tw \geq 2boolw$ bound mentioned above. A drawback of tree-width is that it is always high when edge density is high. In contrast, boolean-width is typically low for dense graphs and our experiments show that within reasonable time we can find decompositions witnessing this. Our results indicate that, for problems like Dominating Set, using boolean-width will outperform dynamic programming by tree-width, at least for graphs of edge density above a certain bound. In view of the amount of previous work on heuristics for tree-width we expect that further work on boolean-width heuristics will substantially increase the class of graphs for which boolean-width outperforms tree-width, also for other problems besides Independent Set and Dominating Set.

The rest of the paper is organized as follows. In Section 2 we define partial and full decomposition trees and boolean-width. In Section 3 we describe the heuristic finding a decomposition of low boolean-width. In Section 4 we describe the experimental results on graphs in TreewidthLIB, and also on small grid graphs. In Section 5 we draw some conclusions.

2   Boolean-Width

We consider undirected graphs $G = (V, E)$ without loops. We denote the neighborhood of a vertex v by $N(v)$ and the union of neighborhoods of a vertex subset A by $N(A) = \cup_{v \in A} N(v)$. The complement of $A \subseteq V$ is denoted by $\bar{A} = V \setminus A$ and we call $(A, \bar{A})$ a cut of G. A partition of a set S consists of non-empty and disjoint subsets of S whose union is S. We follow custom by referring to vertices of a graph and nodes of a tree.

Definition 1 (Full and partial decomposition trees). A partial decomposition tree of a graph $G = (V, E)$ is a pair (T, δ), where T is a full binary tree and δ is a mapping from the nodes of T to non-empty subsets of V, satisfying the following: if x is the root of T then $\delta(x) = V$ and if nodes y and z of T are children of a node x then $(\delta(y), \delta(z))$ is a partition of $\delta(x)$. If a subtree of T rooted at x has $|\delta(x)|$ leaves then it is called a full decomposition subtree. If T has $|V|$ leaves then (T, δ) is called a full decomposition tree.

Note that in a partial decomposition tree (T, δ) of a graph G, if L is the set of leaves of T then $\{\delta(x) : x \in L\}$ is a partition of V. Hence in a full decomposition tree there will for each vertex v of G be a unique leaf x of T with $\delta(x) = \{v\}$. Likewise for each vertex of $\delta(x)$ in a full decomposition subtree rooted at x.

Definition 2 (Unions of neighborhoods and boolean-width). Let (T, δ) be a partial decomposition tree of a graph G. Let $V(T)$ be the nodes of T. Every node $x \in V(T)$ defines a cut $(\delta(x), \overline{\delta(x)})$ of G. The set of unions of neighborhoods of subsets of A across the cut $(A, \overline{A})$ is $UN(A) = \{N(X) \cap \overline{A} : X \subseteq A\}$. The boolean-width of (T, δ) is

$$boolw(T, \delta) = \max_{x \in V(T)} \{log_2 |UN(\delta(x))|\}$$

The boolean-width of a graph G is the minimum boolean-width over all its full decomposition trees $boolw(G) = \min_{\text{full } (T, \delta) \text{ of } G} \{boolw(T, \delta)\}$.

Note that $UN(A)$ are the subsets of $\overline{A}$ for which there exists an $X \subseteq A$ with $N(X) \cap \overline{A}$ being that subset, so we always have $\emptyset \in UN(A)$. It is known from boolean matrix theory [14] that $|UN(A)| = |UN(\overline{A})|$ and this is sometimes used by our code. Let us consider some examples. If $|UN(A)| = 2$ then the set of edges crossing the cut $(A, \overline{A})$ induce a complete bipartite graph. If the set of edges crossing the cut $(A, \overline{A})$ induce a perfect matching of G then $|UN(A)| = 2^{|V/2|}$. In the definition of boolean-width we take the logarithm base 2 of $|UN(A)|$ which ensures that $0 \leq boolw(G) \leq |V|$. If a graph has boolean-width one then it has a full decomposition tree such that, for every cut defined by a node of the tree, the edges crossing the cut, if any, induce a complete bipartite graph. From this it follows that the graphs of boolean-width one are exactly the distance-hereditary graphs [7].

Definition 3 (Split). A split of a set P is a partition into two subsets A and B, with the constraint that $min\{|A|, |B|\} \geq \frac{1}{3}|P|$.

3 Heuristic Algorithm

We present a local search heuristic that given a graph G computes a full decomposition tree of G. The search for new solutions in the space of candidate solutions is based on a fine balance between greedy choices and random choices. The heuristic, given in Algorithm 1, runs for a pre-defined length of time and then returns the best full decomposition found. Each heuristic pass iterates over all decomposition nodes of the current partial decomposition tree, including the children created by this heuristic pass. A newly created tree node always starts out as a leaf node, which δ maps to a set of vertices of G that may be larger than one. We keep track of the best full decomposition subtrees found for each $P \subseteq V$ encountered so far and call it $Best(P)$.

3.1 Greedy Initialization

Step 1 of Algorithm 1 greedily generates a full decomposition tree, to serve as the starting tree for the local search in Step 2. The greedy initialization starts with T containing a single node x (as both root and leaf) with $\delta(x) = V$ and repeatedly calls the **Split** subroutine until we get a full decomposition tree. The **Split**(P) subroutine returns a split (A, B) of P and is given in Algorithm 2. Starting with A being a random half of the vertices of P (unless $P=V$), it adds new vertices to A one by one in a greedy fashion

---

Algorithm 1. Generate a full decomposition of a given graph

Input: a graph G
Output: a full decomposition tree (T, δ) of G
Step 1: /*Greedily generate initial full decomposition tree*/
        Initialize T with $V(T) = \{root\}$, $\delta(root) = V$
        while $\exists$ leaf x of T with $|\delta(x)| > 1$
            $(A, B) = \textbf{Split}(\delta(x))$;
            Add leaves y and z as children of x with $\delta(y) = A$ and $\delta(z) = B$
        for all $x \in V(T)$ store $Best(\delta(x))$, the subtree rooted at x
Step 2: /*Local Search for better trees*/
        for fixed amount of time **do**
            TryToImproveSubtree($root$)
            if (T, δ) is a full decomposition tree **then** $Best(V) = (T, \delta)$
        return $Best(V)$

---

while minimizing $|UN(A)|$ and $|UN(P \setminus A)|$, and returns the best split found along the way complying with the split constraint. The call of **Split**(V) at the root sets the initial conditions for the later splits and for this root-case we start with $A = \emptyset$, rather than a random half of the vertices, to allow the full benefit of the greedy choices. The local search in **TryToImproveSubtree** will for leaves of the current tree make calls to **Split**(P) but not for $P = V$, since the root of T will never again become a leaf and instead the **RandomSwap** subroutine described in the next subsection will be applied to the root.

Algorithm 2. Split(P)

Input: Set of vertices $P \subseteq V$.
Output: a partition (A,B) of P s.t. $min\{|A|, |B|\} \geq \frac{1}{3}|P|$.
if $P = V$ **then** $A_1 \leftarrow \emptyset$
else $A_1 \leftarrow$ random half of the vertices in P
$i = 1$
while $|P \setminus A_i| \geq \frac{1}{3}|P|$ **do**
    find $x \in P \setminus A_i$ s.t. $max\{\textbf{UN}(A_i \cup \{x\}), \textbf{UN}((P \setminus A_i) \setminus \{x\})\}$ is minimized.
    $A_{i+1} = A_i \cup \{x\}$.
    $i = i + 1$.
end while
find i such that $max\{\textbf{UN}(A_i), \textbf{UN}(P \setminus A_i)\}$ is minimized and $|A_i| \geq \frac{1}{3}|P|$.
return $(A_i, P \setminus A_i)$.

The objective function optimized locally in **Split** is $|UN(A)|$, the number of unions of neighborhoods of A, which directly relates to boolean-width, see Definition 2. The computation of $|UN(A)|$ is done in a separate subroutine called **UN**(A) given in Algorithm 3. This subroutine starts by restricting from the cut $(A, \overline{A})$ to the subsets of vertices (S_1, S_2) having an edge going across the cut $(A, \overline{A})$. The list LN is used to accumulate the set $UN(A)$ in a straightforward way. Correctness is easy to show by induction on $|S_1|$. Early termination of the **UN**(A) subroutine is not shown in Algorithm 3 but is done if it is determined that $|LN|$ is too large for the cut $(A, \overline{A})$ to be interesting.

Algorithm 3. UN(A)

Input: Set of vertices $A \subseteq V$.
Output: $|UN(A)|$, the number of unions of neighborhoods of the cut $(A, \overline{A})$
if $|UN(A)|$ has already been computed **return** the stored value
$S_1 = \{v \in A : \exists u \in \overline{A} \wedge (u, v) \in E\}$
$S_2 = \{v \in \overline{A} : \exists u \in A \wedge (u, v) \in E\}$
$LN \leftarrow \{\emptyset\}$ /*neighborhood set accumulator*/
for all $u \in S_1$ **do**
    for all $Y \in LN$ **do**
        $X \leftarrow (N(u) \cap S_2) \cup Y$
        if $X \notin LN$ **then** add X to LN
return The number of elements in LN

3.2 Local Search

The local search used to improve the current decomposition tree is initiated at the root of the tree T, in Step 2 of Algorithm 1. In the subroutine **TryToImproveSubtree(x)**, given in Algorithm 4, x is a node of the current partial decomposition tree (T, δ) and the goal is to improve the subtree of T rooted at x. That subroutine has four main parts.

(1) if x leaf then find candidate for split of its subset
(2) if x non-leaf then find candidate for swap of its two children subsets
(3) conditionally update (T, δ)
(4) for each child of x either use stored subtree or recurse

For (1) we use the **Split** subroutine described earlier. For (2) we use the **RandomSwap(A,B)** subroutine given in Algorithm 5 that randomly swaps vertices between A and B while complying with the split constraint. At the very onset of the local search, the current (T, δ) is the full decomposition tree found by the greedy initialization. However, the current decomposition tree ceases to be full as soon as the split given by **RandomSwap($\delta(y), \delta(z)$)** in (2) is a good one and (3) updates (T, δ) so that y and z become leaves. If the new $\delta(y)$ is a subset of vertices for which a full decomposition subtree has never been stored, or the stored one is not good enough, then in (4) a recursive call is made to **TryToImproveSubtree(y)**, with y a leaf of the current tree. If in that recursive call the split found in (1) is not good then in (3) we will return with y a leaf of the current (T, δ) having $|\delta(y)| > 1$, which explains the if-statement at the very end of Algorithm 1.

Note that the local improvements made in the local search are based on randomly swapping vertices between $\delta(y)$ and $\delta(z)$ for two nodes y and z with the same parent. As usual in local search, there is a fine balance to trying new splits versus sticking with old splits. The goal is to neither get stuck in local minima nor to swap so many nodes that we re-randomize completely and don't get a hill-climbing effect. Note in (4) that we store for each subset P of vertices encountered so far the best found full decomposition subtree $Best(P)$. The decision of when to try new splits and when to use the old splits is tied to the boolean-width of the best subtrees, and to the upper bound on boolean-width of G given by $Best(V)$.

Algorithm 4. TryToImproveSubtree(x)

> **Input:** a node x of T with $|\delta(x)| > 1$
> **(1) if** x is a leaf **then** (A,B) = **Split**($\delta(x)$)
> **(2) else**
> > Let y and z be the children of the node x.
> > (A, B)=**RandomSwap**($\delta(y), \delta(z)$)
>
> **(3) if** $max\{\mathbf{UN}(A), \mathbf{UN}(B)\} < boolw(Best(V))$
> > **then** Set y and z as new leaf children of x with $\delta(y) = A$ and $\delta(z) = B$
> > **else if** x is still a leaf **then return** /* in case we came from (1) */
>
> **(4) if** $max\{\mathbf{UN}(\delta(y)), \mathbf{UN}(\delta(z))\} < boolw(Best(V))$ **then**
> > **for** $w \in \{y, z\}$
> > > **if** subtree for $\delta(w)$ is stored and $boolw(Best(V)) > boolw(Best(\delta(w)))$
> > > > **then** use root of $Best(\delta(w))$ as w.
> > > **else if** $|\delta(w) > 1|$ call **TryToImproveSubtree**(w)
> > **if** the subtree T_x rooted at x is a full subtree of $\delta(x)$
> > > **then** update $Best(\delta(x))$ to T_x

Algorithm 5. RandomSwap($\delta(y), \delta(z)$)

> **Input:** $\delta(y), \delta(z) \subseteq V$ for sibling nodes y and z of T.
> **Output:** split (A, B) of $\delta(y) \cup \delta(z)$.
> Let x be the parent of y and z.
> **choose** randomly i in $0..(|\delta(y)| - \frac{|\delta(x)|}{3})$ and j in $0..(|\delta(z)| - \frac{|\delta(x)|}{3})$.
> **choose** randomly $M_i \subset \delta(y)$ and $M_j \subset \delta(z)$ with $|M_i| = i$ and $|M_j| = j$.
> $A = (\delta(y) \setminus M_i) \cup M_j$
> $B = (\delta(z) \setminus M_j) \cup M_i$
> **return** (A, B).

3.3   Discussion and Implementation Details

We made our implementations in Java. Subsets of vertices are stored as bitvectors of length n, i.e. the number of vertices in the graph. We expect most of the subsets we store to be of size at least $\frac{n}{2}$ so this is an efficient way to store subsets. We also limited the boolean-width to 31, i.e. $|UN(A)| \leq 2^{31}$, but none of the graphs tested reached this limit. The bottleneck is rather the memory available on our machines. Let us explain. Our implementation of subroutine **UN**(A) uses memory proportional to $n * |UN(A)|$ bits. Since $|UN(A)| \leq 2^{min(|A|, |\overline{A}|)}$ the 'boolean-width ≤ 31' becomes a bottleneck only if the graph has at least 64 vertices. In that case the implementation is handling a list of neighborhoods of size $64 * 2^{31}$ bits which is 16 GB of memory and that is more memory than our desktop had. It is part of future research to find memory efficient methods to compute $|UN(A)|$.

As described, we are currently storing the best full decompositions of subtrees. Since bitvectors are easy to compare they are stored in a binary search tree for quick look-up. Storing all these solutions eats up memory, and for some big graphs this is the limiting factor. In the future we will consider more advanced schemes for storing the partial solutions encountered. In particular one should throw out elements that are no longer below the upper bound.

The search for new solutions in the space of candidate solutions is based on a fine balance between greedy choices and random choices, a balance that was arrived at mainly through experimentation. This appears e.g. in the choice of letting the **Split** subroutine start with a random half of the nodes on one side before trying vertices one-by-one in the more costly greedy stage. Similarly for the fully random choice of swapping in subroutine **RandomSwap**, and in the conditional tests in (3) and (4) of **TryToImproveSubtree**.

Although not specified in the pseudocode, for small subtrees we just return an arbitrary one, since if $|\delta(x)| \leq boolw(Best(V))$ then any full subtree at x will have boolean-width at most $boolw(Best(V))$. The **Split**(P) subroutine given in Algorithm 2 could be stopped as soon as a subset A_i with low $|UN(A_i)|$ and $|UN(P \setminus A_i)|$ values has been found. It is not clear that this is always better and currently it is not done. There are many calls of **UN**(A) for many subsets A that only differ in a few vertices. A possible improvement is to store the sets of unions of neighborhoods $UN(A)$ and use these e.g. when computing $UN(A \cup \{v\})$ for a single added vertex v, allthough it is not clear how to do this efficiently. The **UN**(A) subroutine given in Algorithm 3 does not recompute known values, but otherwise it may seem naive. It forms the inner loop of the heuristic and it is the bottleneck for running on graphs with many vertices. We tried different approaches such as randomly sampling subsets to approximate $|UN(A)|$ and exploiting a correlation between the degree of a vertex and its contribution to $|UN(A)|$. These tests led to only insignificant improvements so for the moment we kept the naive algorithm. There are other, similar, improvements to **UN**(A) that can be attempted, and although they may not asymptotically improve the running-time of the heuristic they could potentially be of big help.

The balance between trying new splits and sticking to old splits is guided by the conditional test in (3) of Algorithm 4. We did try imposing stronger conditions in order to arrive at better splits sooner, but only minor improvements were seen, and only in some cases.

The heuristic ran for a predefined amount of time for each graph but there are several ways of experimenting with the stopping criteria, for example based on the size of the input graph, or on the fraction of time since an improved tree was last found.

4 Experimental Results

All presented results have been carried out on a Linux machine with 2.33 GHz Intel Core 2Duo CPU E6550 and 2 GB RAM. Our aim was not fast benchmark results, but to explore heuristics for finding decompositions of low boolean-width. TreewidthLIB is an online depository containing a collection of 710 graphs, to be used as a benchmark for the comparison of algorithms computing treewidth. TreewidthLIB provides selected instance graphs, for which computing the treewidth is relevant, originating from applications like probabilistic networks, vertex coloring, frequency assignment and protein structures [5]. We ran our heuristic on the graphs in TreewidthLIB.

TreewidthLIB contains 710 graphs. For 482 graphs a tree-width bound is given in TreewidthLIB, and for 426 graphs we give a boolean-width bound using our heuristic. For the comparison we concentrate on the 300 graphs for which we have a bound on

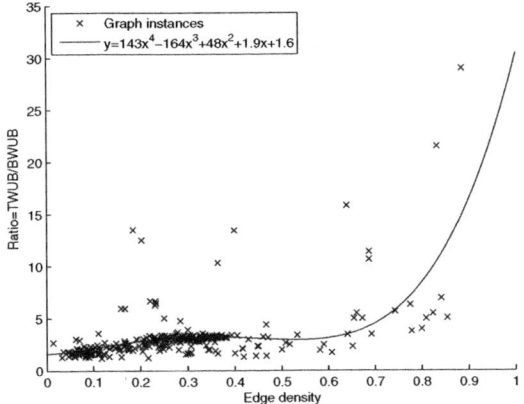

Fig. 1. Ratio (treewidth divided by boolean-width) versus edge density in all the 300 graphs for which heuristically computed upper bounds are known

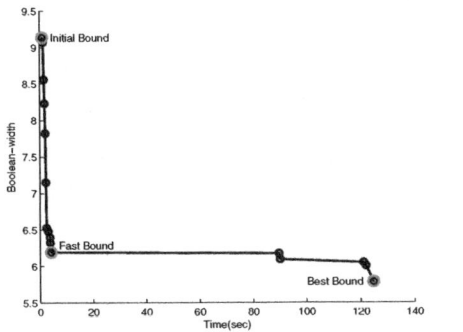

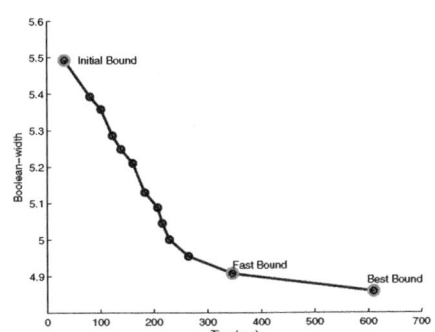

Fig. 2. Improvement of boolean-width upper bound as the local search progresses over time, for the graph *eil51.tsp* (V=51,E=140)

Fig. 3. Improvement of boolean-width upper bound as the local search progresses over time, for the graph *miles1500* (V=128,E=5198)

both tree-width and boolean-width, but let us first discuss the remaining 410 graphs. Among these 410 graphs, there are 126 having only a boolean-width bound, 182 having only a tree-width bound, and 102 having neither. Among the 182 graphs having only a tree-width bound there are some in a graph format not supported by our implementation, but for the majority of these graphs our heuristic simply timed out already at the greedy initialization stage. Note that for these 182 graphs, if we were given the decomposition of low tree-width k, we could easily have produced a decomposition of boolean-width at most k, using the $O(nk^2)$ algorithm which can be deduced from [1].

We now summarize our findings for the 300 graphs having both a tree-width bound and a boolean-width bound. Firstly, the boolean-width bound is always better than the tree-width bound, with the ratio of the tree-width bound divided by the boolean-width bound ranging from 1.15 to 29, with an average of 3.13. Not surprisingly, the ratio increased with higher edge density. In Fig.1 we have plotted this ratio against the edge density of the graphs for a total of 300 graphs. The trend line shows the growth of ratio with edge density.

Our heuristic algorithm starts with greedily finding a full decomposition tree giving an Initial Bound on boolean-width and then improves this bound iteratively. In the experiments we kept track of the decrease in the boolean-width over time. In Fig. 2 and Fig. 3 the upper bounds on boolean-width, i.e. the values of $boolw(BEST(V))$, are shown as they decrease over time, for the two graphs called *eil51.tsp* (V=51 and E=140) and *miles1500* (V=128,E=5198). For the graph *eil51.tsp* the Initial Bound was 9.1 after less than a second, then at the 'knee' of the curve before the improvement decays we found a Fast Bound of 6.2 after 4 seconds, and finally the Best Bound of 5.8 was found after 124 seconds. For each graph, we can likewise speak of three bounds: i) the Initial Bound given by the greedy initialization, ii) a Fast Bound found at the 'knee' of the curve, and iii) the Best Bound found possibly after a long runtime.

In Table 1 we summarize results for 8 selected graphs having a good variety of number of vertices V, edge density $density$, Time in seconds to find Initial Bound, Fast Bound, and Best Bound on boolean-width, its best known treewidth upper bound TWUB, and Ratio=TWUB/BWUB(Best Bound). The graphs are sorted by this Ratio. The *miles1500* graph is translated from the Stanford GraphBase. The *zeroin.i.1* and *mulsol.i.5* graphs originate from the 2nd DIMACS implementation challenge [10] and are generated from a register allocation problem based on real code. The *queen8_12* also comes from the DIMACS[10] graph coloring problems and is an example of n-queens puzzle. The graph *1awd* is from the field of computational biology with each vertex representing a single side chain and each edge representing the existence of a pairwise interaction between the two side chains. The graph *celar06-wpp* is a frequency assignment instance. The graph *BN_28* originates from Bayesian Network from evaluation of probabilistic inference systems at UAI 2006. The graph *eil51.tsp* is a Delauney triangulation of a traveling salesman problem.

Table 1. Results for selected graphs

Graph name	V	Edge $density$	Initial Bound BWUB	Time(s)	Fast Bound BWUB	Time(s)	Best Bound BWUB	Time(s)	TWUB	Ratio
miles1500	128	0.64	5.5	32.6	4.9	345.7	4.8	609.6	77	15.85
zeroin.i.1	211	0.19	4.0	74.1	3.8	116.2	3.7	168.0	50	13.51
mulsol.i.5	186	0.23	6.4	55.3	5.4	130.0	4.9	365.2	31	6.25
queen8_12	96	0.30	16.7	3055	16.7	3055	16.7	3055	65	3.91
1awd	89	0.27	13.3	67.5	11.1	521.1	10.8	702.9	38	3.52
celar06-wpp	34	0.28	4.5	0.1	3.2	0.8	3.0	4.8	11	3.37
BN_28	24	0.18	3.3	0.02	2.3	0.05	2.0	0.3	5	2.50
eil51.tsp	51	0.11	9.1	0.9	6.2	4.1	5.8	124.6	9	1.55

4.1 Small Grid Graphs

We also ran our heuristic on graphs corresponding to the $n \times n$ grid. However, for square grids the current implementation of **UN**(A) is too memory-intensive and we had to limit the size to $n \leq 9$. These are sparse graphs having tree-width n and the upper bound we find on boolean-width is below this. See Figure 4. The boolean-width of square $n \times n$ grids is a topic we are investigating and our current guess is that the optimal upper bound, holding for all n, is about $0.8 * n$. If this is correct, the value computed by the heuristic is close to optimal, which is somewhat interesting as it is our understanding that the heuristics for finding decompositions of low tree-width do not perform well on grid graphs.

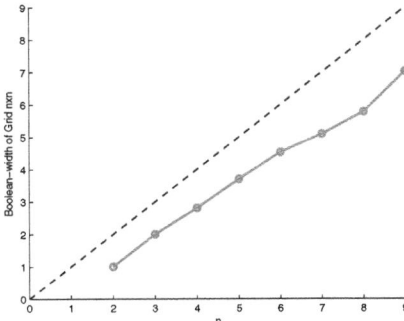

Fig. 4. Upper-bound on boolean-width, as computed by our heuristic, for the $n \times n$ grid, with n ranging from 2 to 9. Tree-width is given by the dotted line $x = y$.

5 Conclusion

We presented the first experimental study on computing a notion of width that works also for non-sparse graphs, based on the boolean-width parameter. Experiments with the graphs in TreewidthLIB show the strength of boolean-width versus tree-width, in a practical setting, in particular for graphs of edge density above a certain value. For more examples of real-world graphs of high edge density and high tree-width we could also look beyond the TreewidthLIB library. There are a number of open problems related to boolean-width heuristics and some have already been discussed in subsection 3.3. Firstly, we need a fast heuristic that directly constructs a reasonable upper bound on the boolean-width for any graph, regardless of how big the graph is or what its edge density is. The main issue will be to give a fast heuristic for the computation of a good upper bound on $|UN(A)|$. Secondly, we need to consider heuristics for computing lower bounds on boolean-width, just as it has been done for tree-width [6]. Thirdly, we should explore pre-processing to simplify the graph instances, again this has been done extensively for tree-width [4]. These problems are of interest since our results indicate that using boolean-width could in the future outperform the use of tree-width in practice for a large class of graphs and problems.

References

1. Adler, I., Bui-Xuan, B.M., Rabinovich, Y., Renault, G., Telle, J.A., Vatshelle, M.: On the Boolean-Width of a Graph: Structure and Applications. In: Thilikos, D.M. (ed.) WG 2010. LNCS, vol. 6410, pp. 159–170. Springer, Heidelberg (2010)
2. Belmonte, R., Vatshelle, M.: Graph classes with structured neighborhoods and algorithmic applications. In: Proceedings of the 37th International Workshop on Graph-Theoretic Concepts in Computer Science, WG 2011 (2011),
   www.ii.uib.no/~martinv/Papers/LogBoolw.pdf
3. Bodlaender, H.L.: A linear time algorithm for finding tree-decompositions of small treewidth. SIAM Journal on Computing 25, 1305–1317 (1996)
4. Bodlaender, H.L.: Treewidth: Characterizations, Applications, and Computations. In: Fomin, F.V. (ed.) WG 2006. LNCS, vol. 4271, pp. 1–14. Springer, Heidelberg (2006)
5. Bodlaender, H.L., Koster, A.M.C.A.: Treewidth computations I. Upper bounds. Information and Computation 208, 259–275 (2010)
6. Bodlaender, H.L., Koster, A.M.C.A.: Treewidth computations II. lower bounds. Technical Report UU-CS-2010-022, Department of Information and Computing Sciences, Utrecht University, Utrecht, The Netherlands (2010) (accepted for publication in Information and Computation)
7. Brandstadt, A.: Personal Communication
8. Bui-Xuan, B.M., Telle, J.A., Vatshelle, M.: Boolean-width of graphs. Theoretical Computer Science (to appear, 2011),
   www.ii.uib.no/~telle/bib/listofpub/BTV11.pdf
9. Chen, H.: Quantified constraint satisfaction and bounded treewidth. In: de Mántaras, R.L., Saitta, L. (eds.) Proceedings of the 17th European Conference on Artificial Intelligence, ECAI 2004, pp. 161–165 (2004)
10. The second DIMACS implementation challenge: NP-Hard Problems: Maximum Clique, Graph Coloring, and Satisfiability (1992-1993),
    http://dimacs.rutgers.edu/Challenges/
11. Gottlob, G., Leone, N., Scarcello, F.: A comparison of structural CSP decomposition methods. Acta Informatica 124, 243–282 (2000)
12. Hicks, I.V., Koster, A.M.C.A., Kolotoğlu, E.: Branch and tree decomposition techniques for discrete optimization. In: Cole Smith, J. (ed.) INFORMS Annual Meeting, TutORials 2005. INFORMS Tutorials in Operations Research Series, ch. 1, pp. 1–29 (2005)
13. Hliněný, P., Oum, S.: Finding branch-decomposition and rank-decomposition. SIAM Journal on Computing 38, 1012–1032 (2008)
14. Kim, K.H.: Boolean matrix theory and its applications. Marcel Dekker (1982)
15. Lauritzen, S.J., Spiegelhalter, D.J.: Local computations with probabilities on graphical structures and their application to expert systems. The Journal of the Royal Statistical Society. Series B (Methodological) 50, 157–224 (1988)
16. Overwijk, A., Penninkx, E., Bodlaender, H.L.: A Local Search Algorithm for Branchwidth. In: Černá, I., Gyimóthy, T., Hromkovič, J., Jefferey, K., Královič, R., Vukolić, M., Wolf, S. (eds.) SOFSEM 2011. LNCS, vol. 6543, pp. 444–454. Springer, Heidelberg (2011)
17. Röhrig, H.: Tree decomposition: A feasibility study. Master's thesis, Max-Planck-Institut für Informatik, Saarbrücken, Germany (1998)
18. Song, Y., Liu, C., Malmberg, R., Pan, F., Cai, L.: Tree decomposition based fast search of RNA structures including pseudoknots in genomes. In: Proceedings of the 2005 IEEE Computational Systems Bioinformatics Conference, CSB 2005, pp. 223–234 (2005)

19. Treewidthlib (2004), `http://www.cs.uu.nl/people/hansb/treewidthlib`
20. van Rooij, J.M.M., Bodlaender, H.L., Rossmanith, P.: Dynamic Programming on Tree Decompositions using Generalised Fast Subset Convolution. In: Fiat, A., Sanders, P. (eds.) ESA 2009. LNCS, vol. 5757, pp. 566–577. Springer, Heidelberg (2009)
21. Zhao, J., Che, D., Cai, L.: Comparative pathway annotation with protein-DNA interaction and operon information via graph tree decomposition. In: Proceedings of Pacific Symposium on Biocomputing, PSB 2007, vol. 12, pp. 496–507 (2007)
22. Zhao, J., Malmberg, R.L., Cai, L.: Rapid ab initio prediction of RNA pseudoknots via graph tree decomposition. Journal of Mathematical Biology 56(1-2), 145–159 (2008)

Parameterized Maximum Path Coloring

Michael Lampis

Graduate Center, City University of New York
mlampis@gc.cuny.edu

Abstract. We study the well-known MAX PATH COLORING problem from a parameterized point of view, focusing on trees and low-treewidth networks. We observe the existence of a variety of reasonable parameters for the problem, such as the maximum degree and treewidth of the network graph, the number of available colors and the number of requests one seeks to satisfy or reject. In an effort to understand the impact of each of these parameters on the problem's complexity we study various parameterized versions of the problem deriving fixed-parameter tractability and hardness results both for undirected and bi-directed graphs.

1 Introduction

The PATH COLORING (PC) and MAXIMUM PATH COLORING (MAXPC) problems are two well-known and widely studied combinatorial problems with applications in the field of optical networks. In PC we are given a graph representing the optical network and a set of paths on that graph and are asked to find a coloring of the paths such that any two paths which share an edge have distinct colors and the number of colors used is minimized. In the MAXPC problem on the other hand we are given a specified number of colors and must select a maximum cardinality set of paths which can be properly colored with the available colors. If the graph contains cycles we may alternatively be given the endpoints of the communication requests only, with the flexibility to choose the most suitable path for each. Then the problem is often called ROUTING AND PATH COLORING (RPC and MAXRPC). Of course, if the underlying graph is a tree the two versions of the problems are equivalent.

PC is unfortunately known to be hard to solve exactly even on very simple topologies and therefore the same holds for MAXPC. As a consequence the vast majority of research on the two problems has focused on coming up with good approximation algorithms for either minimizing the number of colors or maximizing the number of accepted requests. In this paper, however, we investigate the complexity of solving MAXPC on trees and tree-like graphs exactly, from the point of view of parameterized complexity theory. (For an introduction to parameterized complexity theory and the theory of fixed-parameter tractable (FPT) algorithms see [4,16,9]).

The main observation we want to exploit is that MAXPC is a problem rich with reasonable parameters. For example in practical situations one may often expect that the network will have moderate maximum degree and it will be a

D. Marx and P. Rossmanith (Eds.): IPEC 2011, LNCS 7112, pp. 232–245, 2012.

tree or perhaps "tree-like". Furthermore, technological limitations mean that the number of available colors on each edge is also likely to be moderate. Also, as observed in [1], communications networks are often built with maximum capacity in mind, meaning that typically the available resources should be enough to satisfy all or almost all requests. Interestingly, nothing prevents several of these facts from happening together. This motivates the study of the problem through a parameterized lens: one identifies a parameter (or set of parameters) expected to be small and then attempts to design an FPT algorithm for this particular parameterization or prove that none exists. For example, if one is interested in instances with moderate maximum degree Δ, the goal is to design an algorithm with running time $f(\Delta) \cdot n^c$, for some moderately exponential function f or to prove that no such algorithm exists (but perhaps an algorithm running in time $n^{g(\Delta)}$ is possible).

Of course, the observation that Δ or some other parameters may be small in practice is not new; in fact the traditional complexity of PC has been investigated for bounded degree trees for example. The contribution of this paper is that, in addition to giving new results it puts these known results under the light of parameterized complexity, where $f(\Delta) \cdot n^c$ and $n^{g(\Delta)}$ algorithms are considered completely different cases with only the first called tractable, whereas for traditional complexity both are "polynomial for fixed Δ". This allows us to more systematically enter more parameters into the problem and better assess their impact on complexity. Interestingly, it also leads us to discover some interesting gaps between the complexity of PC and MaxPC which were previously overlooked in the literature.

Previous Work. PC and MaxPC are very well-studied problems, starting from the 1980s (see [12]). As mentioned, when the network graph contains cycles one may consider either the case where requests are pre-routed or where routing is part of the problem. Furthermore, the communication network can either be assumed to be undirected or bi-directed, where in the second case every request has a direction and two requests with the same color can share an edge if they use it in opposite directions.

PC is known to be hard even in very restricted topologies, from which fact the hardness of MaxPC also follows trivially. Specifically, PC is NP-hard for undirected stars by equivalence to edge coloring in multi-graphs [6], undirected rings (here the problem is equivalent to coloring circular-arc graphs [10]) and bi-directed binary trees [6,15]. However, it is known to be FPT in undirected trees when parameterized by the maximum degree of the tree Δ [6,15], and also to be FPT in bi-directed trees when parameterized by the maximum number of requests touching any node [6]. A 4/3-approximation algorithm is known for PC in undirected trees [6] and a 5/3-approximation for bi-directed trees [8].

For MaxPC a 2.22-approximation is known for bi-directed trees [5] and a 1.58-approximation is known for undirected trees [17]. For bi-directed trees it is also known that MaxPC is solvable in polynomial time if both the maximum degree and the number of colors are constant [5], a result which can be extended to undirected trees in a straightforward manner. Note though that this is an XP, not an FPT algorithm, a fact that we will return to later. For the special case where only one color is available the problem is also known as MAXIMUM EDGE DISJOINT PATHS (MAXEDP), and is known to be NP-hard for bi-directed trees [7] but in P for undirected trees [11].

To the best of our knowledge PC and MAXPC have not been explicitly studied before from a parameterized perspective, though there are results in the literature which can be translated to the fixed-parameter tractability terminology, such as the algorithm for PC on bounded degree undirected trees mentioned above. For the related CALL CONTROL problem however there has been an investigation of its complexity when parameterized by the number of rejected requests [1], which is one of the parameters we consider in this paper as well. Note though that the situations for CALL CONTROL and MAXPC are quite different for this parameter, as MAXPC is usually hard even when no requests can be rejected (this is the PC problem), while CALL CONTROL is easy in that case. Thus, for MAXPC this parameter can only be useful in combination with other parameters that make PC fixed-parameter tractable.

Contributions of This Paper. In this paper we study several parameterized versions of MAXPC mainly on trees. First, we study parameterizations which do not involve the objective function, that is, the number of requests to be satisfied or rejected. Specifically, for trees the parameters we consider are the maximum degree Δ and the number of available colors W. As mentioned, for undirected trees PC is FPT parameterized by Δ alone while for bi-directed trees PC is FPT parameterized by Δ and W, from the FPT result when parameterized by the maximum number of requests touching a node. However, for MAXPC all that is known is an XP algorithm running in roughly $n^{\Delta W}$ time. From the traditional complexity perspective it is easy to overlook the difference as in both cases we have algorithms polynomial for fixed Δ and W. However, from the parameterized complexity perspective it is natural to ask why no FPT algorithm is known for MAXPC and whether this can be fixed by designing an algorithm that would take at least one or ideally both of the parameters out of the exponent of n. We resolve this question fully by showing that neither parameter can be removed from the exponent of n, under standard complexity assumptions, even if the other is a small constant. This points out the existence of a (previously unknown) gap between the complexity of PC and MAXPC. In particular our results imply that MAXPC is NP-complete even on binary trees (where PC is solvable in polynomial time), which to the best of our knowledge was not known before. They also show that the complexity of MAXPC grows much faster as Δ and W grow than the complexity of PC.

Continuing this line of reasoning we observe that the $n^{\Delta W}$ algorithm can be extended in a straightforward way to graphs of treewidth t^1, running in time roughly $n^{\Delta W t}$. This poses the new problem of whether at least t can be moved out of the exponent which we again resolve negatively.

---

[1] The treewidth of a graph is a measure which estimates how "tree-like" the graph is; for more information see [2].

One intuitive explanation for the complexity gap between PC and MAXPC is that in the instances of our reductions a large fraction of the requests must be rejected, making the situation very different from PC where all requests must be satisfied. Thus, we are led to add as another parameter the number of rejected requests. In this case the complexity gap (at least partially) closes again: we show that for both undirected and bi-directed trees the MAXPC problem is FPT if one considers as parameters Δ, W and the number of requests which can be rejected. Also we show that for undirected binary trees MAXPC is FPT parameterized by the number of rejected requests.

Finally, we consider the naturally parameterized version of MAXPC (that is, parameterized by the size of the solution) and show that it is FPT on any topology where the naturally parameterized version of MAXEDP is FPT, using a color-coding technique. From this, it immediately follows that this parameterization of MAXPC is FPT for undirected trees and for rings, since in those cases MAXEDP is solvable in polynomial time. For bi-directed trees, where MAXEDP is NP-hard, we show that its naturally parameterized version is FPT, a result which may be of independent interest, thus settling the fixed-parameter tractability of MAXPC in this case as well.

2   Definitions and Preliminaries

In this paper we discuss the PATH COLORING problem (PC) and its corresponding maximization problem MAXPC. Our main topic is their restriction to trees. The input we are given in this case consists of an undirected tree $G(V, E)$ and a multi-set of demands $D \subseteq V \times V$, each demand corresponding to the unique path in G that connects its two vertices. We are also given two integers W (the number of colors) and B (the number of demands we seek to satisfy). The question is whether there exist W mutually disjoint subsets $D_1, D_2, \ldots, D_W \subseteq D$ s.t. no set D_i contains two demands that share an edge and $\sum_{i=1}^{W} |D_i| \geq B$.

In other words we are asked if there exists a W-colorable set of at least B paths from the set of the given demands. This problem, where we seek to maximize B is usually called MAXPC, while PC is simply the special case when $B = |D|$. The graph G can either be considered undirected, in which case the ordering of each demand pair is irrelevant, or bi-directed, in which case two satisfied demands with the same color are allowed to use the same edge but only in opposite directions (another way to think of this is as replacing every undirected edge with two parallel arcs of opposite directions). In this paper we will deal with both undirected and bi-directed graphs.

The problems can be generalized to graphs that contain cycles. Here, we will focus on the case where for each demand we are given the path that it must follow on the graph, but also briefly mention how our results can be extended to the case where routing is part of the problem.

We will use Δ to denote the maximum degree of G, n to denote the number of vertices, t to denote the treewidth of G (which is of course 1 if G is a tree) and $T = |D| - B$ to denote the number of demands we are allowed to reject. We will consider various tractable special cases and parameterizations of MAXPC, for example on trees of bounded degree, or in instances with a small number of colors (we assume here that the reader is familiar with the basic definitions of parameterized complexity theory, such as the class FPT). The candidate parameters we are interested in are Δ, t, W, B and T. To keep the presentation short and concise we will use a notation where different parameterizations of MAXPC are denoted by prepending it with the list of variables we consider constant or parameters. For example, the $(p\Delta)$−MAXPC problem is the parameterized version of MAXPC when Δ is our only parameter, while (pB)−MAXPC is the parameterized version where B is the parameter (the "naturally" parameterized version of MAXPC). The reason for this notation is that we will consider various combinations of parameters and also cases where some values are parameters and some others are fixed constants. For example $(pW, c\Delta)$−MAXPC is the special case of (pW)−MAXPC restricted to bounded degree trees. Observe that this is not the same as the problem $(pW, p\Delta)$−MAXPC since a hypothetical algorithm running in time say $2^W n^\Delta$ is FPT for the first problem but not for the second.

Our aim here is to investigate how different parameters (and combinations of parameters) affect the complexity of the problem. Table 1 contains a summary of some of the already known results on the complexity of PC and MAXPC and the results of this paper. Worthy of note is the contrast between some already known tractable cases of PC ($(p\Delta)$-PC for undirected trees and $(p\Delta, pW)$-PC for bi-directed trees) and the hardness we establish for the corresponding cases of MAXPC. Interestingly, tractability returns if we add T as a parameter to $(pW, p\Delta)$-MAXPC in both bi-directed and undirected trees, which makes intuitive sense since T quantifies the "distance" between a PC and a MAXPC instance. We can also prove that (pT)-MAXPC is FPT on undirected binary trees.

Table 1. Summary of results. All results concern trees, except those where the graph's treewidth t is included in the problem description.

Undirected			Bi-Directed		
Problem	Result	Comment	Problem	Result	Comment
(cW)-PC, $W = 3$	NP-h	(edge 3-coloring) [6]	(cW)-PC, $W = 1$	NP-h	[6]
$(p\Delta)$-PC	FPT	[6]	$(c\Delta)$-PC, $\Delta = 3$	NP-h	[6]
			$(pW, p\Delta)$-PC	FPT	[6]
$(c\Delta, ct)$-PC	NP-h	(PC on rings) [10]	$(c\Delta, ct)$-PC	NP-h	[10]
$(cW, c\Delta)$-MAXPC	P	[5]	$(cW, c\Delta)$-MAXPC	P	[5]
$(cW, c\Delta, ct)$-MAXPC	P	Theorem 1	$(cW, c\Delta, ct)$-MAXPC	P	Theorem 1
$(pW, c\Delta)$-MAXPC	W[1]-h	Theorem 2	$(pW, c\Delta)$-MAXPC	W[1]-h	Theorem 2
$(cW, p\Delta)$-MAXPC	W[1]-h	Theorem 3	$(cW, p\Delta)$-MAXPC	W[1]-h	Theorem 3
$(cW, c\Delta, pt)$-MAXPC	W[1]-h	Theorem 4	$(cW, c\Delta, pt)$-MAXPC	W[1]-h	Theorem 4
$(pW, p\Delta, pT)$-MAXPC	FPT	Theorem 5	$(pW, p\Delta, pT)$-MAXPC	FPT	Theorem 5
(pT)-MAXPC, $\Delta = 3$	FPT	Theorem 6			
(pB)-MAXPC	FPT	Corollary 1	(pB)-MAXPC	FPT	Corollary 2

3   Structural Parameterizations

In this section we investigate parameterizations which do not involve the objective function. The candidate parameters will be the maximum degree Δ, the input graph's treewidth t and the number of available colors W. Some fixed-parameter tractability results are known in the case of PC for these cases, but unfortunately for the corresponding cases of MaxPC only XP algorithms are known and as we will show this can probably not be improved.

First, recall that in [5] it was shown that MaxPC can be solved in polynomial time on bi-directed trees if both Δ and W are constant. The basic idea is a bottom-up dynamic programming technique which can be extended in a straightforward way to undirected trees also. Our first observation is that this idea can in fact be extended to graphs of bounded treewidth as well.

Theorem 1. $(cW, c\Delta, ct)$-MaxPC *can be solved in polynomial time for both undirected and bi-directed graphs.*

Theorem 1 essentially applies common dynamic programming techniques associated with treewidth to obtain an XP algorithm. The algorithm is likely to be extremely impractical though, even for small values of the parameters, since the exponent relies on all three. So the natural, and more important question to ask is whether any kind of fixed-parameter tractability result can be obtained.

Ideally, one would like an FPT algorithm running in time $f(W, \Delta, t) \cdot n^c$, that is, an FPT algorithm for $(pW, p\Delta, pt)$-MaxPC. Barring that, it would still be helpful if any one of the three parameters could be moved out of the exponent of n, even by itself. Unfortunately, we resolve this problem in a negative way, showing that even if any two of the parameters are small fixed constants (and are therefore allowed to appear in the exponent of n in an FPT algorithm) it is still impossible to obtain an FPT algorithm for the problem, under standard complexity assumptions. We prove this by using three parameterized reductions.

The reductions presented here will use a slightly more general problem we will call CapMaxPC. In this problem, for each edge $e \in E$ we are given an integer capacity $1 \leq c(e) \leq W$ and have the additional constraint that in a feasible solution at most $c(e)$ satisfied demands may be using e. For parameterizations not involving the objective function this problem is shown FPT-reducible to MaxPC by using a simple trick where limited edge capacity on an edge is simulated by adding an appropriate number of length 1 demands going through the edge.

We will also use another intermediate problem in our reductions, which we will call Disjoint Neighborhoods Packing (DNP). In DNP we are given an undirected graph $G(V, E)$ and are asked to find a maximum cardinality set $V' \subseteq V$ such that $\forall u, v \in V'$ we have $N(u) \cap N(v) = \emptyset$ (we denote by $N(u)$ the set that contains u and all its neighbors, that is, the closed neighborhood of u). The parameter we consider is the size of V'. This problem is sometimes referred to in the literature as 2-Independent Set, see [14].

Overall our strategy is to start from the well-known W[1]-hard problem Independent Set and present reductions to our problems through the two intermediate problems described above, that is, we aim to prove that IS $\leq_{FPT}$ DNP $\leq_{FPT}$ CapMaxPC $\leq_{FPT}$ MaxPC. The trickiest step in this process will be the second reduction, where we will show three different versions, one for each parameterization of MaxPC we are interested in.

Lemma 1. *For both undirected and bi-directed graphs we have*

- $(pW, c\Delta)$-CapMaxPC $\leq_{FPT}$ $(pW, c\Delta)$-MaxPC '
- $(cW, p\Delta)$-CapMaxPC $\leq_{FPT}$ $(cW, p\Delta)$-MaxPC
- $(cW, c\Delta, pt)$-CapMaxPC $\leq_{FPT}$ $(cW, c\Delta, pt)$-MaxPC

Lemma 2. DNP *is W[1]-hard.*

Proof. We present a reduction from the INDEPENDENT SET problem. Given a graph $G(V, E)$ and assuming without loss of generality that it has no isolated vertices and we are looking for an independent set of size $k > 2$ in G, we will construct an equivalent instance of DNP. First, subdivide every edge of G, that is, replace each $(u, v) \in E$ with a path of length 2. Connect all newly added vertices into a clique. We will argue that the new graph has a packing of k disjoint neighborhoods iff the original graph has an independent set of size k.

If the original graph has an independent set of size k this immediately gives us a packing of the same size on the new graph by selecting the same vertices. The packing is valid since the only way two of the original vertices could have a common neighbor in the new graph is if one of the vertices introduced in the subdivisions is connected to both and that can only happen if an edge was connecting them in the original graph.

If the new graph has a packing of $k > 2$ disjoint neighborhoods given by the set of vertices V', then we can immediately infer that V' cannot include two or more of the vertices introduced in the subdivisions, since they are all connected in a clique. If V' contains one of these new vertices, say the one introduced in the subdivision of (u, v) (call that vertex w) then it cannot contain any vertices in $V \setminus \{u, v\}$ because every original vertex is connected to at least one new vertex and that vertex is connected to w. V' may also contain at most one of $\{u, v\}$, so its total size cannot be more than 2 in this case. We conclude that a packing of $k > 2$ disjoint neighborhoods must consist entirely of vertices found in the original graph. To see that these form an independent set in the original graph, observe that if two were originally connected they would have a common neighbor in the new graph, violating the feasibility of the packing.   □

Theorem 2. $(pW, c\Delta)$-MaxPC *is W[1]-hard for both undirected and bi-directed trees.*

Proof. Given Lemma 1 and Lemma 2 the only thing left to prove is that DNP $\leq_{FPT}$ $(pW, c\Delta)$-CapMaxPC. Given an instance of DNP, that is a graph $G(V, E)$ and a target size for the DNP set k, we construct a CapMaxPC instance as described below. We first show the reduction for undirected trees and then describe how it can be made to work for bi-directed trees as well.

First, let $|V| = n$ and we construct a "backbone", which is simply a path on $n + 2$ vertices. We take $n + 2$ disjoint copies of a path on n vertices and attach one of the endpoints of each to one of the vertices of the backbone so that each backbone vertex now has a path hanging from it. Label the backbone vertices $b_i, 0 \leq i \leq n+1$ and the vertices of the other paths $p_{i,j}, 0 \leq i \leq n+1, 1 \leq j \leq n$, so that the path vertex connected to b_i is called $p_{i,1}$, its other neighbor is $p_{i,2}$ and so on. Finally, for each $1 \leq i, j \leq n$ we add three vertices in the graph $v_{i,j}, u_{i,j}$ and $w_{i,j}$ and the edges $(v_{i,j}, w_{i,j}), (u_{i,j}, w_{i,j})$ and $(w_{i,j}, p_{i,j})$. In other words, we construct a path on three vertices and connect the middle vertex to $p_{i,j}$. This completes the description of the graph, which is a tree of maximum degree 3.

Now let us describe the demands. Suppose that the vertices of the original graph are numbered $\{1, 2, \ldots, n\}$. For each $i \in V$ we consider the closed neighborhood $N(i)$ in increasing order and let $N(i) = \{j_0, j_1, j_2, \ldots, j_{d(i)}\}$, where $d(i)$ is the degree of i. We add a demand from $p_{0,i}$ to $u_{j_0,i}$. Then, for each $l, 0 \leq l < d(i)$ we add a demand from $v_{j_l,i}$ to $u_{j_{l+1},i}$. We also add a demand from $v_{j_{d(i)},i}$ to $p_{n+1,i}$. We add all these demands for each $i \in V$ and call these demands global. Finally, for each $1 \leq i, j \leq n$ we add a demand from $v_{i,j}$ to $u_{i,j}$. We call these demands local.

The only thing left is to specify W, which we set to $W = 2k$, and the capacities. We leave all capacities unconstrained except for the edges $(b_i, p_{i,1}), 1 \leq i \leq n$, which have a capacity of 2 and the edges $(u_{i,j}, w_{i,j})$ and $(v_{i,j}, w_{i,j})$ which have a capacity of 1. The construction is now complete.

To give some intuition about this construction, notice the interaction between local and global demands. Each local demand intersects exactly two global demands in edges of capacity 1. Thus, if the local demand is satisfied the global demands are rejected. Furthermore, if exactly one of the global demands is satisfied in a solution we can exchange it with the local demand, therefore this gadget ensures that either both global demands will be taken or both will be rejected in some optimal solution. Observe also that from all the local demands found in a branch attached to the backbone at most one will be rejected, since the edge $(b_i, p_{i,1})$ acts as a bottleneck allowing at most two global demands to go through. The idea will be that if a vertex i is in the neighborhood packing then we will select the global demand starting at $p_{0,i}$ and satisfy one after the other pairs of demands that go into branches that correspond to its neighbors, making these branches unusable for other global demands.

For a more precise argument, suppose that the original graph has a packing V' of size k, we will construct a CAPMAxPC solution of size $n^2 + k$. Start with a solution of size n^2 by selecting all the local demands of the instance and nothing else. Now for each $i \in V'$ we will inrease the size of the solution by 1. We do this by satisfying all the global demands associated with i, that is, all demands touching a vertex $p_{j,i}$ for any j. Each time we perform this improvement step we use two new colors (the two colors are sufficient to color the global demands since they form essentially a path) and remove from the solution all local demands that intersect with these global demands (it is not hard to see that this gives a profit of exactly one demand). Since we are using different colors in each step the only way this process could run into a problem is in an edge where fewer than $2k$ colors can be used. For that to happen we must be trying to satisfy more than two requests going through an edge $(b_j, p_{j,1})$ but that would imply that j is a common neighbor of two vertices of the packing, violating its feasibility.

For the other direction, suppose that a solution of size $n^2 + k$ exists. As mentioned, if in a set of one local and its two intersecting global demands the solution satisfied exactly one of the global demands, we exchange it with the local demand. This means that for each edge $(b_i, p_{i,1})$ we are either satisfying two of the demands crossing it or none and furthermore that if we are satisfying two, one of them is going "left" (that is, its other endpoint is towards b_{i-1}) and the other is going "right" (so its other endpoint is towards b_{i+1}). Therefore, the number of satisfied requests going through each edge (b_i, b_{i+1}) is constant for all i; call this number L. We will establish that $L = k$. Pick an arbitrary satisfied demand which uses a backbone edge and delete it from the solution. This will reduce the size of the solution by one, but it will also allow us to reduce L by one, since by the same arguments used before we can make the number of satisfied demands on each backbone edge the same without affecting the size of the solution[2]. Repeat this process L times and now we have a solution which satisfies only local demands and has size $n^2 + k - L$. Since there are exactly n^2 local demands it must be the case that $k = L$. Now we can conclude that there are k vertices in the branch connected to b_0 whose demands are satisfied and all subsequent global demands associated with them are also satisfied. These give us a neighborhood packing in the original graph because if two of them had a common neighbor the solution would be exceeding some branch's bottleneck capacity of 2.

It is not hard to modify this reduction to also work for bi-directed trees. The only difference in the network is that edges $(b_i, p_{i,1})$ are given a capacity of 1, since they are intended to be traversed twice but in different directions, and that it is now sufficient to have $W = k$ since all the global demands corresponding to a vertex are non-intersecting. Other than that we remain consistent with the ordering that we have implied in our description, that is, every global demand is ordered towards the vertex that lies further to the right (the vertex closer to $p_{n+1,n}$ so to speak). We also make sure that the local demands are directed in such a way that they intersect both global demands with which they share an edge and the rest of the arguments of the reduction go through unchanged.   □

Theorem 3. $(cW, p\Delta)$-MAXPC *is W[1]-hard for both undirected and bi-directed trees. The result holds even for instances where all the vertices but one have degree bounded by 3.*

Proof. Once again we will describe a reduction from DNP, but now the produced instance will have maximum degree depending on k and constant W. We will reuse some of the ideas of Theorem 2, properly adjusted. Again we will first

---

[2] This is implicitly relying on the fact that all global demands must intersect some local demand, which is true if the original graph had no isolated vertices.

describe a construction for undirected graphs and then discuss how it can be modified for bi-directed graphs.

Take k copies of a path on n vertices and label the vertices $S_{i,j}, 1 \leq i \leq k, 1 \leq j \leq n$. Take k more copies and label the vertices $T_{i,j}, 1 \leq i \leq k, 1 \leq j \leq n$. Add a new vertex to the graph, call it C, and connect it to all $S_{i,n}$ and $T_{i,n}$ for $1 \leq i \leq k$. Set the capacities of all edges to 1. Also, for each $i, j, 1 \leq i \leq k, 1 \leq j \leq n$ add a demand from $S(i,j)$ to $T(i,j)$.

Before we go on, let us examine the construction so far. It should be clear that the optimal solution satisfies k paths by selecting k vertices in the S branches and their corresponding vertices in the T branches. The k selected vertices will eventually encode the vertices we will pick for our neighborhood packing. What is of course missing is some machinery to ensure that our selection is indeed a packing in the original graph.

The constraints of a valid packing can be broken down as follows: for each of the $\binom{k}{2}$ pairs of vertices selected for the packing we must make sure that they do not share common neighbors. Thus, our basic tool will be a gadget that takes two of the k choices we have made and checks their compatibility. We will make $\binom{k}{2}$ copies of that gadget, attach them to C and then properly reroute the demands from S to T vertices through these gadgets.

To describe the pairwise consistency gadget, consider the instance constructed in the proof of Theorem 2. We modify it as follows: First, we add local requests gadgets, identical as those used in vertices $p_{i,j}, 1 \leq i, j \leq n$ to the vertices of the paths p_0 and p_{n+1}. We extend all demands which currently had an endpoint in $p_{0,j}$ or $p_{n+1,j}$ for some $j \in \{1, \ldots, n\}$ to the vertices $v_{0,j}$ and $u_{n+1,j}$ respectively, so that they intersect the new local demands. Now we make an exact copy of the branch p_0 and all its connected gadgets (i.e. the vertices $p_{0,j}, u_{0,j}, v_{0,j}, w_{0,j}, 1 \leq j \leq n$). We call the new branch p_0' (and the new vertices respectively $p_{0,j}', u_{0,j}', v_{0,j}', w_{0,j}', 1 \leq j \leq n$) and attach it also to b_0. We also make sure to replicate all demands that existed between the branch p_0 and the rest of the graph so that corresponding demands are placed between the branch p_0' and the rest of the graph. We perform another full copy for the branch p_{n+1} producing the branch p_{n+1}' with identical vertices and demands and attach this to b_{n+1}. Now the whole gadget has $n(n+4)$ local demands overall. We set the capacities of all backbone edges to 4, all edges used by local demands to 1 and all other edges to 2.

To demonstrate the use of this gadget we will connect one such gadget on our initial construction and use it to ensure that in the optimal solution the choices encoded in the paths S_1 and S_2 (i.e. the encoding of the first two choices for the packing) are compatible. Take a gadget as described and connect its b_0 to C by an edge of capacity 4. Recall that for all $j \in \{1, \ldots, n\}$ there is a demand from $S_{1,j}$ to $T_{1,j}$. Remove these n demands and for all $j \in \{1, \ldots, n\}$ add a demand from $S_{1,j}$ to $u_{0,j}$ and a demand from $v_{n+1,j}$ to $T_{1,j}$ (in other words we are rerouting the $S_1 \to T_1$ demands through the gadget). Do the same for demands from S_2 to T_2, only reroute them through the p_0' and p_{n+1}' branches.

A solution of size $n(n+4)+k$ can be achieved now iff the selections for active vertices in S_1 and S_2 are compatible, that is, the corresponding vertices of the initial graph have no common neighbors. This follows from the analysis of the properties of our gadget performed in Theorem 2.

It is now possible to complete the construction by adding more of the consistency gadgets so as to make sure that all $\binom{k}{2}$ pairs of choices are compatible. The final graph consists of the $\binom{k}{2}$ gadgets plus the $2k$ paths all attached to a single vertex of degree $\binom{k}{2} + 2k$. The total number of vertices is $O(n^2k^2)$ and a solution of size $\binom{k}{2}n(n+4)+k$ can be achieved iff the original graph has a packing of size k.

Modifying this construction to bi-directed trees is again strightforward, since a direction was implicit in our description. Again the only major difference is that we change edges with capacities 4 and 2 to capacities 2 and 1 respectively. For the last remark of the theorem, notice that the only vertex of high degree is C. All other vertices have degree at most 3, except the b_0 vertices of the gadgets, but even this can easily be fixed since it is not necessary for the reduction to attach p'_0 and p_0 to the same vertex. We can simply subdivide the (b_0, b_1) edge and attach p'_0 there. $\qquad\square$

Theorem 4. $(cW, c\Delta, pt)$-MaxPC and $(cW, c\Delta, pt)$-MaxRPC are W[1]-hard for both undirected and bi-directed graphs.

As a final note in this section, note that it is known that assuming standard complexity assumptions (specifically the Exponential Time Hypothesis which states that 3-SAT cannot be solved in time $2^{o(n)}$, see [13]) it is not possible to find an independent set of size k on an n-vertex graph in time $n^{o(k)}$. The reductions in Theorems 2 and 4 are linear in the parameter, meaning that assuming the ETH we know there is no $n^{o(W)}$ algorithm for MaxPC even for binary trees and there is no $n^{o(t)}$ algorithm, even when $W = 2, \Delta = 4$. The reduction in Theorem 3 is quadratic in the parameter, meaning that no $n^{o(\sqrt{\Delta})}$ algorithm is possible (see [3]). Putting these results together tells us that no $n^{o(Wt\sqrt{\Delta})}$ algorithm is possible. Contrasting this with the algorithm of Theorem 1 we see that the only small gap left to close here is the complexity as a function of Δ.

4   Parameterizations Involving the Objective Function

In this section we investigate parameterizations of MaxPC where the number of satisfied demands is involved in the parameters. In addition to the parameters of the previous section we consider cases where either one wishes to reject a small number T of requests or one wishes to satisfy at least a small number of requests B. Note that T cannot possibly lead to tractability results if considered as the only parameter as the case $T = 0$ is exactly the PC case which is known to be NP-hard even for simple graph topologies. Thus, T is considered as a parameter together with W and Δ, a combination known to make PC tractable, but for which MaxPC is still intractable after the results of the previous section.

More specifically, for bi-directed trees PC is known to be hard even when $\Delta = 3$ for unbounded W, or for $W = 1$ for unbounded Δ. Therefore, any parameterization involving T would have to include both Δ and W as parameters if it were to be tractable for such trees. Here we show that $(pW, p\Delta, pT)$-MaxPC is indeed fixed-parameter tractable for bi-directed and also for undirected trees.

Theorem 5. $(pW, p\Delta, pT)$-MaxPC *is FPT for both undirected and bi-directed trees.*

Thus, for the three parameters Δ, W, T the problem is now settled for bi-directed trees: if all three are part of the parameter the problem is FPT, if we drop T the problem is W[1]-hard from the results of the previous section and if we drop any of the other two the problem is NP-hard. For undirected trees it is an interesting question what happens if one drops only W from the list of parameters (the problem $(p\Delta, pT)$-MaxPC). Here we will resolve a special case of this problem by showing that (pT)-MaxPC is FPT when restricted to undirected binary trees.

Theorem 6. (pT)-MaxPC *is FPT on undirected trees of maximum degree 3.*

Proof. (Sketch) The algorithm relies on the fact that PC on undirected trees can be decomposed into PC on stars. We first apply this step and locate good (i.e. locally colorable) and bad stars, pruning away parts of the tree where everything is good. Now, a kernelization-like argument shows that the resulting tree cannot have more than $O(T)$ leaves, otherwise it will be impossible to touch all bad stars by dropping only T requests. By extension, there can be no more than $O(T)$ internal vertices of degree 3 without attached leaves. So, we are left with a graph such that if we remove all leaves most vertices have degree 2 and there is a small number ($O(T)$) of "special" other vertices (degree 3 or 1).

Now, to select the first endpoint of a request to be dropped we simply pick one of the bad leaves. The last crucial ingredient is that for the second endpoint we can either guess the other endpoint among the $O(T)$ "special" vertices, or if the other endpoint is a non-special vertex we can use a provably optimal greedy criterion of picking the endpoint that is furthest away.

We remark that this last step is the only part of the algorithm that crucially relies on $\Delta = 3$, as all previous arguments work generally when Δ is a parameter.
□

Let us now move on to consider the "natural" parameterization of MaxPC, that is, the case where the parameter is simply the number of demands B one seeks to satisfy. In this case the hardness results for PC are of course irrelevant. Here we solve the problem for any topology where the naturally parameterized version of MaxEDP is FPT using a randomized color-coding technique.

Theorem 7. *In any graph topology where* (pB)-MaxEDP *is FPT,* (pB)-MaxPC *is also FPT.*

Corollary 1. (pB)-MAXPC *is FPT on undirected trees and rings.*

For bi-directed trees it is known that MAXEDP is NP-hard ([7]), so we cannot immediately apply Theorem 7. Here we will prove that its naturally parameterized version is FPT, a result which may be of independent interest.

Theorem 8. (pB)-MAXEDP *is FPT on bi-directed trees.*

Corollary 2. (pB)-MAXPC *is FPT on bi-directed trees.*

5   Conclusions and Open Problems

A short way to summarize the results of this paper is the following: it was known that having small (or moderate) Δ and W can help solve PC on trees. We showed that in general this cannot help us much to solve MAXPC, but it does still help if we only want to reject a small (or moderate) number of requests. This short summary captures to a large extent our results for bi-directed trees, while for the undirected case, where it is known that small Δ alone suffices to make PC tractable, we have left the complexity of $(p\Delta, pT)$-MAXPC as an interesting open problem, though settling the special case of $\Delta = 3$ (recall that even this is known to be intractable for the bi-directed case).

Much else could be done in the general direction of this work by experimenting with more parameters for the MAXPC problem and their combinations. In particular, all the structural parameters we considered here have to do with the network only. It would be nice to also explore parameters that have to do with the structure of the demands, for example limiting the maximum number of demands touching a vertex, or the maximum length of a demand. Also, many variations of MAXPC have been proposed in the past (e.g. multi-fiber networks, networks with limited hops where color conversion is allowed) and each is likely to have its own reasonable parameters to be exploited.

References

1. Anand, R.S., Erlebach, T., Hall, A., Stefanakos, S.: Call control with k rejections. Journal of Computer and System Sciences 67(4), 707–722 (2003)
2. Bodlaender, H.L., Koster, A.M.C.A.: Combinatorial optimization on graphs of bounded treewidth. Comput. J. 51(3), 255–269 (2008)
3. Chen, J., Huang, X., Kanj, I.A., Xia, G.: Linear FPT reductions and computational lower bounds. In: Babai, L. (ed.) STOC, pp. 212–221. ACM (2004)
4. Downey, R.G., Fellows, M.R.: Parameterized complexity. Springer, New York (1999)
5. Erlebach, T., Jansen, K.: Maximizing the Number of Connections in Optical Tree Networks. In: Chwa, K.-Y., Ibarra, O.H. (eds.) ISAAC 1998. LNCS, vol. 1533, pp. 179–188. Springer, Heidelberg (1998)
6. Erlebach, T., Jansen, K.: The complexity of path coloring and call scheduling. Theoretical Computer Science 255(1-2), 33–50 (2001)

7. Erlebach, T., Jansen, K.: The maximum edge-disjoint paths problem in bidirected trees. SIAM Journal on Discrete Mathematics 14(3), 326–355 (2001)
8. Erlebach, T., Jansen, K., Kaklamanis, C., Mihail, M., Persiano, P.: Optimal wavelength routing on directed fiber trees. Theoretical Computer Science 221(1-2), 119–137 (1999)
9. Flum, J., Grohe, M.: Parameterized complexity theory. Springer-Verlag New York Inc. (2006)
10. Garey, M.R., Johnson, D.S., Miller, G.L., Papadimitriou, C.H.: The complexity of coloring circular arcs and chords. SIAM Journal on Algebraic and Discrete Methods 1, 216 (1980)
11. Garg, N., Vazirani, V.V., Yannakakis, M.: Primal-dual approximation algorithms for integral flow and multicut in trees. Algorithmica 18(1), 3–20 (1997)
12. Golumbic, M.C., Jamison, R.E.: Edge and vertex intersection of paths in a tree. Discrete Mathematics 55(2), 151–159 (1985)
13. Impagliazzo, R., Paturi, R.: On the complexity of k-sat. J. Comput. Syst. Sci. 62(2), 367–375 (2001)
14. Kong, M.C., Zhao, Y.: On computing maximum k-independent sets. In: Congressus Numerantium, pp. 47–60 (1993)
15. Kumar, S.R., Panigrahy, R., Russell, A., Sundaram, R.: A note on optical routing on trees. Information Processing Letters 62(6), 295–300 (1997)
16. Niedermeier, R.: Invitation to fixed-parameter algorithms. Oxford University Press, USA (2006)
17. Wan, P.J., Liu, L.: Maximal throughput in wavelength-routed optical networks. Multichannel Optical Networks: Theory and Practice 46, 15–26 (1998)

On Cutwidth Parameterized by Vertex Cover

Marek Cygan[1], Daniel Lokshtanov[2], Marcin Pilipczuk[1],
Michał Pilipczuk[1], and Saket Saurabh[3]

[1] Institute of Informatics, University of Warsaw, Poland
{cygan@,malcin@,mp248287@students.}mimuw.edu.pl
[2] University of California, San Diego, La Jolla, CA 92093-0404, USA
dlokshtanov@cs.ucsd.edu
[3] The Institute of Mathematical Sciences, Chennai - 600113, India
saket@imsc.res.in

Abstract. We study the CUTWIDTH problem, where input is a graph G, and the objective is find a linear layout of the vertices that minimizes the maximum number of edges intersected by any vertical line inserted between two consecutive vertices. We give an algorithm for CUTWIDTH with running time $O(2^k n^{O(1)})$. Here k is the size of a minimum vertex cover of the input graph G, and n is the number of vertices in G. Our algorithm gives an $O(2^{n/2} n^{O(1)})$ time algorithm for CUTWIDTH on bipartite graphs as a corollary. This is the first non-trivial exact exponential time algorithm for CUTWIDTH on a graph class where the problem remains NP-complete. Additionally, we show that CUTWIDTH parameterized by the size of the minimum vertex cover of the input graph does not admit a polynomial kernel unless NP $\subseteq$ coNP/poly. Our kernelization lower bound contrasts the recent result of Bodlaender et al.[ICALP 2011] that TREEWIDTH parameterized by vertex cover does admit a polynomial kernel.

1 Introduction

In the CUTWIDTH problem we are given an n-vertex graph G together with an integer w. The task is to determine whether there exists a linear layout of the vertices of G such that any vertical line inserted between two consecutive vertices of the layout intersects with at most w edges. The *cutwidth* $(cw(G))$ of G is the smallest w for which such a layout exists. The problem has numerous applications [8,19,20,25], ranging from circuit design [1,23] to protein engineering [4]. Unfortunately CUTWIDTH is NP-complete [14], and remains so even when the input is restricted to subcubic planar bipartite graphs [24,10] or split graphs [16] where all independent set vertices have degree 2. On the other hand, the problem has a factor $O(\log^2(n))$-approximation on general graphs [22] and is polynomial time solvable on trees [28,9], graphs of constant treewidth and constant degree [27], threshold graphs [16], proper interval graphs [29] and bipartite permutation graphs [15].

In this article we study the complexity of computing cutwidth exactly on general graphs, where the running time is measured in terms of the size of the smallest vertex cover of the input graph G. A *vertex cover* of G is a vertex

D. Marx and P. Rossmanith (Eds.): IPEC 2011, LNCS 7112, pp. 246–258, 2012.

set S such that every edge of G has at least one endpoint in S. We show that CUTWIDTH can be solved in time $2^k n^{O(1)}$ where k is the size of the smallest vertex cover of G. An immediate consequence of our algorithm is that CUTWIDTH can be solved in time $2^{n/2} n^{O(1)}$ on bipartite graphs. This is the first non-trivial exact exponential time algorithm for CUTWIDTH on a graph class where the problem is NP-complete. Furthermore, our algorithm improves considerably over the previous best algorithm for CUTWIDTH parameterized by vertex cover [11], whose running time is $O(2^{2^{O(k)}} n^{O(1)})$ (however, it was not the focus of [11] to optimize the running time dependence on k).

Additionally, we show that CUTWIDTH parameterized by vertex cover does not admit a polynomial kernel unless $CoNP \subseteq NP/poly$. A *polynomial kernel* for CUTWIDTH parameterized by vertex cover is a polynomial time algorithm that takes as input a CUTWIDTH instance (G, w), where G has a vertex cover of size at most k and outputs an equivalent instance (G', w') of CUTWIDTH such that G' has at most $k^{O(1)}$ vertices. We show that unless $NP \subseteq coNP/poly$ such a kernelization algorithm can not exist. This contrasts a recent result of Bodlaender et al. [6] that TREEWIDTH parameterized by the vertex cover number of the input graph does admit a polynomial size kernel.

Context of our work. The CUTWIDTH problem is one of many *graph layout* problems, where the task is to find a permutation of the vertices of the input graph that optimizes a problem specific objective function. Graph layout problems, such as TREEWIDTH, BANDWIDTH and HAMILTONIAN PATH are not amenable to "branching" techniques, and hence the design of faster exact exponential time algorithms for these problems has resulted in several new and useful tools. For example, Karps *inclusion-exclusion* based algorithm [21] for HAMILTONIAN PATH was the first application of inclusion-exclusion in exact algorithms. Another example is the introduction of *potential maximal cliques* as a tool for the computation of treewidth. Most graph layout problems (with the exception of BANDWIDTH) admit an $O(2^n n^{O(1)})$ time dynamic programming algorithm [2,17]. For several of these problems, faster algorithms with running time below $O(2^n)$ have been found [3,12,26], a stellar example is the recent algorithm by Björklund [3] for HAMILTONIAN PATH. The CUTWIDTH problem is perhaps the best known graph layout problem for which a $O(2^n n^{O(1)})$ time algorithm is known, yet no better algorithm has been found. Hence, whether such an improved algorithm exists is a tantalizing open problem. While we do not resolve this problem in this article, we make considerable progress; hard instances of CUTWIDTH can not contain any independent set of size cn for any $c > 0$.

The study of kernelization for problems parameterized by vertex cover has recently received considerable attention [5,6,18]. The existence of a $O(2^{k^{O(1)}} n^{O(1)})$ time algorithm is a necessary, but not sufficient condition for CUTWIDTH parameterized by vertex cover to have a polynomial kernel. Hence our $O(2^k n^{O(1)})$ time algorithm makes it natural to ask whether such a kernel exists. In particular, the recent result of Bodlaender et al. [6], that TREEWIDTH parameterized by vertex cover admits a polynomial kernel suggests that CUTWIDTH might have one as well. We show that this is not the case, unless $NP \subseteq coNP/poly$.

Organization of the paper. In Section 2 we present a dynamic programming algorithm which computes cutwidth in time $O(2^k n^{O(1)})$ for a given vertex cover of size k, whereas in Section 3 we show that CUTWIDTH parameterized by vertex cover does not admit a polynomial kernel unless $NP \subseteq coNP/poly$. Section 4 is devoted to concluding remarks.

Notation. All graphs in this paper are undirected and simple. For a vertex $v \in V$ we define its neighbourhood $N_G(v) = \{u : uv \in E(G)\}$ and closed neighbourhood $N_G[v] = N_G(v) \cup \{v\}$. If G is clear from the context, we might omit the subscript. For $X \subseteq V$ we denote $N_G[X] = \bigcup_{v \in X} N_G(v) \setminus X$.

2   Faster Cutwidth Parameterized by Vertex Cover

In this section we show that given a graph $G = (C \cup I, E)$ such that C is a vertex cover of G of size k, we can compute the cutwidth of G in time $O(2^k n^{O(1)})$, using a dynamic programming approach. We start by showing that there always exists an optimal ordering of a specific form.

For an ordering $\sigma = v_1 \ldots v_n$ of $V = C \cup I$ we define $V_i = \{v_j : j \leq i\}$. For vertices u and $v \in V$ we say that $u \leq_\sigma v$ if u occurs before v in σ. Denote by $\delta(V_i)$ the number of edges between V_i and $V \setminus V_i$. The cutwidth of the ordering, $cw_\sigma(G)$, is defined as the maximum of $\delta(V_i)$ for $i = 1, 2, \ldots, |V| - 1$. The *rank* of a vertex v_i with respect to σ is denoted by $rank_\sigma(v_i)$ and it is equal to $|N(v_i) \setminus V_i| - |N(v_i) \cap V_i|$. Notice that $\delta(V_{i+1}) = \delta(V_i) + rank_\sigma(v_{i+1})$ and hence $\delta(V_i) = \sum_{j \leq i} rank_\sigma(v_j)$. Moving a vertex v_p *backward* to position q with $q < p$ results in the ordering

$$\sigma' = v_1 v_2 \ldots v_{q-2} v_{q-1} \mathbf{v_p} v_q v_{q+1} \ldots v_{p-2} v_{p-1} v_{p+1} v_{p+2} \ldots v_n.$$

Moving v_p *forward* to a position q with $q > p$ results in the ordering

$$\sigma' = v_1 v_2 \ldots v_{p-2} v_{p-1} v_{p+1} v_{p+2} \ldots v_{q-2} v_{q-1} \mathbf{v_p} v_q v_{q+1} \ldots v_n.$$

Notice that any vertex with odd degree must have (nonzero) odd rank.

Lemma 1. *If moving v_p backward to position q results in an ordering σ' such that $rank_{\sigma'}(v_p) \leq 0$ then $cw_{\sigma'}(G) \leq cw_\sigma(G)$. If moving v_p forward to position q results in an ordering σ' such that $rank_{\sigma'}(v_p) \geq 0$ then $cw_{\sigma'}(G) \leq cw_\sigma(G)$.*

Proof. Suppose moving v_p backward to position q results in an ordering σ' such that $rank_{\sigma'}(v_p) \leq 0$. For every non-negative integer i define V_i' to contain the first i vertices of σ'. Then, for every $i < q$ and $i \geq p$ we have $V_i' = V_i$ and hence $\delta(V_i') = \delta(V_i)$. For every i such that $q \leq i < p$ we have that $V_i' = V_{i-1} \cup \{v_p\}$. Observe that for any other vertex v_j, $j \neq p$, $rank_{\sigma'}(v_j) \leq rank_\sigma(v_j)$, while $rank_{\sigma'}(v_p) \leq 0$. Thus $\delta(V_i') = rank_{\sigma'}(v_p) + \sum_{j \leq i-1} rank_{\sigma'}(v_j) \leq \delta(V_{i-1})$ and $cw_{\sigma'}(G) \leq cw_\sigma(G)$. The proof that if moving v_p forward to position q results in an ordering σ' such that $rank_{\sigma'}(v_p) \geq 0$ then $cw_{\sigma'}(G) \leq cw_\sigma(G)$ is analogous. $\square$

Lemma 1 allows us to rearrange optimal orderings. Let σ be an optimal cutwidth ordering of G, $c_1 c_2 \dots c_k$ be the ordering which σ imposes on C and $C_i = \{c_1, \dots, c_i\}$ for every i. Observe that if u and v are both in I then moving u does not affect the rank of v. In particular, if moving u yields the ordering σ', then $rank_{\sigma'}(v) = rank_{\sigma}(v)$. For every vertex $u \in I$ with odd degree and $rank_{\sigma}(u) < 0$ we move u backward to the leftmost position where u has rank -1. For every vertex $u \in I$ with odd degree and $rank_{\sigma}(u) > 0$ we move u forward to the rightmost position where u has rank 1. For every vertex of the set I with even degree we move it (forward or backward) to the rightmost position where u has rank 0. This results in an optimal cutwidth ordering σ' with the following properties.

1. For every vertex $v \in I$ of even degree $rank_{\sigma'}(v) = 0$ and every vertex $v \in I$ of odd degree $rank_{\sigma'}(v) \in \{-1, 1\}$.
2. For every vertex $v \in I$ such that $rank_{\sigma'}(v) \geq 0$ and $c_i \in C$ we have $c_i \leq_{\sigma'} v$ if and only if $|N(v) \cap C_i| \leq |N(v) \setminus C_i|$.
3. For every vertex $v \in I$ such that $rank_{\sigma'}(v) < 0$ and $c_i \in C$ we have $c_i \leq_{\sigma'} v$ if and only if $|N(v) \cap C_{i-1}| < |N(v) \setminus C_{i-1}|$.

Define I'_0 and I'_k to be the set of vertices in I appearing before c_1 and after c_k in σ', respectively. For i between 1 and $k-1$ we denote I'_i the set of vertices in I appearing between c_i and c_{i+1} in σ'. For any i, if I'_i contains any vertices of rank -1, we move them backward to the position right after c_i. This results in an ordering σ'' where for every i, all the vertices of I'_i with negative rank appear before all the vertices of I'_i with non-negative rank. By Lemma 1 and the fact that moving a vertex from independent set does not affect the rank of another vertex from the independent set we have that σ'' is still an optimal cutwidth ordering. Also, σ'' satisfies the properties $1 - 3$. We say that an ordering σ is C-good if it satisfies properties $1 - 3$ and orders the vertices between vertices of C in such a way that all vertices of negative rank appear before all vertices of non-negative rank. The construction of σ'' from an optimal ordering σ proves the following lemma.

Lemma 2. *Let $G = (C \cup I, E)$ be a graph and C be a vertex cover of G. There exists an optimal cutwidth ordering σ of G which is C-good.*

In a C-good ordering σ, consider a position i such that $c_i \in C$. Because of the properties of a C-good ordering we can essentially deduce $V_i \cap I$ from $V_i \cap C$ and the vertex c_i. We will now formalize this idea. For a set $S \subseteq C$ and vertex $v \in S$ we define the set $X(S, v) \subseteq I$ as follows. A vertex $u \in I$ of even degree is in $X(S, v)$ if $|N(u) \cap S| > |N(u) \setminus S|$. A vertex $u \in I$ of odd degree is in $X(S, v)$ if $|N(u) \cap (S \setminus \{v\})| > |N(u) \setminus (S \setminus \{v\})|$. Now we define the set $Y(S, v)$. A vertex $u \in I$ is in $Y(S, v)$ if $uv \in E$ and $|(N(u) \setminus \{v\}) \cap S| = |(N(u) \setminus \{v\}) \setminus S|$. Note that the vertices in $Y(S, v)$ have odd degrees and $Y(S, v)$ is disjoint with $X(S, v)$. The following observation follows directly from the properties of a C-good ordering.

Observation 3. *In a C-good ordering σ let i be an integer such that $c_i \in C$ and let $S = V_i \cap C$. Then $X(S, c_i) \subseteq V_i \cap I \subseteq X(S, c_i) \cup Y(S, c_i)$.*

A *prefix ordering* ϕ is a set $V_\phi \subseteq C \cup I$ together with an ordering of V_ϕ. The *size* of the prefix ordering ϕ is just $|V_\phi|$. Similarly to normal orderings we define $V_i^\phi = \{v_1 \ldots v_i\}$. Let $c_1 c_2 \ldots c_{|V_\phi \cap C|}$ be the ordering imposed on $V_\phi \cap C$ by ϕ, and for every $i \leq |V_\phi \cap C|$ we set $C_i^\phi = \{c_1, \ldots, c_i\}$. The *rank* of a vertex $v \in V_\phi$ with respect to ϕ is defined as $rank_\phi(v_i) = |N(v_i) \setminus V_i^\phi| - |N(v_i) \cap V_i^\phi|$. We now extend the notion of being C-good from orderings of G to prefix orderings of G in such a way that that the restriction of any C-good ordering σ of G to the first t vertices, where $v_t \in C$, must be C-good. We say that a prefix ordering $\phi = v_1 \ldots v_t$ of size t with $v_t \in C$ is *C-good* if the following conditions are satisfied.

1. For every vertex $v \in V_\phi \cap I$ of even degree, $rank_\phi(v) = 0$ and for every vertex $v \in V_\phi \cap I$ of odd degree, $rank_\phi(v) \in \{-1, 1\}$.
2. $X(V_\phi \cap C, v_i) \subseteq V_\phi \cap I \subseteq X(V_\phi \cap C, v_i) \cup Y(V_\phi \cap C, v_i)$
3. For every vertex $v \in X(V_\phi \cap C, c_i)$ such that $rank_\phi(v) \geq 0$ and $c_i \in V_\phi \cap C$ we have $c_i \leq_\phi v$ if and only if $|N(v) \cap C_i^\phi| \leq |N(v) \setminus C_i^\phi|$.
4. For every vertex $v \in X(V_\phi \cap C, c_i)$ such that $rank_\phi(v) < 0$ and $c_i \in V_\phi \cap C$ we have $c_i \leq_\phi v$ if and only if $|N(v) \cap C_{i-1}^\phi| < |N(v) \setminus C_{i-1}^\phi|$.
5. Between two vertices $c_i, c_{i+1} \in C \cap V_\phi$, all vertices with $rank_\phi(v) < 0$ come before all vertices with $rank_\phi(v) \geq 0$.

Comparing the properties of C-good orderings and C-good prefix orderings it is easy to see that the following lemma holds.

Lemma 4. *Let $\sigma = v_1 \ldots v_n$ be a C-good ordering and let ϕ be the restriction of σ to the first t vertices, such that $v_t \in C$. Then ϕ is a C-good prefix ordering.*

For a prefix ordering ϕ define the cutwidth of G with respect to ϕ to be $cw_\phi(G) = \max_{i \leq |V_\phi|} \delta(V_i^\phi)$. For a subset S of C and vertex $v \in S$, define $T(S, v)$ to be the minimum value of $cw_\phi(G)$ where the minimum is taken over all C-good prefix orderings ϕ with $V_\phi \cap C = S$ and v being the last vertex of ϕ. Notice that property 5 of C-good prefix orderings implies that in a C-good prefix ordering ϕ there must be some i with $v_i \in C \cap V_\phi$ such that $cw_\phi(G) = \delta(V_i)$ or $cw_\phi(G) = \delta(V_{i-1})$. Also, notice that for any set $S \subseteq C$ and vertices $u, v \in S$ we have that $X(S \setminus \{v\}, u) \subseteq X(S, v)$ and $Y(S \setminus \{v\}, u) \subseteq X(S, v)$. Finally, observe that for any set $S \subseteq C$ and vertex $v \in S$, every vertex $u \in Y(S, v)$ is adjacent to v and satisfies $|(N(u) \setminus \{v\}) \cap S| = |(N(u) \setminus \{v\}) \setminus S|$. Thus, for any set $I' \subseteq I$ with $X(S, v) \subseteq I' \subseteq X(S, v) \cup Y(S, v)$ the value of $\delta(S \cup I')$ depends only on $|Y(S, v) \cap I'|$ and not on $Y(S, v) \cap I'$ in general. We let $Y_i(S, v)$ be an arbitrary subset of $Y(S, v)$ of size i. The discussion above yields that the following recurrence holds for $T(S, v)$, where $S \subseteq C$ and $v \in S$.

$$T(S, v) = \min_{u \in S} \min_{0 \leq i \leq |Y(S, v)|} \max \left\{ \begin{array}{c} \delta(S \cup X(S, v) \cup Y_i(S, v)) \\ \delta((S \setminus \{v\}) \cup X(S, v) \cup Y_i(S, v)) \\ T(S \setminus \{v\}, u) \end{array} \right\}.$$

Observe that $cw(G) = \min_{v \in C} T(S, v)$ because in any ordering σ all vertices of I appearing after the last vertex of C must have negative rank. Thus the recurrence above naturally leads to a dynamic programming algorithm for CUTWIDTH running in time $O(2^k n^{O(1)})$. This proves the main theorem of this section.

Theorem 5. *There is an algorithm that given a graph $G = (C \cup I, E)$ such that C is a vertex cover of G, computes the cutwidth of G in running time $O(2^{|C|}(|C| + |I|)^{O(1)})$. Thus, MINIMUM CUTWIDTH on bipartite graphs can be solved in time $O(2^{n/2} n^{O(1)})$, where n is the number of vertices of the input graph.*

3   Kernelization Lower Bound

In this section we show that CUTWIDTH parameterized by vertex cover does not admit a polynomial kernel unless $\text{NP} \subseteq \text{coNP/poly}$.

3.1   The Auxiliary Problem

We begin with introducing an auxiliary problem, namely HYPERGRAPH MINIMUM BISECTION. Let $H = (V, E)$ be a multihypergraph with $|V| = n$, where n is even. A *bisection* of V is a colouring $\mathcal{B} : V \to \{0, 1\}$ such that $|\mathcal{B}^{-1}(0)| = |\mathcal{B}^{-1}(1)| = n/2$. For a hyperedge e let us define the cost of e with respect to a bisection $\mathcal{B}$ as $cost(e, \mathcal{B}) = \min\left(\left|e \cap \mathcal{B}^{-1}(0)\right|, \left|e \cap \mathcal{B}^{-1}(1)\right|\right)$. The cost of a bisection is defined as the sum of the contributions of the hyperedges, i.e., $cost(\mathcal{B}) = \sum_{e \in E} cost(e, \mathcal{B})$.

HYPERGRAPH MINIMUM BISECTION	**Parameter:** n
Input: Multihypergraph H with n vertices, where n is even; an integer k	
Question: Does there exist a bisection of H with cost at most k?	

In the case when all the hyperedges are in fact edges (have cardinalities 2) and there are no multiedges, the problem is equivalent to the classical MINIMUM BISECTION problem. As MINIMUM BISECTION is NP-hard, HYPERGRAPH MINIMUM BISECTION is also NP-hard, so NP-complete as well.

The goal now is to prove that CUTWIDTH parameterized by the size of vertex cover does not admit a polynomial kernel, unless $\text{NP} \subseteq \text{coNP/poly}$. We do it in two steps. First, using the OR-distillation technique of Fortnow and Santhanam [13] we prove that HYPERGRAPH MINIMUM BISECTION does not admit a polynomial kernel, unless $\text{NP} \subseteq \text{coNP/poly}$. Second, we present a parameterized reduction of HYPERGRAPH MINIMUM BISECTION to CUTWIDTH parameterized by vertex cover.

3.2   No Polykernel for HYPERGRAPH MINIMUM BISECTION

We use the OR-distillation technique by Fortnow and Santhanam [13], put into framework called *cross-composition* by Bodlaender et al. [5]. Let us recall the crucial definitions.

Definition 6 (Polynomial equivalence relation [5]). *An equivalence relation $\mathcal{R}$ on Σ^* is called a* polynomial equivalence relation *if (1) there is an algorithm that given two strings $x, y \in \Sigma^*$ decides whether $\mathcal{R}(x,y)$ in $(|x| + |y|)^{O(1)}$ time; (2) for any finite set $S \subseteq \Sigma^*$ the equivalence relation $\mathcal{R}$ partitions the elements of S into at most $(\max_{x \in S} |x|)^{O(1)}$ classes.*

Definition 7 (Cross-composition [5]). *Let $L \subseteq \Sigma^*$ and let $Q \subseteq \Sigma^* \times \mathbb{N}$ be a parameterized problem. We say that L* cross-composes *into Q if there is a polynomial equivalence relation $\mathcal{R}$ and an algorithm which, given t strings $x_1, x_2, \ldots x_t$ belonging to the same equivalence class of $\mathcal{R}$, computes an instance $(x^*, k^*) \in \Sigma^* \times \mathbb{N}$ in time polynomial in $\sum_{i=1}^{t} |x_i|$ such that (1) $(x^*, k^*) \in Q$ iff $x_i \in L$ for some $1 \leq i \leq t$; (2) k^* is bounded polynomially in $\max_{i=1}^{t} |x_i| + \log t$.*

Theorem 8 ([5], Theorem 9). *If $L \subseteq \Sigma^*$ is NP-hard under Karp reductions and L cross-composes into the parameterized problem Q that has a polynomial kernel, then $NP \subseteq coNP/poly$.*

Lemma 9. HYPERGRAPH MINIMUM BISECTION *does not admit a polynomial kernel, unless $NP \subseteq coNP/poly$.*

Proof. As MINIMUM BISECTION is NP-hard under Karp reductions, it suffices to prove that it cross-composes into the HYPERGRAPH MINIMUM BISECTION problem. Let R be an equivalence relation on Σ^* defined as follows:

- all words that do not correspond to instances of MINIMUM BISECTION form one equivalence class;
- all the well-formed instances are partitioned into equivalence classes having the same number of vertices, the same number of edges and the same demanded cost of the bisection.

It is straightforward, that R is a polynomial equivalence relation. Therefore, we can assume that the composition algorithm is given a sequence of instances $(G_0, k), (G_1, k), \ldots, (G_{t-1}, k)$ of MINIMUM BISECTION with $|V(G_i)| = n$ and $|E(G_i)| = m$ for all $i = 0, 1, \ldots, t - 1$ (n is even). Moreover, by copying some instances if necessary we can assume without losing generality that $t = 2^l$ for some integer l. Note that in this manner we do not increase the order of $\log t$.

We now proceed to the construction of the composed HYPERGRAPH MINIMUM BISECTION instance (H, K). Let $N = 2m \cdot ((l + 2)2^{l-1} - 1) + 2k + 1$ and $M = N(l^2 - l) + N$. We begin with creating two sets of vertices A_0 and A_1, each of size $2nl$. We introduce each set A_0, A_1 as a hyperedge of the constructed hypergraph M times.

Then, we introduce $2l$ vertices s_i^0, s_i^1 for $i = 0, 1, \ldots, l - 1$ and denote the set of all these vertices by S. For every $i \neq j$ we put N times each hyperedge $\{s_i^0, s_j^0\}$, $\{s_i^0, s_j^1\}$, $\{s_i^1, s_j^0\}$, $\{s_i^1, s_j^1\}$. Thus, the hypergraph induced by S is a clique without a matching, repeated N times. Furthermore, for every $p = 0, 1$ and $i = 0, 1, \ldots, l - 1$ we construct a set S_i^p of $n - 1$ vertices and put $S_i^p \cup \{s_i^p\}$ as a hyperedge of the constructed hypergraph M times.

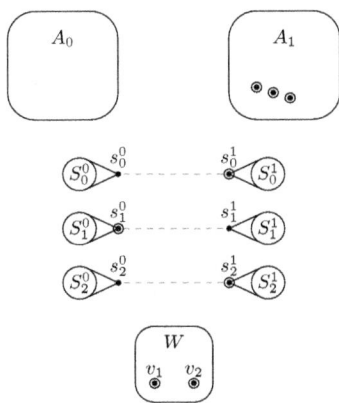

Fig. 1. The constructed hypergraph H for $l = 3$; encircled vertices indicate the hyperedge e^0 constructed for $e = v_1^5 v_2^5 \in E(G_5)$

Now, we construct a set of n vertices $v_1, v_2, \ldots, v_n$ and denote it by W. For every instance G_a we arbitrarily choose an ordering of its vertices $v_1^a, v_2^a, \ldots, v_n^a$. Let $b_{l-1}b_{l-2}\ldots b_1 b_0$ be the binary representation of a, with trailing zeroes added so that its length is equal to l. For every edge $e = v_g^a v_h^a \in E(G_a)$ we create two hyperedges:

- e^0, consisting of vertices v_g, v_h, $s_i^{b_i}$ for all $i = 0, 1, \ldots, l-1$ and l vertices from A_1, chosen arbitrarily;
- e^1, consisting of vertices v_g, v_h, $s_i^{1-b_i}$ for all $i = 0, 1, \ldots, l-1$ and l vertices from A_0, chosen arbitrarily.

Finally, we set the expected cost of the bisection to $K = M - 1 = N(l^2 - l) + 2m \cdot ((l+2)2^{l-1} - 1) + 2k$.

Now, assume that some graph G_a has a bisection $\mathcal{B}$ having cost at most k. Let $b_{l-1}b_{l-2}\ldots b_1 b_0$ be the binary representation of a, as in the previous paragraph. We now construct a bisection $\mathcal{B}'$ of H as follows:

- for each $u \in A_0$ we set $\mathcal{B}'(u) = 0$, for each $u \in A_1$ we set $\mathcal{B}'(u) = 1$;
- for each $u \in S_i^p \cup \{s_i^p\}$ for $p = 0, 1$, $i = 0, 1, \ldots, l-1$ we set $\mathcal{B}'(u) = p + b_i$ (mod 2);
- for each $v_j \in W$ we set $\mathcal{B}'(v_j) = \mathcal{B}(v_j^a)$.

Observe that $\mathcal{B}'$ bisects each of the sets $A_0 \cup A_1$, S and W, so it is a bisection. We now prove that its cost is at most K. Let us count the contribution to the cost from every hyperedge of H.

Each copy of hyperedges A_0, A_1 and $S_i^p \cup \{s_i^p\}$ for $p = 0, 1$, $i = 0, 1, \ldots, l-1$ has zero contribution, as it is monochromatic. The edges of $H[S]$ have contribution 0 or 1, depending whether the endpoints are coloured in the same or in a different way in $\mathcal{B}'$. There are l vertices s_i^p that map to 0 in $\mathcal{B}'$ and l that map to 1, so there are l^2 pairs of vertices coloured in a different way. Between every pair of vertices there are N edges, apart from pairs (s_i^0, s_i^1). Note that all these pairs are coloured differently; therefore, there are exactly $N(l^2 - l)$ edges in $H[S]$ contributing 1 to the cost.

Take $c \in \{0, 1, \ldots, t - 1\}$ such that $c \neq a$. Let $d_{l-1}d_{l-2} \ldots d_0$ be the binary representation of c. For $e \in E(G_c)$ let us count the contribution to $cost(\mathcal{B}')$ of hyperedges e^0 and e^1. Suppose that $q = |\{i : b_i \neq d_i\}| > 0$. Among vertices of e^0, l from A_1 are coloured 1, q from S are coloured 1 as well and $l - q$ from S are coloured 0. In total, we have $l + q$ vertices coloured 1 and $l - q$ coloured 0, so regardless of the colouring of the remaining two vertices from W, the contribution is equal to the number of vertices coloured 0 in e^0, namely $l - q + |e^0 \cap W \cap \mathcal{B}'^{-1}(0)|$. Analogously, the contribution of the hyperedge e^1 is equal to the number of vertices of e^1 coloured 1, namely $l - q + |e^1 \cap W \cap \mathcal{B}'^{-1}(1)|$. As there are exactly two vertices in $e^0 \cap W = e^1 \cap W$, $cost(e^0, \mathcal{B}') + cost(e^1, \mathcal{B}') = 2(l - q) + 2$. Thus, the total contribution of hyperedges e^0, e^1 for $e \in E(G_c)$ is equal to $2m(l - q) + 2m$.

Now we count the contribution of edges e^0 and e^1 for $e \in E(G_a)$. Analogously as in the previous paragraph, both edges e^0, e^1 contain l vertices coloured 0, l vertices coloured 1 plus two vertices from W. If both these vertices are coloured in the same way, the sum of contributions of e^0 and e^1 is equal to $2l$; however, if the vertices are coloured differently, the sum is equal to $2l + 2$. As the cost of bisection $\mathcal{B}$ was at most k, the total contribution of edges e^0, e^1 for $e \in E(G_a)$ is at most $2ml + 2k$.

Finally, we sum up the contributions:

$$cost(\mathcal{B}') \leq N(l^2 - l) + 2m \sum_{q=1}^{l}(l - q + 1)\binom{l}{q} + 2ml + 2k$$

$$= N(l^2 - l) + 2m \cdot (2^l - 1) + 2m \sum_{q=0}^{l}(l - q)\binom{l}{q} + 2k$$

$$= N(l^2 - l) + 2m \cdot (2^l - 1) + 2ml2^{l-1} + 2k$$

$$= N(l^2 - l) + N - 1 = K.$$

We proceed to the second direction. Assume that we have a bisection $\mathcal{B}'$ of H such that $cost(\mathcal{B}') \leq K$. Observe that as $M > K$, both the sets A_0, A_1 are monochromatic with respect to $\mathcal{B}'$. Moreover, they have to be coloured differently, as they contain more than half of the vertices of the graph in total. Without losing generality we can assume that A_0 is coloured in colour 0, while A_1 is coloured in colour 1, by flipping the colours if necessary.

Now consider the set $S_p^i \cup \{s_p^i\}$ for $p = 0, 1$, $i = 0, 1, \ldots, l - 1$. Analogously as in the previous paragraph, $S_p^i \cup \{s_p^i\}$ has to be monochromatic. Furthermore, observe that exactly l such sets have to be coloured 0 in $\mathcal{B}'$ and the same number have to be coloured 1, as every set $S_p^i \cup \{s_i^p\}$ contains the same number of vertices as the set W and $\mathcal{B}'$ is a bisection. Therefore, $\mathcal{B}'$ has to bisect each of the sets $A_0 \cup A_1$, S and W.

Exactly l vertices s_i^p are coloured 0 in $\mathcal{B}'$ and exactly l are coloured 1. Let r be the number of indices i, such that s_i^0 and s_i^1 are coloured differently. Observe that analogously as previously, the contribution of the edges of $H[S]$ to $cost(\mathcal{B}')$ is equal to $N(l^2-r) = N(l^2-l)+N(l-r)$. If $r < l$, then $cost(\mathcal{B}') \geq N(l^2-l)+N > K$, a contradiction. Therefore, all the pairs (s_i^0, s_i^1) are coloured differently.

Let a be a number with binary representation $\mathcal{B}'(s_{l-1}^0)\mathcal{B}'(s_{l-2}^1)\ldots\mathcal{B}'(s_0^1)$. Consider a bisection $\mathcal{B}$ of G_a defined as follows: $\mathcal{B}(v_i^a) = \mathcal{B}'(v_i)$. We claim that the cost of $\mathcal{B}$ is at most k. Indeed, the same computations as in the previous part of the proof show that

$$cost(\mathcal{B}') = N(l^2 - l) + 2m((l+2)2^{l-1} - 1) + 2cost(\mathcal{B})$$

Therefore, as $cost(\mathcal{B}') \leq K$, then $cost(\mathcal{B}) \leq k$. □

3.3   From HYPERGRAPH MINIMUM BISECTION to CUTWIDTH

Let us briefly recall the notion of polynomial parameter transformations.

Definition 10 ([7]). *Let P and Q be parameterized problems. We say that P is polynomial parameter reducible to Q, written $P \leq_p Q$, if there exists a polynomial time computable function $f : \Sigma^* \times \mathbb{N} \to \Sigma^* \times \mathbb{N}$ and a polynomial p, such that for all $(x, k) \in \Sigma^* \times \mathbb{N}$ the following holds: $(x,k) \in P$ iff $(x',k') = f(x,k) \in Q$ and $k' \leq p(k)$. The function f is called a polynomial parameter transformation.*

Theorem 11 ([7]). *Let P and Q be parameterized problems and $\tilde{P}$ and $\tilde{Q}$ be the unparameterized versions of P and Q respectively. Suppose that $\tilde{P}$ is NP-hard and $\tilde{Q}$ is in NP. Assume there is a polynomial parameter transformation from P to Q. Then if Q admits a polynomial kernel, so does P.*

We apply this notion to our case.

Lemma 12. *There exists a polynomial-time algorithm that, given an instance of HYPERGRAPH MINIMUM BISECTION problem with n vertices, outputs an equivalent instance of CUTWIDTH problem along with its vertex cover of size n.*

Proof. Let $(H = (V, E), k)$ be an instance of HYPERGRAPH MINIMUM BISECTION given in the input, where $|V| = n$ (n is even) and $|E| = m$. We construct a graph G as follows.

Let us denote $N = mn + 1$. We begin with taking the whole set V to the set of vertices of G. For every distinct $u, v \in V$ we introduce N new vertices $x_{u,v}^i$ for $i = 1, 2, \ldots, N$, each connected only to u and v. Then, for every $e \in E$ we introduce a new vertex y_e connected to all $v \in e$. Denote the set of all vertices $x_{u,v}^i$ by X and the set of all vertices y_e by Y. This concludes the construction. Observe that V is a vertex cover of G of size n. We now prove that H has a bisection with cost at most k if and only if G has cutwidth at most $n^2N/4 + k$.

Assume that H has a bisection $\mathcal{B}$ with cost at most k. Let us order the vertices of the graph G as follows. Firstly, we order the vertices from V: we place $\mathcal{B}^{-1}(0)$ first, in any order, and then $\mathcal{B}^{-1}(1)$, in any order. Then, we place every $x_{u,v}^i$ anywhere between u and v. At the end, for every $e \in E$ we place y_e

at the beginning if at least half of the vertices of e is in $\mathcal{B}^{-1}(0)$, and in the end otherwise. Vertices y_e at the beginning and at the end are arranged in any order.

Now, we prove that the cutwidth of the constructed ordering is at most $n^2N/4 + k$. Consider any cut C, dividing the order on $V(G)$ into the first part V_1 and second V_2. Suppose that $|V_1 \cap V| = n/2 - l$ for some $-n/2 \le l \le n/2$, thus $|V_2 \cap V| = n/2 + l$. Observe that C cuts exactly $N(n/2-l)(n/2+l) = n^2N/4 - l^2N$ edges between V and X. Note that there are not more than $nm < N$ edges between V and Y. Therefore, if $l \ne 0$, then C can cut at most $n^2N/4 - N + nm < n^2N/4 + k$ edges.

We are left with the case when $l = 0$. Observe that $V_1 \cap V = \mathcal{B}^{-1}(0)$ and $V_2 \cap V = \mathcal{B}^{-1}(1)$. Moreover, the cut C cuts exactly $n^2N/4$ edges between sets V and X. As far as edges between V and Y are concerned, for every hyperedge $e \in E$ cut C cuts exactly $cost(e, \mathcal{B})$ edges incident with y_e. As $cost(\mathcal{B}) \le k$, the cut C cuts at most $n^2N/4 + k$ edges.

Now assume that there is an ordering of vertices of G that has cutwidth at most $n^2N/4 + k$. We construct a bisection $\mathcal{B}$ of H as follows. Let $\mathcal{B}(v) = 0$ for every v appearing among the first $n/2$ vertices from V with respect to the ordering, and $\mathcal{B}(v) = 1$ for v among the second $n/2$ vertices. We now prove that the cost of this bisection is at most k.

Let C be any cut dividing the order into the first part V_1 and the second part V_2, such that $V_1 \cap V = \mathcal{B}^{-1}(0)$ and $V_2 \cap V = \mathcal{B}^{-1}(1)$. As the cutwidth of the ordering is at most $n^2N/4 + k$, C cuts at most $n^2N/4 + k$ edges. Observe that C needs to cut at least $n^2N/4$ edges between sets V and X, therefore it cuts at most k edges between sets V and Y. For every hyperedge $e \in E$, C cuts at least $cost(e, \mathcal{B})$ edges incident to y_e, thus $cost(\mathcal{B}) \le k$. □

From Lemmata 9, 12 and Theorem 11 we can easily conclude the following.

Theorem 13. CUTWIDTH *parameterized by the size of vertex cover does not admit a polynomial kernel, unless* $NP \subseteq coNP/poly$.

4   Conclusions

In this paper we studied the complexity of computing cutwidth of the graph in terms of the size of given vertex cover. We have shown an algorithm with running time $O(2^k n^{O(1)})$, where k is the cardinality of the vertex cover and n is the number of vertices of the graph. Moreover, we have proven that polynomial kernelization of the problem is unlikely, thus counterpoising the recent result of Bodlaender et al. [6].

The thrilling and natural question is whether the insight we gave into the problem can be a starting point to breaking the 2^n barrier for an exact algorithm computing cutwidth. Our result implies that one can assume that in the hard instance all the independent sets are small, i.e., of size not larger than cn for arbitrarily small constant $c > 0$.

References

1. Adolphson, D., Hu, T.C.: Optimal linear ordering. SIAM J. Appl. Math. 25, 403–423 (1973)
2. Bellman, R.: Dynamic programming treatment of the travelling salesman problem. J. ACM 9(1), 61–63 (1962)
3. Björklund, A.: Determinant sums for undirected hamiltonicity. In: FOCS, pp. 173–182 (2010)
4. Blin, G., Fertin, G., Hermelin, D., Vialette, S.: Fixed-parameter algorithms for protein similarity search under mrna structure constraints. Journal of Discrete Algorithms 6, 618–626 (2008)
5. Bodlaender, H.L., Jansen, B.M.P., Kratsch, S.: Cross-composition: A new technique for kernelization lower bounds. In: Schwentick, T., Dürr, C. (eds.) STACS. LIPIcs, vol. 9, pp. 165–176. Schloss Dagstuhl - Leibniz-Zentrum fuer Informatik (2011)
6. Bodlaender, H.L., Jansen, B.M.P., Kratsch, S.: Preprocessing for Treewidth: A Combinatorial Analysis through Kernelization. In: Aceto, L., Henzinger, M., Sgall, J. (eds.) ICALP 2011. LNCS, vol. 6755, pp. 437–448. Springer, Heidelberg (2011)
7. Bodlaender, H.L., Thomasse, S., Yeo, A.: Analysis of data reduction: Transformations give evidence for non-existence of polynomial kernels, technical Report UU-CS-2008-030, Institute of Information and Computing Sciences, Utrecht University, Netherlands (2008)
8. Botafogo, R.A.: Cluster analysis for hypertext systems. In: SIGIR, pp. 116–125 (1993)
9. Chung, M., Makedon, F., Sudborough, I., Turner, J.: Polynomial time algorithms for the min cut problem on degree restricted trees. SIAM Journal on Computing 14, 158–177 (1985)
10. Diaz, J., Penrose, M., Petit, J., Serna, M.: Approximating layout problems on random geometric graphs. Journal of Algorithms 39, 78–117 (2001)
11. Fellows, M.R., Lokshtanov, D., Misra, N., Rosamond, F.A., Saurabh, S.: Graph Layout Problems Parameterized by Vertex Cover. In: Hong, S.-H., Nagamochi, H., Fukunaga, T. (eds.) ISAAC 2008. LNCS, vol. 5369, pp. 294–305. Springer, Heidelberg (2008)
12. Fomin, F.V., Kratsch, D., Todinca, I., Villanger, Y.: Exact algorithms for treewidth and minimum fill-in. SIAM J. Comput. 38(3), 1058–1079 (2008)
13. Fortnow, L., Santhanam, R.: Infeasibility of instance compression and succinct PCPs for NP. In: STOC 2008: Proceedings of the 40th Annual ACM Symposium on Theory of Computing, pp. 133–142. ACM (2008)
14. Gavril, F.: Some np-complete problems on graphs, pp. 91–95 (1977)
15. Heggernes, P., van 't Hof, P., Lokshtanov, D., Nederlof, J.: Computing the Cutwidth of Bipartite Permutation Graphs in Linear Time. In: Thilikos, D.M. (ed.) WG 2010. LNCS, vol. 6410, pp. 75–87. Springer, Heidelberg (2010)
16. Heggernes, P., Lokshtanov, D., Mihai, R., Papadopoulos, C.: Cutwidth of Split Graphs, Threshold Graphs, and Proper Interval Graphs. In: Broersma, H., Erlebach, T., Friedetzky, T., Paulusma, D. (eds.) WG 2008. LNCS, vol. 5344, pp. 218–229. Springer, Heidelberg (2008)
17. Held, M., Karp, R.M.: A dynamic programming approach to sequencing problems. Journal of the Society for Industrial and Applied Mathematics 10(1), 196–210 (1962)
18. Jansen, B.M.P., Kratsch, S.: Data reduction for graph coloring problems. CoRR abs/1104.4229 (2011)

19. Junguer, M., Reinelt, G., Rinaldi, G.: The travelling salesman problem. In: Handbook on Operations Research and Management Sciences, pp. 225–330 (1995)
20. Karger, D.R.: A randomized fully polynomial time approximation scheme for the all-terminal network reliability problem. SIAM J. Comput. 29(2), 492–514 (1999)
21. Karp, R.M.: Dynamic programming meets the principle of inclusion and exclusion. Oper. Res. Lett. 1, 49–51 (1982)
22. Leighton, F., Rao, S.: Multicommodity max-flow min-cut theorems and their use in designing approximation algorithms. Journal of the ACM 46, 787–832 (1999)
23. Makedon, F., Sudborough, I.H.: On minimizing width in linear layouts. Discrete Applied Mathematics 23, 243–265 (1989)
24. Monien, B., Sudborough, I.H.: Min cut is np-complete for edge weighted trees. Theoretical Computer Science 58, 209–229 (1988)
25. Mutzel, P.: A Polyhedral Approach to Planar Augmentation and Related Problems. In: Spirakis, P.G. (ed.) ESA 1995. LNCS, vol. 979, pp. 494–507. Springer, Heidelberg (1995)
26. Suchan, K., Villanger, Y.: Computing Pathwidth Faster than 2. In: Chen, J., Fomin, F.V. (eds.) IWPEC 2009. LNCS, vol. 5917, pp. 324–335. Springer, Heidelberg (2009)
27. Thilikos, D.M., Serna, M.J., Bodlaender, H.L.: Cutwidth ii: Algorithms for partial w-trees of bounded degree. Journal of Algorithms 56, 24–49 (2005)
28. Yannakakis, M.: A polynomial algorithm for the min cut linear arrangement of trees. Journal of the ACM 32, 950–988 (1985)
29. Yuan, J., Zhou, S.: Optimal labelling of unit interval graphs. Appl. Math. J. Chinese Univ. Ser. B (English edition) 10, 337–344 (1995)

Twin-Cover: Beyond Vertex Cover
in Parameterized Algorithmics

Robert Ganian[*]

Faculty of Informatics, Masaryk University
Botanická 68a, Brno, Czech Republic
ganian@mail.muni.cz

Abstract. Parameterized algorithms are a very useful tool for dealing with NP-hard problems on graphs. In this context, vertex cover is used as a powerful parameter for dealing with problems which are hard to solve even on graphs of bounded tree-width. The drawback of vertex cover is that bounding it severely restricts admissible graph classes. We introduce a new parameter called twin-cover and show that it is capable of solving a wide range of hard problems while also being much less restrictive than vertex cover and attaining low values even on dense graphs.

The article begins by introducing a new FPT algorithm for Graph Motif on graphs of bounded vertex cover. This is the first algorithm of this kind for Graph Motif. We continue by defining twin-cover and providing some related results and notions. The next section contains a number of new FPT algorithms on graphs of bounded twin-cover, with a special emphasis on solving problems which are hard even on graphs of bounded tree-width. Finally, section five generalizes the recent results of Michael Lampis for MS_1 model checking from vertex cover to twin-cover.

1 Introduction

One very successful approach to dealing with NP-hard problems on graphs is the use of parameterized algorithms. The idea is that in real-life applications it is usually not necessary to solve problems on general graphs, but rather on graphs with some kind of structure present. It is then possible to use a structural parameter k to describe this structure and use it to design algorithms which run in polynomial time as long as k is bounded. Specifically, we are interested in so-called Fixed Parameter Tractable (FPT) algorithms, i.e. those with a runtime of $O(f(k) \cdot poly(n))$. We refer to [6] for an introduction to parameterized complexity.

It is well known that FPT algorithms exist for a large number of NP-hard problems on graphs of bounded tree-width. However, for those problems where tree-width does not help, it is necessary to use more powerful parameters – vertex cover being the most successful one to date. A wide range of such problems have been solved in FPT time on graphs of bounded vertex cover, including Equitable Coloring[12], Equitable Connected Partition[7], Boxicity[1], Precoloring

---

[*] This research has been supported by the Czech research grants 202/11/0196 and MUNI/E/0059/2009.

D. Marx and P. Rossmanith (Eds.): IPEC 2011, LNCS 7112, pp. 259–271, 2012.

Extension[12], and various graph layout problems such as Imbalance, Cutwidth, Distortion and Bandwidth [10].

So, if vertex cover is so powerful in parameterized algorithmics, why even bother with other parameters at all? The answer is that any good structural parameter needs to balance between being as powerful as possible (i.e. useful in designing algorithms) while also being as little restrictive as possible (the class of graphs with the parameter bounded should be rich). The greatest disadvantage of vertex cover is that it is very restrictive, and this severely limits its practical usefulness.

The main contribution of the article lies in introducing twin-cover as a much more general alternative to vertex cover. Almost all interesting problems solvable on graphs of bounded vertex cover also become easy on cliques, and so it was a natural question to ask if one can take the best of the two worlds. We show that twin-cover is capable of dealing with a very wide range of problems of practical interest, some of which are hard even on graphs of bounded tree-width. In fact, the contribution of twin-cover is twofold: it directly generalizes the best known parameterized algorithms for problems where vertex cover is the most general parameter for which the problem is FPT, and at the same time provides an alternative to tree-width so that problems hard on clique-width can be solved on dense graphs.

Additionally, the article contains a new parameterized algorithm for solving the Graph Motif problem of graphs of bounded vertex cover. The algorithm is located in the second section of the article, and we use it to familiarize the reader with parameterized algorithm design on graphs of bounded vertex cover. Section three introduces and discusses twin-covers, and includes a comparison with other popular parameters as well as an algorithm for computing a twin-cover when its size is bounded – a very useful property for structural parameters to have.

Sections four and five deal with algorithmic applications of the new parameter. Specifically, section four contains several parameterized algorithms for graphs of bounded twin-cover. Four of the solved problems are hard or open on graphs of bounded tree-width, and vertex cover was the best (and sometimes only) known parameter which could be used – until now. The remaining five problems are hard on graphs of bounded clique-width, and in most cases the presented parameterized algorithms allow them to be solved on dense graphs for the first time. Section five shows how to extend a recent result for MS_1 on graphs of bounded vertex cover to twin-cover. The concluding notes also discuss alternative ways of defining a similar parameter, and why they fail in algorithmic design.

2 An FPT Algorithm for Graph Motif

Definition 2.1 (Graph Motif)
Input: A vertex-colored[1] undirected graph G and a multiset M of colors.
Question: Does there exist a connected subgraph H of G such that the multiset of colors $col(H)$ occurring in H is identical to M?

---

[1] This coloring need not be proper – neighboring vertices may have the same color.

The Graph Motif problem was introduced in [16] and arises naturally in bioinformatics, especially in the context of metabolic network analysis. Its complexity has been studied in [8] and [2], the latter proving that the problem remains NP-hard even on graphs of path-width 2, superstar graphs and other very restricted classes of graphs. Our first new result is an FPT algorithm for Graph Motif on graphs of bounded vertex cover. This is the first parameterized algorithm for Graph Motif which only bounds the structure of the graph and not M. We will later show how to extend this result to the much more general class of graphs of bounded twin-cover.

Theorem 2.2. *The Graph Motif problem can be solved in time* $O(2^{k+2^k} \cdot (\sqrt{|V|}|E|))$ *on graphs of vertex cover number at most* k.

Proof: We begin by finding a vertex cover C in time $O(1.2738^k + k|V|)$ [4]. For any non-cover vertex v we say its type $t(v) \subseteq C$ is equal to its neighborhood in G. Notice there are 2^k possible types and that all non-cover vertices are partitioned into sets $T_{c \subseteq 2^k}$ containing all vertices of the same type.

Next, we run over all possible 2^k subsets C' of C and try to find a Motif H which intersects with C exactly in C'. If this is done correctly for all C' and no admissible H is found, it is clear H does not exist. What remains is to decide whether there exists a motif H for a given C'. This is easy if C' is already connected, since we may simply add any adjacent non-cover vertex to H if their color is in $M - col(H)$ (non-existence of a vertex with this color indicates that no H exists for given C').

On the other hand, if C' is not connected, then it is necessary to add up to $k-1$ non-cover vertices to H before it becomes connected. One needs to be careful here, since trying $|V|^{k-1}$ possible vertices requires too much time, and e.g. using some red vertex to connect two components may prevent the use of red vertices elsewhere. However, notice that it suffices to select at most one vertex from each type to make H connected. So, we may run over all possible (at most 2^{2^k}) sets of types which are used to make H connected. Each type contains vertices with various colors, and we need to make sure to select a color in each type so that its occurrence in H does not exceed that in M.

This final subproblem may be solved by simply finding a maximum matching between all colors remaining in the multiset M and all types in the selected set – with edges between colors and types containing that color. Such a bipartite graph may be constructed in time $|E|$, has at most $|V|$ vertices and allows a maximum matching to be found in time $O(\sqrt{|V|}|E|)$. If the resulting matching does not include all type-vertices we try the next set of types.

So, the whole algorithm constructs H by first trying all possible subsets of C in time $O(2^k)$, then trying all possible ways to make H connected by selecting at most $O(2^{2^k})$ possible sets of types, and then deciding which vertices to use from these types based on a maximum matching algorithm in time $O(\sqrt{|V|}|E|)$ – if the matching does not exist, we skip to the next set of types. At this point we have some "skeleton" of H such that $col(H) \subseteq M$, and we may simply run through all neighbours of H and add everything that is missing from M.

This can be done in time $O(|V|)$ and either results in a yes answer or we skip to the next selection of C'.   ∎

3   Introducing Twin-Covers

It has already been said that the usefulness of vertex cover in practice is severely limited by how restrictive it is. However, practically all interesting problems where vertex cover is used are easy to solve on cliques, in spite of cliques having a high vertex cover number. This suggests that there should exist a parameter which

1. attains low values on a significantly more general class of graphs, including cliques and graphs of bounded vertex cover, and
2. is capable of solving "most" problems which are solvable by vertex cover.

Note that the sought parameter cannot solve *all* of these problems, since Courcelle, Makowski and Rotics have shown the existence of problems which are hard on cliques and yet solvable in FPT time on graphs of bounded tree-width [5]. However, known problems of this kind are of mostly theoretical interest and have little practical importance.

Definition 3.1. $X \subseteq V(G)$ *is a twin-cover of G if for every edge $e = \{a, b\} \in E(G)$ either*

1. *$a \in X$ or $b \in X$, or*
2. *a and b are twins, i.e. all other vertices are either adjacent to both a and b, or none.*

We then say that G has twin-cover number k if k is the minimum possible size of a twin-cover of G.

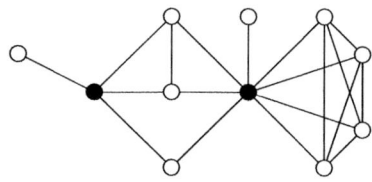

Fig. 1. The twin-cover of a graph

The underlying idea is to allow large cliques in graphs of bounded twin-cover, but restrict the number of vertices with out-edges from each clique. Notice that the relation of "being twins" is transitive, and that twin vertices form cliques in a graph. In fact, there exists another, perhaps more intuitive definition for twin-cover:

Definition 3.2. $X \subseteq V(G)$ *is a twin-cover of G if there exists a subgraph G' of G such that*

1. *$X \subseteq V(G')$ and X is a vertex cover of G'.*
2. *G can be obtained by iteratively adding twins to non-cover vertices in G'.*

We note that complete graphs indeed have a twin-cover of zero. Section 6 discusses an alternative – less restrictive – approach to defining a similar parameter, and why it fails. Let us conclude this section with a brief comparison of twin-cover to other graph parameters and an algorithm to compute twin-cover when its size is bounded. The latter result is especially important, since otherwise we would need to rely on an oracle to provide the twin-covers before running our parameterized algorithms (as is the case with clique-width [11]).

Proposition 3.3 (cf. full version for proof)

1. *The vertex cover of graphs of bounded twin-cover may be arbitrarily large.*
2. *There exist graphs with arbitrarily large twin-cover and bounded tree-width, and vice-versa.*
3. *The clique-width of graphs of twin-cover k is at most $k + 2$.*
4. *The rank-width [14] and linear rank-width [13] of graphs of twin-cover k are at most $k + 1$.*

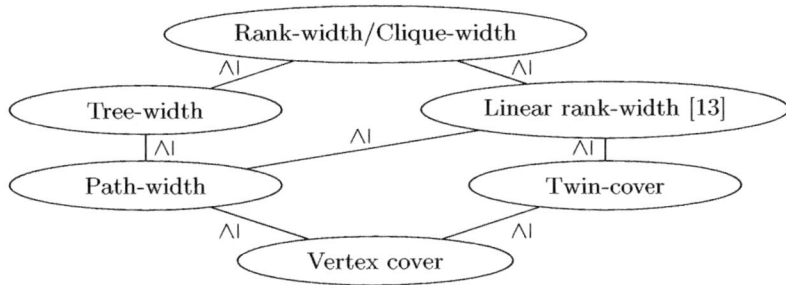

Fig. 2. Relationships between selected graph parameters

Theorem 3.4. *It is possible to find a twin-cover of size k in time $O(|E||V| + k|V| + 1.2738^k)$ (c.f. [4]).*

Proof: To compute the twin-cover, we remove all edges between twins (which need not be covered) and then compute the vertex cover of the remaining graph. The first step simply requires running through all $|E|$ edges and marking those which form twins (i.e. both incident vertices have the same neighborhood minus each other), and then deleting all marked edges. The second step can be done in time $O(1.2738^k + k|V|)$ [4]. ∎

4   Algorithms on Twin-Cover

In this section we intend to show how twin-cover can be used to solve various NP-hard problems. In some cases it suffices to slightly adjust the parameterized algorithms for graphs of bounded vertex cover, whilst in others the situation is more complicated. We first focus on problems where FPT algorithms are not known even when the tree-width is bounded: Graph Motif, Boxicity, Equitable Coloring and Precoloring Extension. In the last part of this section we also sketch a few algorithms for problems which can be solved in FPT time on tree-width, but which are hard on clique-width.

4.1   Graph Motif

Theorem 4.1. *The Graph Motif problem can be solved in time $O(|V| \cdot |M| + 2^{k+2^k} \cdot (\sqrt{|V|}|E| + |V|))$ on graphs of twin-cover at most k.*

Proof: Given a twin-cover X, we first try whether it is possible to find H in $V - X$; this is easy to do since all that remains are disconnected cliques. Specifically, we compute the multiset of colors contained in each clique and check whether M is a subset of this multiset, with a total time requirement of $|V| \cdot |M|$.

The important finding is that if any solution H contains at least one cover vertex, then removing all the edges between non-cover vertices in G does not affect the existence of a solution. To see this, consider any vertex v in H which is not a cover vertex nor a neighbor of a cover vertex in H. Since H is connected, it contains at least one cover vertex x such that there is a path of non-cover vertices between x and v in H. Every edge on this path (except the first) is an edge between twins, and since one non-cover vertex on this path is adjacent to x, all the other vertices need to be adjacent to x as well. This means that there exists another solution H' which uses $\{x, v\}$ instead of the original edge incident to v.

So, if we find no solution in G which does not contain any cover vertices, we simply remove all the edges between non-cover vertices and run the algorithm from Theorem 2.2. If this algorithm finds a solution, it will also apply to the original G before the edge deletions, and if no solution is found then no solution exists in the original G either.   ∎

4.2   Equitable Coloring

Definition 4.2 (Equitable Coloring)
Input: A graph $G = (V, E)$ and a positive integer r.
Question: Is there a proper vertex coloring c using at most r colors, with the property that the sizes of any two color classes differ by at most one?

The notion of Equitable Coloring first appeared in 1973 as part of an application for scheduling garbage trucks [18], and has since appeared as a subproblem in various scheduling applications. Fellows, Fomin, Lokshtanov et al. proved that it remains W[1]-hard on graphs of bounded tree-width [9], and Kratochvíl, Fiala and Golovach introduced an FPT algorithm solving the problem on graphs of bounded vertex cover [12].

Theorem 4.3. *The Equitable Coloring problem can be solved in time* $O(|V|^3 r \cdot (\frac{0.792k}{ln(k+1)})^k \cdot 2^k)$ *on graphs of twin-cover at most* k.

Proof: Unfortunately, the presence of large cliques in our graph class requires a new approach to solve the problem. Notice that if the number of colors c is fixed, then the numbers of color classes of size s and of size $s+1$ become fixed as well. Unlike coloring, it is not true that equitable colorability by r color implies colorability by $r+1$ colors – for example $K_{n,n}$ is equitably 2-colorable but not 3-colorable. To account for this we run our algorithm for all r possible values of c, and so we may assume that the numbers of s- and $s+1$-classes are fixed.

The algorithm begins by considering all possible ways of coloring the k cover vertices. However, we do not care about which colors are actually used to color the cover vertices, only whether they are colored by distinct colors and whether that color class has size s or $s+1$ – specifically, we will account for all possible partitions of the twin-cover into color classes and all possible total sizes of these color classes, without distinguishing between cases which are isomorphic up to color swapping. The number of different partitions of k vertices into sets is called the Bell number B_k. A recent result of Berend and Tassa provides an upper bound of $B_k < (\frac{0.792k}{ln(k+1)})^k$ [3]. For each such partition it remains to decide which of the (at most k) color sets containing a cover vertex have size s and which have size $s+1$. This amounts to at most 2^k possibilities per partition. Note that we discard all precolorings which are not proper.

Now that we have a precoloring of the cover vertices, all that needs to be colored in G are disconnected cliques of various sizes, and all the vertices in each uncolored clique have the same neighbors in the cover. This means that for each clique, we know exactly which colors may or may not be used in that clique, and that each color may be used there only once. This information allows us to construct a network flow instance to color the remaining vertices in G, as follows.

We have a source which is connected by arcs to r vertices representing colors, and the capacity of these arcs is equal to the number of vertices which still need to be colored by that color – for example $s-2$ if the class was of size s and 2 cover vertices already have this color. Then we create a vertex to represent every uncolored clique in G, and arcs from these vertices to the sink with capacities equal to the size of the clique. Finally, we add arcs of capacity 1 from color vertices to clique vertices if the color may be used in that clique. After computing the maximum flow, we check whether it is equal to the number of uncolored vertices in G (also equal to the sum of arcs outgoing from the source), and if this is the case we immediately obtain a solution in G. This flow subproblem can be solved in time $O(|V(G)|^3)$.

To recapitulate, we run through all $c \leq r$ allowed numbers of colors for G. Then we run through all possible ways of partitioning the cover vertices into color classes of sizes s and $s+1$. Finally, we use network flow to decide whether the remaining vertices can be equitably colored with respect to this selection of c, partitioning of cover vertices, and sizes of color classes in this partitioning. ∎

4.3   Precoloring Extension

Precoloring Extension is a natural problem where we are given a partial proper coloring of a graph G and the task is to extend it into a proper coloring of G with r colors. Similarly to Equitable coloring, it is W[1]-hard on graphs of bounded tree-width [9] and FPT when the vertex cover is bounded [12]. Unfortunately, the algorithm of [12] cannot be directly applied to graphs of bounded twin-cover due to them containing large cliques. Instead, several tricks are required to obtain an FPT algorithm – showcasing the kind of tools available for solving similar problems on graphs of bounded twin-cover.

Theorem 4.4. *The Precoloring Extension problem can be solved in time* $2^{O(k^3 2^k)} \cdot |V|$ *on graphs of twin-cover at most* k.

Proof: Recall that two non-cover vertices have the same type if they are adjacent to the same vertices in the cover. The vertices in each type T_i are organized into cliques of various sizes and some vertices are precolored. Note that we will consider isolated vertices in the type as cliques of size 1. Any color which is precolored at least in one vertex of this type cannot occur in any cover vertex adjacent to this type, and so it can never hurt the optimality of the solution to add this color into some non-precolored vertex in every clique of this type (i.e. use it as much as possible within the type). In this way we can extend the precoloring so that all cliques of the same type are precolored with the same colors; the smaller ones may be fully precolored with a subset of the colors. Cliques now only differ by the number of non-precolored vertices still available. Since their precolorings match, the coloring of any of the largest cliques in the optimal solution can simply be taken and copied onto smaller or equal cliques of the same type. So, if any vertices remain uncolored in the given type, we can delete everything except for one clique of maximum size. All of this may be done in $O(|V|)$ time.

Next we will need *color types*. While the number of colors r may be large, we will divide all colors into color types based on the set of types they are precolored in (additionally, the at most k colors precoloring cover vertices may be handled as separate types). There can be at most $2^{2^k} + k$ color types, and for each color type we remember the number of colors in that type (which may be large). Notice that colors of the same type are completely symmetric on the cover vertices, i.e. may be arbitrarily swapped in the cover without creating conflicts.

This allows us to use a similar trick to the one in precoloring extension – we consider all possible colorings of the cover vertices, but do not distinguish between colors of the same color type. There are $2^{O(k 2^k)}$ ways of dividing cover vertices into color types, and for each we need to consider all possible pairs of cover vertices which will be assigned the same color if they are in the same color type – at most 2^{k^2} possibilities.

Now the algorithm is clear. The first step runs through all the $2^{O(k^3 2^k)}$ possibilities for color types on the cover vertices. The second step then assigns any colors in the appropriate type to cover vertices with respect to the selected types and color equalities, and checks whether any conflicts occur (non-proper coloring or not enough colors of the given type) and whether it is possible to color the remaining vertices in the various types, all of which can be done by a greedy algorithm in $O(|V|)$ time.   ∎

4.4 Boxicity

Definition 4.5 ([1]). *A k-box is a Cartesian product of closed intervals $[a_1, b_1] \times [a_2, b_2] \ldots [a_k, b_k]$. A k-box representation of a graph G is a mapping of the vertices of G to k-boxes in the k-dimensional Euclidean space such that two vertices in G are adjacent iff their corresponding k-boxes have a non-empty intersection. The boxicity of a graph G is the minimum integer k such that G has a k-box representation.*

Alternatively, boxicity may be defined as the minimum positive integer b such that the graph can be represented as the intersection of b interval graphs. The notion was first introduced by Roberts [19] in 1969 and has found applications in social sciences and biology. Determining whether the boxicity of a given graph is at most 2 is already NP-hard [15], and it is believed that computing the Boxicity of graphs of bounded tree-width is NP-hard, however the problem remains open [1]. An FPT algorithm for Boxicity on graphs of bounded vertex cover was recently introduced in [1] and this may be easily used to prove:

Theorem 4.6. *The Boxicity problem can be solved in time $2^{O(2^k k^2)}|V|$ on graphs of twin-cover at most k.*

Proof: We first prove the following statement: For any graph G and any pair of twins $a, b \in V(G)$, deleting b and all of its incident edges does not change the boxicity of G.

Consider a minimal k-box representation of a graph G' which was obtained by removing b from G. We may simply add b and all of its incident edges back into G', and map b to a k-box identical to the k-box of a in G'. Thus the boxicity of G is equal to the boxicity of G'.

This allows us to simply delete non-cover twin vertices until we obtain a graph of vertex cover k with no edges between non-cover vertices. Afterwards it suffices to run the algorithm from [1], which has a runtime of $2^{O(2^k k^2)}|V|$ if the vertex cover is at most k.   ∎

4.5 Additional algorithms

So far we have only considered problems which are hard on trees or graphs of bounded tree-width. Additionally, any parameterized algorithm on rank-width/clique-width will work on graphs of bounded twin-cover (since their clique-width is bounded). We include a few algorithms for problems which are known to be hard on clique-width and yet can be solved in FPT time on twin-cover bounded graphs.

Proposition 4.7 (cf. full version for proof). *Hamiltonian Path, Chromatic Number, Vertex-Disjoint Paths, Edge-Disjoint Paths and Max-Cut can be solved in FPT time on graphs of twin-cover at most k.*

5   Monadic Second-Order Logic

Monadic second-order logic is often used in algorithmic graph theory to describe a wide range of problems on graphs. We distinguish between the so-called MS_2 logic, which allows quantification over (sets of) vertices and edges, and the weaker MS_1 logic, which only allows quantification over (sets of) vertices. It is well known that all MS_2-definable problems can be solved in FPT time on graphs of bounded tree-width, and the same is true for MS_1-definable problems on graphs of bounded rank-width.

Unfortunately, in both of these cases the height of the tower of exponents grows with the number of quantifier alternations in the formula. This means that while complicated formulas are still solvable in FPT time on these graph classes, the time complexity would be enormous in practice.

The work of Michael Lampis in ESA 2010 [17] explains the problem in detail and, more importantly, shows how to overcome it for MS_1 logic on graphs of bounded vertex cover (note that MS_2 can also be solved in FPT time on graphs of bounded vertex cover, but the tower of exponents may grow arbitrarily). We will sketch how the approach of Lampis can be extended to graphs of bounded twin-cover.

5.1 MS_1 on Graphs of Bounded Vertex Cover

First, let us briefly recapitulate Lampis' results. Given is a first-order sentence (an MS_1 sentence without set quantification) with q quantifiers and a graph with vertex cover k. It is not hard to prove that for any vertex variable x and any two vertices of the same type a, b which have not been assigned any vertex variable yet, the evaluations of the sentence with x assigned to a and x assigned to b are identical. What this means is that as long as vertices are variable-free and of the same type, they are completely indistinguishable. Additionally, since the logical sentence cannot distinguish between vertices in any other way than by assigning quantified vertex variables to them, it can only "count" up to q. Thus, if there are over q vertices of any type in G, we may delete them and obtain a kernel of size at most $k + 2^k \cdot q$. To evaluate satisfiability of the first-order sentence it now suffices to try all combinations of assigning these vertices to q variables, which results in a time complexity of $2^{O(kq+q \log q)}$ [17].

For MS_1 logic the idea is similar, however now we must also deal with set quantifiers. Consider an MS_1 sentence with q vertex quantifiers, s set quantifiers and a graph with vertex cover k. The indistinguishability argument still holds, however each set quantifier now multiplies the number of "countable" vertices by two – for all yet-unassigned vertex variables it is necessary to allow for them being assigned to vertices in the set or to vertices outside of the set. Still, this allows us to limit the number of vertices of each type to $q \cdot 2^s$, resulting in a kernel of size at most $k + 2^k \cdot q \cdot 2^s$. Satisfiability evaluation of the MS_1 sentence requires trying all possible assignments of set and vertex quantifiers on this graph; it is not hard to verify that this can be done in time $2^{2^{O(s+q+k)}}$ [17].

5.2 MS_1 on Graphs of Bounded Twin-Cover

The difficulty with adapting these results to twin-cover stems from the fact that vertices of the same type (i.e. adjacent to the same cover vertices) may be divided into many cliques of various sizes. While it is still true that two vertices of the same type and in the same clique are indistinguishable with respect to MS_1, this does not hold for vertices of the same type in different cliques. Luckily, this may be dealt with by properly utilizing the "counting" limitations of first-order and MS_1 logic.

Consider a first-order sentence with q quantifiers and a graph with twin-cover k. Dividing the vertices into $(k + 2^k)$ indistinguishable classes will not work this time due to the presence of cliques. However, since it only has q variables available, a first-order sentence cannot distinguish between a graph which has q cliques of type A and size i and one with *more than* q cliques of type A and size i. In other words, it can only "count" to up to q as far as the number of cliques of a certain type and size goes. The same argument also holds for the size of the cliques – a first-order sentence with q variables cannot distinguish between cliques of size q and larger cliques as long as they contain vertices of the same type. We use this to obtain a kernel of size at most $k + 2^k \cdot (q + 2q + \cdots + q^2) = O(2^k \cdot q^3)$, for which we need to try all options of assigning q vertex variables.

Corollary 5.1. *There exists an algorithm which, given a first-order sentence ϕ with q variables and a graph G with twin-cover at most k, decides if $G \models \phi$ in time $2^{O(kq+q \log q)}$.*

MS_1 can be dealt with in much the same way. We will need 2^k types, in each type the sentence can distinguish between up to $q \cdot 2^s$ cliques of the same size, and the maximum size of a clique we need to consider is $q \cdot 2^s$. All in all the required kernel has a size of at most $k + 2^k \cdot (q2^s + 2q2^s + \cdots + (q2^s)^2) = O(q^3 2^{k+3s})$. Again, all options of assigning vertex and set variables need to be tried, with a runtime of $O\big(\big(2^{q^3 2^{k+3s}}\big)^s \cdot \big(q^3 2^{k+3s}\big)^q\big)$.

Corollary 5.2. *There exists an algorithm which, given an MS_1 sentence ϕ with q vertex variables and s set variables and a graph G with twin-cover at most k, decides if $G \models \phi$ in time $2^{2^{O(s+q+k)}}$.*

6 Concluding notes

For a new parameter to be accepted by the computer science community, it needs to make a strong case. We firmly believe that the presented algorithmic results provide more than enough evidence that twin-cover is a useful and powerful parameter.

Potential applications of twin-cover include:

1. Replacing vertex cover in parameterized algorithms for various problems hard on tree-width and path-width.
2. Providing a more viable alternative to tree-width for dense graph classes and problems which are hard on clique-width.
3. Significantly faster MS_1 model checking on certain graph classes.

Future work should focus on the practipal aspects of twin-cover. How much smaller is the twin-cover of various graphs compared to their vertex cover, and how does it compare to tree-width? How much of a speed-up does it actually provide over vertex cover? One would expect the runtime differences to be huge, however the actual numbers are yet to be seen.

Lastly, we would like to comment on the definition of twin-cover. Michael Lampis uses an auxiliary parameter called *neighborhood diversity* to obtain his results for vertex cover [17]. While the parameter is also low on cliques, its size can be exponentially higher than the size of the vertex cover. Additionally, it is open whether neighborhood diversity can actually be used to solve various problems instead of vertex cover.

An inquisitive reader might be wondering why it is necessary to bound the vertices with outgoing edges from cliques. Why not just bound the number of cliques, and require that all edges are incident to some clique in the cover? The problem with this approach is that such a parameter would be low on split graphs (graphs which may be partitioned into a clique and an independent set). It is known that several interesting problems are already NP-hard on split graphs, and as a final result we show that Graph Motif is one such problem.

Theorem 6.1 (cf. full version for proof). *The Graph Motif problem is NP-hard on split graphs.*

References

1. Adiga, A., Chitnis, R., Saurabh, S.: Parameterized Algorithms for Boxicity. In: Cheong, O., Chwa, K.-Y., Park, K. (eds.) ISAAC 2010, Part I. LNCS, vol. 6506, pp. 366–377. Springer, Heidelberg (2010)
2. Ambalath, A.M., Balasundaram, R., Chintan Rao, H., Koppula, V., Misra, N., Philip, G., Ramanujan, M.S.: On the Kernelization Complexity of Colorful Motifs. In: Raman, V., Saurabh, S. (eds.) IPEC 2010. LNCS, vol. 6478, pp. 14–25. Springer, Heidelberg (2010)
3. Berend, D., Tassa, T.: Improved bounds on bell numbers and on moments of sums of random variables. Probability and Mathematical Statistics 30, 185–205 (2010)
4. Chen, J., Kanj, I.A., Xia, G.: Improved upper bounds for vertex cover. Theor. Comput. Sci. 411, 3736–3756 (2010)
5. Courcelle, B., Makowsky, J.A., Rotics, U.: Linear time solvable optimization problems on graphs of bounded clique-width. Theory Comput. Syst. 33(2), 125–150 (2000)
6. Downey, R.G., Fellows, M.R.: Parameterized complexity. Monographs in Computer Science. Springer, Heidelberg (1999)

7. Enciso, R., Fellows, M.R., Guo, J., Kanj, I., Rosamond, F., Suchý, O.: What Makes Equitable Connected Partition Easy. In: Chen, J., Fomin, F.V. (eds.) IWPEC 2009. LNCS, vol. 5917, pp. 122–133. Springer, Heidelberg (2009)

8. Fellows, M.R., Fertin, G., Hermelin, D., Vialette, S.: Upper and lower bounds for finding connected motifs in vertex-colored graphs. J. Comput. Syst. Sci. 77, 799–811 (2011)

9. Fellows, M.R., Fomin, F.V., Lokshtanov, D., Rosamond, F., Saurabh, S., Szeider, S., Thomassen, C.: On the complexity of some colorful problems parameterized by treewidth. Inf. Comput. 209, 143–153 (2011)

10. Fellows, M.R., Lokshtanov, D., Misra, N., Rosamond, F.A., Saurabh, S.: Graph Layout Problems Parameterized by Vertex Cover. In: Hong, S.-H., Nagamochi, H., Fukunaga, T. (eds.) ISAAC 2008. LNCS, vol. 5369, pp. 294–305. Springer, Heidelberg (2008)

11. Fellows, M.R., Rosamond, F.A., Rotics, U., Szeider, S.: Clique-width is NP-complete. SIAM J. Discret. Math. 23, 909–939 (2009)

12. Fiala, J., Golovach, P.A., Kratochvíl, J.: Parameterized complexity of coloring problems: Treewidth versus vertex cover. Theoretical Computer Science (2010) (in Press)

13. Ganian, R.: Thread Graphs, Linear Rank-Width and their Algorithmic Applications. In: Iliopoulos, C.S., Smyth, W.F. (eds.) IWOCA 2010. LNCS, vol. 6460, pp. 38–42. Springer, Heidelberg (2011)

14. Ganian, R., Hliněný, P.: On parse trees and Myhill–Nerode–type tools for handling graphs of bounded rank-width. Discrete Appl. Math. (2009) (to appear)

15. Kratochvíl, J.: A special planar satisfiability problem and a consequence of its NP-completeness. Discrete Appl. Math. 52, 233–252 (1994)

16. Lacroix, V., Fernandes, C.G., Sagot, M.-F.F.: Motif search in graphs: application to metabolic networks. IEEE/ACM Transactions on Computational Biology and Bioinformatics 3(4), 360–368 (2006)

17. Lampis, M.: Algorithmic Meta-Theorems for Restrictions of Treewidth. In: de Berg, M., Meyer, U. (eds.) ESA 2010, Part I. LNCS, vol. 6346, pp. 549–560. Springer, Heidelberg (2010)

18. Meyer, W.: Equitable coloring. American Mathematical Monthly 80, 920–922 (1973)

19. Roberts, F.S.: On the boxicity and cubicity of a graph. In: Recent Progresses in Combinatorics. Academic Press (1969)

Author Index

GPSR Compliance

The European Union's (EU) General Product Safety Regulation (GPSR) is a set of rules that requires consumer products to be safe and our obligations to ensure this.

If you have any concerns about our products, you can contact us on ProductSafety@springernature.com

In case Publisher is established outside the EU, the EU authorized representative is:

Springer Nature Customer Service Center GmbH
Europaplatz 3
69115 Heidelberg, Germany

Batch number: 09478804

Printed by Printforce, the Netherlands